Neuroscience in Space

Gilles Clément · Millard F. Reschke

Neuroscience in Space

 Springer

Gilles Clément
Centre de Recherche Cerveau et Cognition
UMR 5549 CNRS-Université Paul Sabatier
Faculté de Médecine de Rangueil
31062 Toulouse Cedex 9
France
gilles.clement@cerco.ups-tlse.fr

Millard F. Reschke
Neuroscience Laboratories
NASA Johnson Space Center
Mail Code SK-272
2101 NASA Parkway
Houston, Texas 77058-3696
USA
millard.f.reschke@nasa.gov

ISBN: 978-0-387-78949-1   e-ISBN: 978-0-387-78950-7
DOI: 10.1007/978-0-387-78950-7

Library of Congress Control Number: 2008927058

Cover illustrations: Design by Sean Collins, NASA Johnson Space Center. Images: Nebula, courtesy of NASA;
Neuron Forest, provided by Alzheimer's Association, artist: Stacy Janis. Authors' photograph: Angie Bukley.
Background image: Aurora Programme, courtesy of ESA, AOES Medialab.

Printed on acid-free paper

9 8 7 6 5 4 3 2 1

springer.com

PREFACE

Why write this book? Of all the intricate components of the human body, the central nervous system is the most responsive to the environment, detecting and responding to changes immediately. Its complexity, however, also means that it is still one of nature's best-kept secrets. Considering that the exploration of space is often thought of as the final frontier in the discovery of our origin and the preparation for our future, *Neuroscience in Space* is a book addressing the last, and greatest, scientific frontier.

All living things on Earth have evolved in the presence of gravity and all of their biological systems have anatomical and physiological mechanisms designed to interpret and measure the force of gravity. However, in the near weightlessness of space, the sensory systems that provide basic information regarding linear acceleration no longer function as they did on Earth. As a result, most if not all, physiological systems dependent on the body's central nervous system are in flux until a new microgravity state is realized. This includes adaptation of basic life sustaining functions such as blood pressure control and cardiac function, as well as other critical functions for everyday activities including balance, coordinated movement in three-dimensional space, and the regulation of sleep. Bones that supported body weight on the ground no longer have that load to bear. They begin to lose mass and strength, as do weight-bearing and postural muscles in the legs. Reduced physical activity and a shift of fluids into the upper body combine to reduce cardiovascular capacity. While in space, cardiovascular, bone and muscle deconditioning does not present a serious problem. However, whether returning to Earth or landing on some other planet, the body's adaptation to microgravity increases the risk of bone fractures, reduces work capacity, and can result in severe balance disorders and even blackouts when standing.

Other significant changes take place in the central nervous systems of astronauts during and following exposure to microgravity. Space travelers are transported in vehicles that move in three-dimensional space and generate inertial forces that create environmental factors to which they are not accustomed, either by evolution or experience. The responses of the vestibular organs in the inner ear, as well the kinesthetic, pressure and touch receptors, may be altered by hyper- or hypogravity. These altered responses to inertial stimulation outside their normal physiological range, or, even within this range, signal appropriately for the force environment, but inappropriately for the other sensory systems. These changes can modify situational awareness, induce spatial disorientation, result in illusions of self-motion, trigger dizziness and vertigo, and bring about motion sickness. However, the plasticity of the central nervous system allows individuals to adapt to these altered sensory stimulus conditions, and after a few days in space the symptoms disappear. The price paid for this in-flight adaptation (what has become known as "space normal") is a deconditioning of antigravity responses necessary for effective living following a return to Earth or landing on Mars. The duration of these altered responses is function of the time spent in space. In order to minimize the impact of adaptation to microgravity on crew health and performance following long-duration space flight, effective countermeasures must be developed.

Since the first human space flight in 1961, extensive experimental and operational research has been performed to investigate these adaptive processes by looking at electrophysiological changes in neural activity, behavioral changes including movements of the eye or body segments compensating for head or visual surround, as well as changes in perception and spatial orientation. The results obtained during and after space flight have contributed to a better understanding of the functioning and adaptation of multi-sensory interaction within the central nervous system. It could be said that the microgravity environment of space flight provides the ideal laboratory to study the underlying function and interactions among physiological systems. This environment can only be improved by an ability to switch gravity on and off during flight. New concepts and questions about the functioning and adaptation of the balance system have been raised directly from results of studies conducted in space. For example, new knowledge of neuronal plasticity, the way nerve cells "re-wire" to compensate for disease or injury, has been gained from animal studies during space flight, and will allow insights into treatment of nervous system disorders.

If one were interested in studying space travel, one would have little difficulty finding descriptions of the early developments in the Soviet space activities, as well as of those who helped establish NASA and their efforts. Should one be more serious about studying the development of space programs, one can find libraries of information addressing the bureaucracy of space flight full of tomes written by mission managers and project engineers. A student of space studies can find a plethora of technical books describing the principals of propulsion and rocket development, orbital mechanics and astrodynamics. as well as books detailing the design of spacecraft and the ground stations required to control them and collect their data. Entire museums dedicated to the progression of flight technology, from the first brief aircraft flights to the development and assembly of the International Space Station, have been established world-wide. However, it is truly challenging for anyone to find a comprehensive history of the life sciences experiments that have been performed in space and the role that neuroscience has played in our quest for space flight.

Our intent and purpose of compiling this historical overview of neuroscience and its role in space flight serves two purposes. The first is to equip researchers with a single reference document compiling a representation of those neuroscience experiments that have been flown in space. The second is to highlight the accomplishments of many scientists who have contributed to the history of space neuroscience. It is our hope that insights generated by reading this book will greatly contribute to the future agenda of space neuroscience.

In a sense, this book originated in a small office in Paris, France when the authors were first introduced. From that initial meeting a shared interest for sensorimotor and vestibular function in space flight would come to define a collaboration that has lasted over 25 years.

We are indebted to all of those astronauts and cosmonauts who became the subjects for much of the work detailed in this book. In particular, we would like to acknowledge Patrick Baudry, Sonny Carter, Owen Garriott, Claudie Haigneré, Joe Kerwin, Bob Parker, Rhea Seddon, and William Thornton who have provided both guidance and inspiration.

We also acknowledge our colleagues Drs. Owen Black, Alain Berthoz, Jacob Bloomberg, Bernard Cohen, Fred Guedry, Deborah Harm, Jerry Homick, Makoto Igarashi, Inessa Kozlovskaya, R. John Leigh, Francis Lestienne, William Paloski, Scott Wood, and Larry Young for their counsel and support. We would also like to express our gratitude to Jody Krnavek and Liz Fisher for the early mornings, late nights and missed holidays they gladly sacrificed to collect data at remote landing sites.

We are particularly grateful to Dr. Donald E. Parker for his support and guidance, as well as his contribution to Chapter 7 for the refinement of the otolith tilt-translation reinterpretation hypothesis. We are also grateful to Dr. Robert Welch who gave us permission to use the material from his contribution to the *"Space Human Factors Engineering Gap Analysis Project Final Report"* (Hudy & Woolford 2005). And finally, thanks to Angie Bukley for her time spent editing this book.

Gilles Clément & Millard Reschke

Houston, 24 January 2008

Contents

Chapter 1

SPACE NEUROSCIENCE: WHAT IS IT?

Neuroscience is a biological discipline whose goal is to describe and understand how the brain controls behavior. Behavior controlled by the brain ranges from the regulation of hormonal secretions that control body functions as varied as physiological and emotional responses to stress and the regulation of blood pressure, to simple reflexive controls, such as spinal and brainstem reflexes, which provide automatic behavioral responses to a variety of environmental stimuli. It also includes the performance of complex sensorimotor behavioral tasks, such as locomotion, posture, or the eye-hand-head coordination required to pilot an aircraft or a spacecraft, the perception of the body's orientation in three-dimensional space, and the control of learning and memory.

For this control, receptors sensitive to light, sound, blood pressure, muscle length, and acceleration, among others, send inputs to the *central nervous system* (CNS). This information is integrated and evaluated to produce reactions to environmental conditions that are appropriate for the survival of the organism. Feedback and feed forward loops, as well as the prediction of our own actions based on our knowledge and experience (cognition), complete this control mechanism.

Space Neuroscience studies attempt to understand how the brain controls responses to the special environment conditions of space travel, such as weightlessness, reduced gravity, unusual combinations of acceleration, radiation, and the stress induced by long confinement in isolated quarters and potentially hazardous activities.

A number of important central nervous system functions are affected by space flight. Among them are spatial orientation, posture, vestibular reflexes, central nervous system processing, and autonomic control. All have received attention in ground-based studies, but most are incompletely understood in space flight.

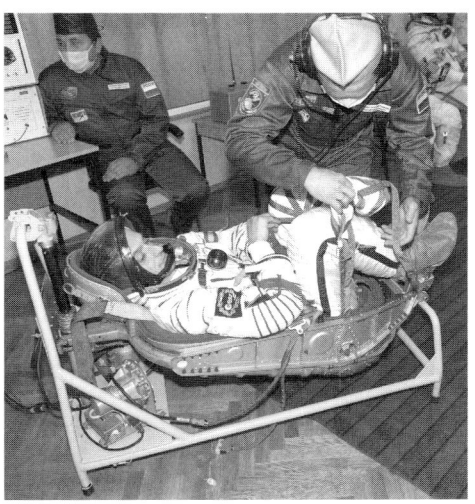

Figure 1-01. The astronaut's couch during launch on a Soyuz vehicle is constructed of a crushable honeycomb material bonded to a fiberglass shell and lined with rubber padding. Each astronaut has a couch contoured to his specific shape. The couch is designed to protect the astronaut's body from the g loads during launch and re-entry. Photo courtesy of NASA.

G. Clément, M.F. Reschke, *Neuroscience in Space*.
DOI: 10.1007/978-0-387-78950-7_1, © Springer Science+Business Media, LLC 2008

1   THE SPACE ENVIRONMENT

The space environment is characterized by the absence of atmosphere (vacuum), harmful levels of ionizing radiation, the absence of a 24-hour day and night cycle, confinement of crewmembers in cramped spacecraft, isolation from family and friends, and weightlessness. All these conditions have a potential impact on the central nervous system. However, we will not review here the effects of the absence of atmosphere because we will assume that the basic life support system functions (air, humidity, pressure, water, food) are provided to the astronauts during space missions. The human needs and the engineering techniques to sustain these needs have been extensively reviewed in Peter Eckart's book *"Space Flight Life Support and Biospherics"* (Kluwer, Dordrecht 1996).

1.1   Microgravity

The large mass of Earth creates a gravitational field that acts to attract objects with a force inversely proportional to the square of the distance between the center of the object and the center of Earth. When we measure the acceleration of an object acted upon only by Earth gravity on the surface of the Earth, we commonly refer to it as '1 g' or 'one-g'. This acceleration is approximately 9.81 m/sec^2.

Isaac Newton (1687) envisioned a cannon at the top of a very tall mountain extending above Earth's atmosphere so that friction with the air would not be a factor, firing cannonballs parallel to the ground. Newton demonstrated how additional cannonballs would travel farther from the mountain each time if the cannon fired using more black powder. With each shot, the path would lengthen and soon the cannonballs would disappear over the horizon. Eventually, if one fired a cannon with enough energy, the cannonball would fall entirely around Earth and come back to its starting point. The cannonball would begin to orbit Earth. Provided no force other than gravity interfered with the cannonball motion, it would continue circling Earth in that orbit (Figure 1-02).

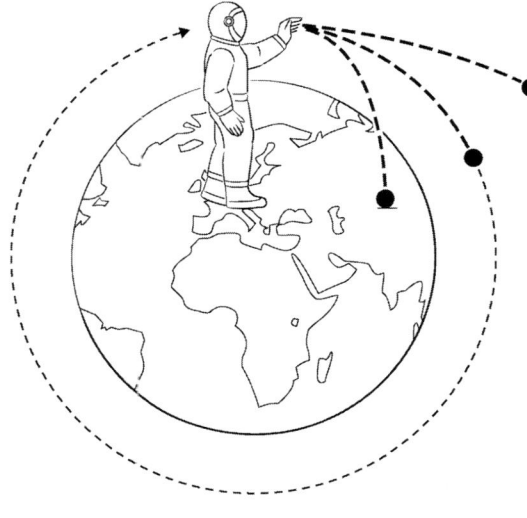

Figure 1-02. To stay in an orbit within a bounded distance from the Earth, artificial satellites or spacecraft must have a speed equal to 7.8 km/sec. At this velocity, gravity causes it to fall around the Earth without hitting the ground. With a lower velocity, the spacecraft would fall back on Earth; with a higher velocity it would eventually leave Earth orbit (Clément 2005).

This is how the Space Shuttle stays in orbit. It launches into a parabolic trajectory above Earth so that the spacecraft reaches the right speed to keep it falling while maintaining a constant altitude above the surface. For example, if the Space Shuttle climbs to a 320-km high orbit, it must travel at a speed of about 27,740 km/h to achieve a stable orbit. At that speed and altitude, due to the extremely low friction of the upper atmosphere, the Space Shuttle executes a falling path parallel to the curvature of Earth. In other words, the spacecraft generates a centripetal acceleration that counterbalances Earth's gravitational acceleration at that vehicle's center of mass. The spacecraft is therefore in a state of free-fall around Earth, and its occupants are in a weightless environment. Gravity *per se* is only reduced by about 10% at the altitude of low Earth orbit, but the more relevant fact is that gravitational acceleration is essentially cancelled out by the centrifugal acceleration of the spacecraft.

The term *microgravity* generally appears in science texts as a substitute for *weightlessness*. However, we can interpret 'microgravity' in a number of ways, depending upon the context:

a. The prefix 'micro' derives from the original Greek *mikros,* meaning 'small'. By this definition, a microgravity environment is one that imparts to an object a net acceleration that is small compared with that produced by Earth at its surface. We can achieve such an environment by using various methods including Earth-based drop towers, parabolic aircraft flights, and Earth-orbiting laboratories. In practice, such accelerations will range from about one percent of Earth's gravitational acceleration (on board an aircraft in parabolic flight) to better than one part in a million (on board a space station). Earth-based drop towers create microgravity environments with intermediate values of residual acceleration.

b. Quantitative systems of measurement, such as the metric system, commonly use the term *micro* to mean one part in a million. By this second definition, the acceleration imparted to an object in microgravity will be 10^{-6} of that measured at Earth's surface.

The use of the terms *microgravity* or *zero-g* in this book corresponds to the first definition: small gravity levels or low gravity. Microgravity can be created in two ways. Because gravitational pull diminishes with distance, one way to create a microgravity environment is to travel away from Earth. To reach a point where Earth's gravitational pull is reduced to one-millionth of that at the surface, we would have to travel into space a distance of 6.37 million kilometers from Earth (almost 17 times farther away than the Moon). This approach is impractical, except for automated spacecraft.

However, the act of free fall can create a more practical microgravity environment. Although airplane, drop tower facilities, and small rockets can establish a microgravity environment, all of these laboratories share a common problem. After a few seconds or minutes of reduced gravity, Earth gets in the way and the free-fall stops. To establish microgravity conditions for long periods of time, one must use spacecraft in orbit. Spacecraft are launched into a trajectory that arcs above Earth at the right speed to keep them falling while maintaining a constant altitude above the surface.

1.2   Accelerations

To reach a speed of nearly 28,000 km/h in about eight minutes requires quite some level of acceleration. Because overall human tolerance to sustained g forces in the chest-to-back direction is effectively twice that in the head-to-toe direction, all manned space vehicles launches and entries before the Space Shuttle have oriented thrust near the crew's chest-to-back direction. Maximum peak acceleration for the Apollo spacecraft reached approximately 6 g on re-entry with lesser values for launch and orbital maneuvers. Mercury and Gemini spacecraft operated at slightly higher values, with acceleration up to 6.4 g for near one minute. On board today's Soyuz and the Space Shuttle, the astronauts and cosmonauts experience peak forces of 3.4 g in the chest-to-back direction during launch (Figure 1-03A). On board the Soyuz, re-entry forces are even higher, reaching a value greater than 6 g (still in the chest-to-back direction) during the period of deceleration during atmospheric re-entry, and another shock during parachute deployment. In the current scenario for the *Crew Exploration Vehicle* to be used for missions to the Moon and Mars, peak forces during launch and re-entry may be as high as 5 g (chest-to-back).

Because it is a piloted vehicle during re-entry, the Space Shuttle imposes quite a different acceleration environment. Visibility requirements during landing necessitate an orientation of the crew couches that results in g forces during re-entry that are primarily in the head-to-foot direction. The g forces are lower than in the other vehicles, but these forces are present for a longer duration. The astronauts in the Space Shuttle are exposed to nominal forces of about 1.3 g for a period ranging from 17 to 20 minutes (Figure 1-03B). A cardiovascular counter pressure garment covering the legs and lower torso is being made available for use during Space Shuttle re-entry to reduce the effects of this vertical acceleration. This g-suit is a pair of tight fitting trousers containing built-in air bladders that are inflated or deflated by a manually operated air valve. By preventing the pooling of fluids in the body below the level of the heart during the deceleration of re-entry, this garment forestalls possible orthostatic hypotension.

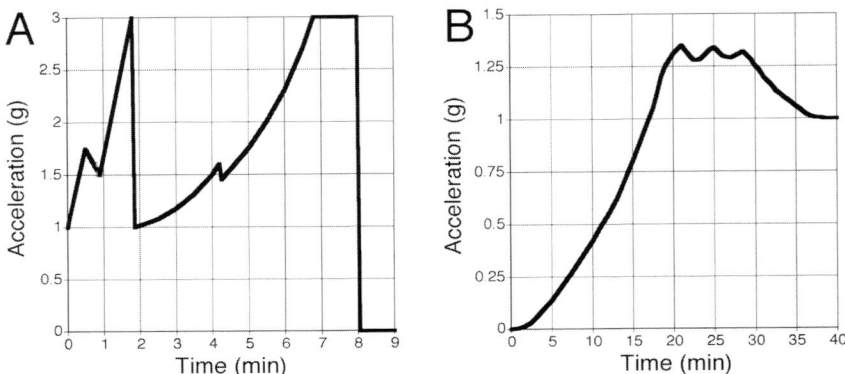

Figure 1-03. Typical acceleration profile of the Space Shuttle during launch and re-entry. A. The two Solid Rocket Boosters *deliver most of the acceleration during the first two minutes of flight. When they are jettisoned away from the vehicle the main Shuttle engines continue to fire until about 8 minutes after lift-off. The acceleration is designed to stay below 3 g and the g forces are exerted in the chest-to-back direction. B. Approximately 40 minutes are required to reach the point of wheel-stop from the beginning of re-entry. Considerable differences between missions exist because of many variations, including energy parameters, wind conditions, Orbiter weight, orbital inclination, and landing site. Re-entry is accomplished under a high relative velocity and varying levels of vertical and lateral acceleration. Nevertheless, the peak g force in the head-to-toe direction is designed to stay below 1.4 g. Adapted from Holland & Vander Ark (1993).*

1.3 Light

The absence of natural light in spacecraft may have significant effects on humans, too. A typical person spends his days outdoor, exposed to light provided by the Sun's rays (filtered through the ozone layer), including a small but important amount of mid- and near-ultraviolet light, and approximately equal portions of the various colors of visible light. Indoor lighting in most offices and in spacecraft is of a much lower intensity and, if emitted by fluorescent "daylight" or "cool-white" bulbs, is deficient in ultraviolet light (and the blues and reds) and excessive in the light colors (yellow-green) that are best perceived as brightness by the retina.

If the only effect of light on humans was to generate subjective brightness, then this artificial light spectrum might be adequate. It has become clear, however, that light has numerous additional physiological and behavioral effects. For example, light exerts direct effects on chemicals near the surface of the body, photoactivating vitamin D precursors and destroying circulating photoabsorbent compounds (melanin). It also exerts indirect effects via the eye and brain on neuroendocrine functions, circadian rhythms, secretion from the pineal organ, and, most clearly, on mood. Many people exhibit major swings in mood seasonally, in particular toward depression in the fall and winter when the hours of daylight are the shortest. When pathological, the *seasonal affective disorder syndrome* is a disease related to excessive secretion of the pineal hormone, melatonin, which also may be treatable with several hours per day of supplemental light. While not yet proved, it seems highly likely that prolonged exposure to inadequate lighting (that is, the wrong spectrum, or too low an intensity, or too few hours per day of light) may adversely affect mood and performance.

1.4 Ionizing Radiation

The space environment also exposes animals and individuals to high-energy radiation unlike anything they experience on Earth. Beyond the shield of Earth's magnetosphere, solar and galactic radiation can cause severe cellular damage or even cancer. Galactic cosmic rays are constantly present, coming from all directions. The radiation consists of heavy, slow moving atomic nuclei that can do far more damage to more cells than the alpha and beta particles coming from solar flares. This radiation requires several meters of shielding for complete blockage, and since the nuclei come from all directions at all times, unlike the brief solar flares that last only a few hours or days, a storm shelter is insufficient to protect the crew.

Several independent factors contribute to the overall risk to astronauts exposed to the complex radiation the space environment. Of primary concern is the induction of late-occurring cancers. But there is also the possibility of cell loss from radiation damage affecting the functional integrity of the CNS. Recent studies also point to radiation-induced cellular pathologies based on the communication between damaged and undamaged cells and the induction of unstable states that lead to late expression of genetic damage (Azzam *et al.* 2001). Space radiation seems to be uniquely effective in causing such cellular changes.

1.5 Confinement and Isolation

Confinement in cramped spaces with the same small group of people, separation from family and friends, as well as possible cultural, isolation from limited communication with Earth, loss of privacy due to habitability constraints, are all stressors that could affect mood and performance. Additional neurobehavioral risks are posed by prolonged exposure to microgravity, radiation, and life support system equipment failure in space. Judging from current evidence, language, culture, gender, and differences in work role will also pose challenges to crew communication and effectiveness (Santy 1994).

Without mitigation, these stressors can impose a burden on astronaut behavioral capability and health, both individually and collectively. They have the potential to erode cognitive performance, change neuroendocrine, cardiovascular and immune responses; disrupt appetite, sleep and other basic regulatory physiology, lead to neuropsychiatric impairment through anxiety and depression, and potentially cause serious interpersonal problems among crewmembers (White & Arener 2001).

In order to ease this harsh condition the habitability features and ergonomics of the crew habitats must be thoroughly understood for minimizing the discomfort of space explorers. In fact, equipment and habitat design, supplies, training materials and crew operations must be planned and developed on the basis of best available information from numerous disciplines, including human factors, biomechanics, cognitive and social psychology, and physiology. This information is required for the design, integration, and support of human, machine, mission and environmental elements that promote optimal performance, physical and psychological health, and safety in long duration space flight.

2   OPPORTUNITIES FOR SPACE RESEARCH

2.1   Space Missions in Low Earth Orbit

Currently the following spacecraft are used for human space flight in *low Earth orbit* (LEO): Soyuz, Space Shuttle, *International Space Station* (ISS), and Shenzhou. Historically, Vostok, Mercury, Voskhod, and Gemini spacecraft, as well as the Salyut, Skylab, and Mir space stations have also been used for human space flight. Unmanned Bion/Foton capsules provide unique opportunity for flying biological specimen such as animals, cells, and plants when no crew activity is needed.

A typical Soyuz or Space Shuttle mission lasts about 10-12 days, whereas increments of expedition crews on board the ISS range from 1-6 months. There are typically two taïkonauts on board ShenZhou, three cosmonauts on board Soyuz, seven astronauts on board the Shuttle, and three astronauts on board the ISS. The characteristics and constraints of these missions for space life sciences research are discussed in greater detail in the book *"Fundamentals in Space Medicine"* by Clément (2005).

The major difference between a life sciences flight experiment and the same experiment in a standard laboratory on Earth is the smaller number of subjects or specimens and observations in the flight experiment. Experimental sample size has been and will continue to be small. Due to limited space and power, a finite number of animals, specimens, or test subjects is available for in-flight research. This limited number often requires the development of elaborate and detailed sharing plans to maximize their use.

Also, in Earth laboratories, it is common to repeat experiments. This is even the basis for a scientifically sound investigation. Every published scientific manuscript contains a Method section detailed enough to allow other scientists to repeat the experiments, to verify and confirm the proposed hypothesis or interpretation of results. For space experiments, it is very rare when the exact same experiment is repeated on a second or a third mission, because of the financial and time constraints. The investigators are doomed to success in the first trial.

In order to limit the risk of failure in scientific return from a mission, and to give more opportunities to investigators, it has become common practice to "integrate" several experiments. In this process, multiple investigators must share the same equipment. The inconvenient is that some common procedures or conditions must be "negotiated" between the investigators (for example, a given centrifuge velocity, or a given temperature for animal habitats), and that a malfunction of the equipment might alter several experiments.

For a life sciences mission, crew time is the most precious resource. The availability of the crew for training prior to the mission is also limited. Training requirements depend on the complexity of both the individual instruments and the integrated payload.

Also, during the flight, the investigators have limited access to real-time data. In an Earth laboratory, a flaw in one experimental protocol is immediately detected and corrected before the experiment continues. During a space flight experiment, it is difficult to assess the exact situation remotely, and to suggest changes in an experimental protocol that has been designed over several years. The suggested changes could also have an impact on other experiments. Perhaps for these reasons, the results of flight experiments are mainly unexpected results. In some cases, results of space investigations have confirmed classical or generally held hypotheses. However, most results have been startling and unexpected, requiring researchers to reexamine their assumptions about the intricate relationship between gravity and life.

2.2 Interplanetary Missions

Interplanetary missions raise some issues concerning human health, particularly due to the long-duration of reduced gravity and the ionizing radiation environment. Although manned missions to the Moon are in reach, many believe that more research is needed before humans could safely go for on a mission to Mars. Research is needed in radiation shielding, space suit systems, in situ fabrication and repair, fire suppression and detection, remote instrumentation technology, as well as in nutrition, immunology, cardiovascular, and neurovestibular studies. A solution to a problem of long-term exposure to weightlessness is artificial gravity, i.e., exposing the crew to centripetal or linear acceleration continuously or intermittently throughout the mission. For a complete description of the principle, design options, and research in this area, the reader could refer to the book on "*Artificial Gravity*" by Clément & Bukley (2007).

2.2.1 Moon Missions

The only experience of manned planetary missions comes from the Apollo program, during which twelve astronauts walked on the Moon surface between 1969 and 1972. The Apollo program demonstrated effective human geological exploration in the hostile environment of another planet. During the initial Apollo-11 lunar landing mission, the crew remained in the one-sixth g environment of the Moon for less than one day and conducted a single excursion of less than three hours, during which they ventured only about 50 m from the Lunar Module. By the sixth and final lunar exploration mission, the distance traveled on the lunar surface had greatly increased. By the last mission to the Moon, Apollo crews had traversed a total distance of more than 97 km on the lunar surface and spent over 160 man-hours outside the Lunar Module.

The scientific experiments carried out on the Apollo missions provided important information about the Moon as well as the solar system. A total of 381.7 kg of lunar material was returned from six unique and scientifically significant lunar locations. Additionally, almost 30,000 high-resolution photographs were taken on the surface and from orbit during Apollo missions, recording the characteristics and features of the Moon in great detail.

NASA and its international partners are currently focusing on extending the human presence beyond Earth orbit in support of human exploration and scientific discovery. Two critical milestones in this "Vision for Space Exploration" program include initially exploiting the capabilities of the International Space Station and then building a long-term outpost on the Moon. These activities are in preparation for traveling to Mars (NASA 2004).

It is NASA's plan that by 2020, four-person crews will be making one-week visits to the Moon. By 2024 it is anticipated that a permanent base will be established with astronauts executing rotating six-month tours of duty. In an effort to narrow the time gap between the retirement of the Space Shuttle, scheduled for 2010, and the next manned flights, the new spacecraft that will take astronauts to the moon will be built using off-the-shelf technology. The new craft was recently named *Orion* and is a large space capsule. It has been nicknamed "Apollo on steroids" because of its striking resemblance to this heritage vehicle along with the fact that it will have more than double the habitable internal volume of the Apollo capsules. This will allow for the transport of six astronauts to the ISS and four to lunar orbit (Figure 1-04) where it will stay for up to six months before returning the crew to Earth. Another spacecraft, the *Altair* lunar lander, will transport the astronauts between the Orion and the surface of the Moon and back (Noland 2007).

Figure 1-04. The crew exploration vehicle, called Orion, *will carry a crew of four to six astronauts, and will be launched by the new* Ares I *launch vehicle. Both Orion and Ares I are elements of NASA's* Project Constellation, *which plans to send human explorers back to the Moon by 2020, and then onward to Mars and other destinations in the solar system. Photo courtesy of NASA.*

Exploiting in-situ resources to the extent possible, the lunar base will include a living area and laboratories. To take advantage of near constant sunlight for solar power, the lunar base will be located in the vicinity of one of the poles. It is postulated that frozen water may exist at the poles. If this turns out to be the case, then the ice could theoretically be processed into oxygen for the crew and liquid oxygen and hydrogen to fuel launch vehicles. The Altair lander has the capability to haul cargo and operate robotically. Astronaut explorers will employ long-range pressurized roving vehicles to explore the surface of the Moon. Robotic diggers and processors will be used to convert lunar soil into usable resources like water, oxygen, and fuel. It is envisioned that by 2030 or 2040, a lunar fuel processing plant to support missions to Mars will be online. All of the lessons-learned through the new lunar exploration program can also be applied to the construction of explorations bases on the surface of Mars (Noland 2007).

2.2.2 Mars Missions

While the Moon will serve as a first step towards the eventual exploration of Mars, the diverse environments of Mars will no doubt prove to be significantly more challenging. The geological features of Mars are gargantuan in scale. The north polar ice cap is itself as large at Antarctica. There are deserts on Mars far more vast than the Sahara. Mars boasts the highest mountain in the entire Solar System, Olympus Mons, which is twice the height of Mount Everest. Our Mars explorers will require extraordinary expeditionary skills and significant physical fitness to cross the forbidding landscapes they will encounter. Their situation will be further complicated by the fact that they will be required to work in a pressurized environment at all times because the atmosphere of Mars is lethal to humans.

The profiles for a manned Mars mission are based on a number of factors including the orbital characteristics of Earth and Mars, possible flight trajectories, energy requirements, travel times, and surface stay times. Two main mission profiles are being examined: [1]

a. The first type of mission is referred to as an *opposition-class mission.* This short-stay mission profile provides only 30 to 90 days on the surface of Mars. The total round-trip time ranges from 400 to 650 days. The opposition-class mission requires a large amount of energy in transit, even when taking advantage of either a Venus swing-by (on either the inbound or outbound leg) or a deep space propulsive maneuver in order to limit Mars and Earth re-entry speeds. This scenario may not provide sufficient time for the astronauts to recover in the 0.38 g Martian gravity from the approximately 250 days spent in microgravity on the outbound trip. In fact, over 90% of the total mission time would be spent in microgravity traveling to and from Mars. Concerns for the safety of the crew arise from the cumulative exposure to the microgravity environment.

b. The second type of mission under consideration is referred to as *a conjunction-class mission.* This long-stay mission minimum energy

---

[1] A third mission profile is also under consideration. This profile is similar to the one described in section *b* (long-stay, conjunction-class mission), but with faster transit between Earth and Mars (one-way travel time ranges from 120 to 180 days). However, the energy requirements for this mission profile are unrealistic with the current propulsion technology.

profile provides for stays on Mars of up to 500 days with a round trip total time of about 900 days. The one-way travel time is about the same as for the opposition-class missions, approximately 250 days. This mission profile maximizes surface stay time while keeping energy requirements within reason (Figure 1-05). Furthermore, the astronauts have much more time to recover from the prolonged exposure to microgravity.

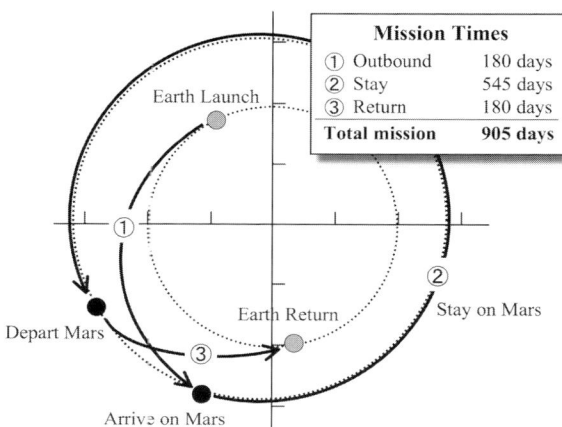

Mission Times	
① Outbound	180 days
② Stay	545 days
③ Return	180 days
Total mission	**905 days**

Figure 1-05. This drawing illustrates one feasible scenario for a human mission to Mars. Total mission time is 905 days away from Earth. This conjunction-class mission profile includes a 180-day transit to Mars, a 545-day stay on the surface, and a 180-day return flight. This mission profile was favored by a recent NASA study (Hoffman & Kaplan 1997) and is referred to as the Mars Design Reference Mission (Clément & Bukley 2007).

The latter mission profile (long-stay, conjunction-class mission) is the favored option at this point, both for the Mars Exploration Study Team at NASA (Hoffman & Kaplan 1997) and numerous authors outside NASA (Oberg 1982, Collins 1990, Zubrin 1996). The first Mars crew will most likely comprise six individuals of diverse backgrounds in multiple disciplines. These disciplines will include geology, biology, engineering, and medicine. All crewmembers will be required to have skills in the areas of communications, information technology, navigation, management, and public relations. A significant portion of the return on investment for the Mars missions will be the reports on surface operations.

The human missions to Mars will be significantly more risky and demanding than the Apollo missions. On the plus side, the Mars astronauts will have a higher level of gravity in which to work, approximately one-third of that on Earth (0.38 g), as well as reduced solar particle exposure as compared to the space environment. This is due to presence of a day-night cycle and some protection provided by the Martian atmosphere. Furthermore, astronaut exposure to the harmful effects of *galactic cosmic rays* (GCRs) will be 75% less than in LEO and transit. GCRs come from all directions in interplanetary space. Mars and its atmosphere provide protection from the GCRs, particularly those coming from the other side of the planet. The minus side of the equation is that it is unknown if the reduced gravity environment on Mars will provide an appropriate environment in which the astronauts can at least partially recover from the effects of the prolonged microgravity exposure during the transit phase of the mission.

In the course of the Mars missions, the crew will be exposed to four g-level transitions and two episodes of high g-load exposure. The transitions are from 1 g to 0 g, 0 g to 0.38 g, 0.38 g to 0 g, and finally 0 g to 1 g upon return to Earth. The high-g episodes are experienced during the Mars and Earth aerobraking maneuvers. Currently, astronauts on board the ISS experience periods of high physical demand only once a month during extra-vehicular activities. However, activities on the surface of Mars will impose such demands on a near daily basis. Also, the experience from the Mir and ISS missions thus far indicates that about 50% of crewmembers who have executed long-duration missions are ambulatory with assistance immediately after landing. After a number of hours have passed, nearly 100% are ambulatory, albeit with assistance. We can therefore reasonably conclude from these experiences that only three of the six crewmembers will be ambulatory immediately after landing on Mars.

Many hold the opinion that the human element is the most complex of the Mars mission design. Significant ground-based and specialized flight research will be required in support of a Mars mission because the ISS provides a platform that can only indirectly address the physiological and psychological issues associated with these long-duration exploration missions.

2.3 Ground-Based Facilities

Many ground-based procedures and devices are currently being used to assess the effects of microgravity and partial gravity on astronauts and train them how to perform in these environments, including the following.

a. Parabolic flight in aircraft executing Keplerian trajectories provides a means to expose individuals to short periods of microgravity, lunar gravity and Martian gravity. These maneuvers reproduce the unloading and neurovestibular effects of these sub-terrestrial gravity environments. This procedure is limited, however, because the exposures to reduced gravity are quite brief, lasting approximately 20, 30 and 40 seconds, respectively. The situation is further complicated by the presence of alternating phases of hypergravity.

b. Underwater immersion in neutral buoyancy water tanks also produces the unloading effects of microgravity, lunar and Martian gravity levels by varying the ballast from none, for microgravity, to the appropriate amount to produce the appropriate g-level loading.

c. Virtual environments are used to induce preflight adaptation training.

d. Bed rest facilities simulate body unloading and fluid shift effects of chronic weightlessness when participants are positioned in a six-degree head-down position.

e. Suspension systems employing vertical cables to suspend the major segments of the body simulate partial gravity by relieving some of the weight exerted by the participant (Figure 1-06).

f. Centrifuges and slow rotating rooms in which participants are exposed to hypergravity for various time durations also provide useful data.

3   NEUROSCIENCE

3.1   Definitions

Neuroscience is the study of the brain. In the 19th century, researchers like Benjamin Franklin, Descartes, Gall, Broca, and Darwin investigated the anatomy and structure of the brain. However, a revolution in neuroscience arose when it was realized that the best hope for understanding of the brain came from an interdisciplinary approach. Today, the scientists that investigate the nervous system generally come from different disciplines: medicine, biology, psychology, physics, chemistry, and mathematics.

Understanding how the brain works is a big challenge. To reduce the complexity of the problem, neuroscientists experiment on several levels: molecular, cellular, systems, behavioral, and cognitive. The most elementary level, *molecular neuroscience*, studies the basic mechanisms by which neurons express and respond to molecular signals and how axons form complex connectivity patterns. The next level is *cellular neuroscience*, which studies the mechanisms by which neurons process signals physiologically and electrochemically. Neurons are assembled in networks and circuits performing physiological functions such as vision, sensory integration, voluntary movement, learning, and memory. How these neural circuits analyze sensory information, forms perception of the environment, make decision, and execute movements is the field of *systems neuroscience*. The mechanisms through which behaviors are then generated from these neural systems are studied in *behavioral neuroscience*. Finally, research of the neural mechanisms responsible for the higher level of human mental activity, such as mental imagery, emotion, is called *cognitive neuroscience* (Bear *et al.* 1996).

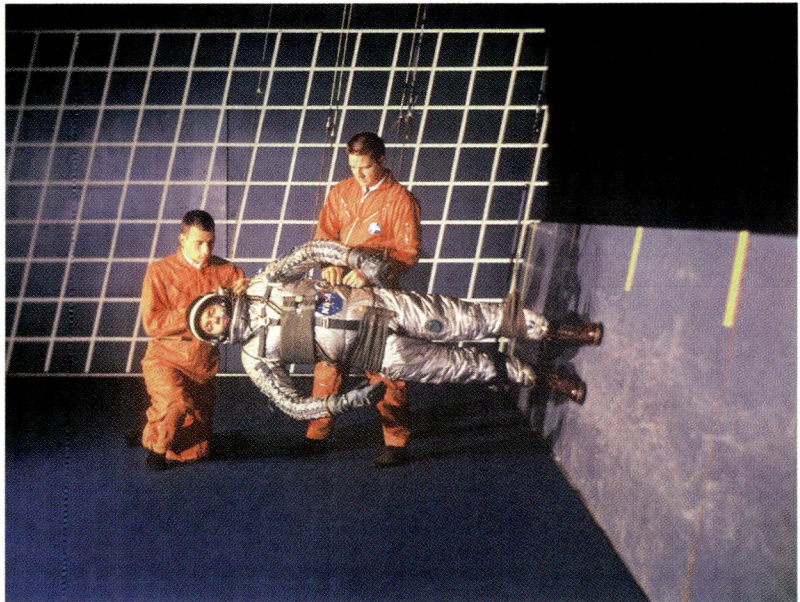

Figure 1-06. Researchers at NASA's Langley Research Center conducted studies in the 1960s to determine the types of difficulties astronauts might encounter walking on the lunar surface. They designed a system of slings to support the weight of a lying subject so that his body axis made an angle of 9.5 deg with the horizontal. When he stood on a platform perpendicular to his body axis, the component of the Earth's gravity forcing him toward the platform was one times the sine of 9.5 deg or approximately one-sixth of the Earth's normal gravity. The force that he would exert on the platform was then the same as if he were standing upright on the lunar surface (Hansen 1995). Photo courtesy of NASA.

The prevention and treatment of brain disorders, such as Alzheimer's disease, cerebral palsy, multiple sclerosis, Parkinson's disease, and depression among others, require an understanding of normal brain function, and this basic understanding is the goal of neuroscience. Similarly, the goal of space neuroscience is the basic understanding of how gravity affects brain functions. It also includes the prevention of medical issues that originates from dysfunction or maladaptation in the nervous system during space flight, such as space motion sickness and spatial disorientation.

3.2   The Human Brain

The human brain is now three times as big as it was two and a half million years ago, which denotes a very rapid progress in evolutionary terms. With each generation, the brain has added about 150,000 nerve cells. The reason for this quick growth dates back to when humans decided to stand up and walk. This freed up their hands, which developed from being rough instruments used to grasp implements and walk to being finely tuned tools that could perform a wide variety of tasks.

The success of these functions depends on all the many units of the brain, i.e., neurons, synapses, and networks, working in harmony. The brain has to adapt to new tasks. Like a child, an astronaut has to learn to move and function in a newly experienced, weightless world. One large part of the brain is devoted to vision and its coordination with other senses for balance, movement, and spatial orientation. This part is undoubtedly most affected by the weightless conditions.

It is commonly admitted that one fifth of the energy generated by the body is used by the brain. Aristotle, the first person to think about the functioning of the brain, believed that it regulated body temperature while the heart controlled the emotions. Now, with powerful experimental techniques such as *functional magnetic resonance imagery* (fMRI), we can actually see what different parts of the human brain become active under different conditions, and how they deal with the processing of sensory information, memory, and cognition.

Such computer-assisted imaging techniques are not yet available on board the International Space Station. The engineers must first solve the issues of shielding for the electromagnetic emissions from these apparatus that could interfere with the onboard avionics. Nevertheless, many important insights can be gained from a vantage point outside the head. From the recording of neuron activity, electrophysiologists can investigate the relationship between a stimulus and its response, and study the adaptation of this response under different conditions. Also, because the brain's activity is reflected in behavior, careful behavioral measurements can inform us about the capabilities and limitations of brain functions.

4   EFFECTS OF GRAVITY ON BRAIN FUNCTIONS

To control the body usefully, the CNS must constantly by aware all the details of the surrounding environment. The body *senses* or *perceives* the environment by the interactions of specialized end organs with some aspect or another of the environment. In common speech, five different senses are usually recognized: sight, hearing, taste, smell, and touch. These senses reach us through special sensory organs, or *receptors*: the eye, ear, tongue, nose, and skin, respectively. There are also sensations arising from organs within the body, from muscles, tendons, ligaments, and joints. These are the *proprioceptive* sensations, by contrast with the *interoceptive* sensations (in Latin, 'received from inside') received from the viscerae, and the five *exteroceptive* sensations (in Latin, 'received from outside') described above. The detection of the head orientation relative to gravity by the specialized vestibular organs in the inner ear can be considered a part of the exteroceptive sensations. In fact, where the proprioceptive senses tell us the position of one part of the body with relation to another, the vestibular sense tells us the position of the body as a whole with respect to its environment, especially with regard to the direction of the pull of gravity.

The stimulation of these sensory end organs gives rise to specific impulses in the nerve endings, which travel through ascending pathways to the brainstem. The lower portions of the brain integrate information from all over the body to coordinate and organize muscular movements to maintain equilibrium and adjust our positions. Although the routine work is done at the low levels and we are not consciously aware of what is going on while we stand, walk, or run, certain information does eventually reach the cerebrum, and through them we remain consciously aware at all times of the relative positions of our body parts. This information is also used to determine the sense of motion of the body relative to the environment and to construct spatial maps or internal representations of our surroundings. How far this sense is dependent on the gravitational force in the long run is the main question addressed by space neuroscientists.

4.1   Gravity-Sensing Receptors

Since the beginning of life on Earth some four billion years ago gravity has remained a fairly stable environmental factor affecting the development of all living organisms. On the one hand, gravity has forced all living organisms to develop a skeleton, from actin to bone, which help to retain form and overcome gravity-enforced size limits. On the other hand, gravity has been used as an appropriate cue for orientation and postural control, because it is constant and has a fixed direction.

Even the most archaic unicellular organisms have been shown to perceive gravity via particular organelles, called *Müller's bodies* or *statoliths*. In an analogous way some fungi and most higher order plants have *amyloplasts* in each cell of the root cap that orient towards the direction of Earth's gravity (Clément & Klenzka 2006). In vertebrates, gravity sensors are not located in each cell of a given tissue, but are found in specialized organs, such as the *vestibular organs* in the inner ear. The vestibular organs, or labyrinths, comprise the non-acoustic portion of the inner ear and consist of three semicircular canals, one utricle, and one saccule in each ear (Figure 1-07).

The semicircular canals are angular accelerometers that sense angular accelerations in any direction as the head is rotated. They do not react to the body's position with respect to gravity, but to a *change* in the body's position. By contrast, the utricle and the saccule constitute multidirectional linear accelerometers, which react to translation and tilt of the body. The saccule and the utricle are tiny sacs filled with fluid and lined along their inner surface with sensory hair cells of various lengths. Overlying the sensory hair cells is a gelatinous matrix, the otoconia, containing solid calcium carbonates crystals. Collectively, these calcium carbonate crystals are called *otoliths*, meaning 'ear stone' in Greek.

Electrophysiological studies have shown that the utricle and saccule, i.e., the otolith organs, respond only to linear acceleration (Fernández & Golberg 1976). Because of the way they are situated within the vestibular apparatus, the saccule is more sensitive to vertical acceleration and the utricle is more sensitive to horizontal acceleration. The otolith organs don't respond to gravito-inertial acceleration perpendicular to the macular plane.

Linear accelerations of the head are sensed by the otoliths. In fact, both head tilt with respect to gravity and linear accelerations resulting from translational motion are sensed by the otoliths; however, these motions are transduced ambiguously. This means that a constant tilt of the head is indistinguishable from a constant translational acceleration. Because the responses required to maintain orientation and equilibrium are different for tilt and translation, the inherent ambiguity in the way these two different motions are transduced must be resolved for normal spatial behavior and postural stability. The fact that we successfully maintain balance, clear vision, and accurate orientation throughout our normal activities, which include both translation and tilt of the head, indicates that this ambiguity is resolved sufficiently for natural behavior.

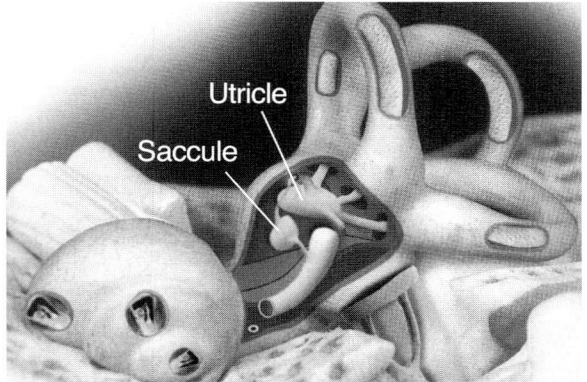

Figure 1-07. The vestibular system located in the inner ear is composed of the three semicircular canals, which are sensitive to angular accelerations, and the otolith organs, the utricle and saccule, which are sensitive to linear accelerations, including head translation and head tilt relative to gravity.

The general explanation for our natural behavior is that over the course of otogenesis, the CNS learns that the gravitational acceleration has a constant magnitude and direction as opposed to accelerations resulting from locomotion generally vary. With this information, the CNS is able to subtract the gravitational component from the total acceleration resulting from head movements, thus ascribing the remaining components of acceleration to self-motion. The neural mechanisms underlying this separation of acceleration components are not yet understood, even thought the scheme has been quite successfully characterized by various mathematical models (Merfeld *et al.* 1993, Bos & Bles 2002, Merfeld *et al.* 2005). Investigators have proposed that the CNS uses internal models to discriminate gravity and inertia based on representations of the body's dynamics and higher order perceptual systems, which presumably use available extra-vestibular information, to supplement the response at high and low frequency (Merfeld *et al.* 1999, Wood 2002).

If the head tilts or translates, the vestibular nuclei in the brainstem will automatically relay the information from the otolith organs via the vestibulo-spinal pathways to activate the muscles necessary to correct our posture. The neural inputs from the inner ear are integrated in the brainstem or cerebellum with those from the peripheral vision, skin, tendons, joint efferents, and gastrointestinal receptors. This integration forms the physiological basis for sensorimotor functions, such as the maintenance of equilibrium, and stabilization of images on the retina as the head and body move, as well as spatial orientation and representation of the environment. On the one hand, basic reflexes are activated by inputs from the vestibular organs. For example, the vestibulo-ocular reflex controls the eye velocity of relative to the head, the vestibulo-collic reflex controls the position of head relative to the trunk, and vestibulo-spinal reflexes control the position of the trunk relative to the ground. On the other hand, our sense of motion and location is determined by processing of vestibular inputs in the hippocampus, thalamus, and cortex cerebral centers.

4.2 Central Processing

During daily activities, the CNS must deal with many effects induced by gravity on the human body. As a primary effect, gravity dictates the law of motion of our body and extremities. For example, to avoid falling, the maintenance of a standing posture requires constant intervention from the neuromuscular system. A simple arm movement needs a different muscle command whether the movement is directed upwards or downwards. The dynamic interactions between our body and the physical environment clearly depend on gravity. Even within our body, gravity plays an important role. For example, fluid shift is strongly influenced by the constant pressure exerted by gravity. Using this property, orthostatic regulation is one the most important functions of the autonomic nervous system.

The regulation and control of body functions are not the only domains where the CNS must interact with gravity. Cerebral functions have also developed during evolution in taking into account the gravity constraints imposed to cognition. For example, the world around us is primarily two-dimensional, i.e., flat, particularly for bipedal creatures such as humans. When massive three-dimensional structures were built, such as skyscrapers, they were designed as multiple layers of bi-directional surfaces. The neuronal processes that allow us to navigate in this world are specialized for the representation of two-dimensional spatial maps. On Earth, we also expect to see objects in a particular orientation. For example, objects on a table will be placed horizontally, such as the silverware, or vertically, such as a vase; people and buildings we see are upright (Figure 1-08); during free-fall, objects accelerate downwards. Using these models predetermined by gravity, we can essentially predict the behavior of objects and people and optimize the performance of cognitive tasks. The success of these tasks can be determinant in survival, such as escaping a predator or avoiding the fall of a brick. In summary, cognitive, motor, and autonomic functions depend on the CNS ability to integrate the gravitational information (Hubbard 1995).

Figure 1-08. This photo was taken in one of San Francisco's steep, hilly streets. The image is tilted to the left so that the road (ground) is shown horizontal. From this point of view, the buildings no longer look horizontal but tilted to the left by the same amount.

Gravity can be sensed by several receptors. The otolith organs in the inner ear measure linear acceleration of gravity. We can sense the gravitational force on our limbs and internal organs. When sitting, we perceive the gravitational force through the tactile contact with the seat. We can also "see" the direction of gravity simply by looking at the horizon, vertical building, trees, and the direction of falling rain. On Earth, it is difficult to separate the influence of these various sensory modalities on neuro-regulative functions, because the manipulation of each of these sources can induce parallel effects on the others. For example, tilting the head generates the perception of a change in the orientation of the head relative to gravity by stimulation of both the otolith organs and neck proprioceptors. Visual stimuli are not only interpreted relative to the body and eyes, but also in relation to the direction of gaze. We also know that objects have a mass and a weight. In weightlessness, we can suppress the reference point provided by the gravitational force and examine separately the effects of the other sensory sources.

Of course, many of these questions can be and have been addressed by well-designed ground-based studies. The addition of mass on the limbs, performance of tasks under water, or experiments in centrifuges are methods that allow changing the gravity constraints. However, each of these conditions is different from weightlessness. Adding a mass on the arm increases the inertia during motion as well as the angular momentum at joint level. Movements under water are also affected by viscosity that resists motion. A scuba diver can come up to the surface without having to figure out where is up or down because gravity is still perceived. Experiments performed in centrifuges are exposed to an increased gravito-inertial force, but the rotational effect and the Coriolis forces can confound these effects during movement. Finally, all these conditions can change the amplitude and direction of gravitational force, but do not remove the gravitational reference from the equation. Weightlessness encountered during space flight is a unique environment that cannot be reproduced on Earth (Gerathewohl 1959).

The physiology of the nervous system is a fascinating mystery that keeps defying our scientific curiosity. Weightlessness provides an opportunity to challenge the CNS in a unique way, which allows investigating new aspects of central processing. Experiments performed in weightlessness help to better understand how the brain uses gravity in normal conditions on Earth. More importantly, knowledge gained from this research helps understand the consequences of an alteration of the gravitational account because of disease, accident, or malformation at birth. Finally, a better understanding of the adaptation of brain functions to a reduced gravity environment will eventually contribute to the success of human exploration missions beyond low Earth orbit.

As will be reviewed in the next chapter of this book, during the early years of manned space flight program, efforts in the life sciences were driven by operational medicine and biomedical support of short-duration space missions. During these missions, no significant problems occurred with regard to neuroscience, and in particular to sensory systems function. However, during subsequent missions, a number of astronauts and cosmonauts reported severe nausea and disorientation. Space neuroscience studies were then initiated to understand the basic etiology of this condition, called *space motion sickness*. Studies were also undertaken to develop tests that would predict susceptibility and enhance development of suitable countermeasures.

Note that in space medicine, a *countermeasure* is a procedure, device, or therapy used to prevent or minimize adverse health and medical events resulting from exposure to short- (< 30 days) or long-duration (> 30 days) space flight. A countermeasure prescription is a direction for a countermeasure including the modality (e.g., hardware device, drug, exercise), duration, intensity, and frequency, as well the physiological monitoring equipment and parameters necessary to gauge the countermeasure effectiveness (see Clément 2005 for review).

Unfortunately, with the exception of drugs, research approaches toward prevention and control of space motion sickness have not led to practical countermeasures. During today's space missions, space motion sickness remains a significant and unpredictable problem. One of the major difficulties in understanding the etiology of space motion sickness has been a lack of understanding of its neuroanatomical and physiological substrates. Only recently has it been realized that attention must be devoted to other anatomical, neurophysiological, and perceptual changes that occur during the process of adaptation to altered gravity, and that may have implications for re-adaptation on return to a gravitational environment. An overview of these changes is given in the following sections. The design and results of the space experiments that investigated these various functions are described in larger details in the following chapters of this book.

4.3   Motor Responses

The properly functioning vestibular system is responsible for a number of reflexes and reactions. The reflexes and reactions are of two types: those involving body posture and muscle tone, and those involving compensatory movement of the eyes. Muscle tone refers to a state of tension of the body musculature, which is maintained without voluntary innervation and which serves to control body posture and the relative positions of the various parts of the body.

4.3.1   Otolith Reflexes

A stimulation of the otolith organs elicits tonic labyrinthine reflexes, righting reflexes, and compensatory eye movements. These reflexes occur in animals even when the cerebrum has been completely removed, and hence are independent of conscious sensations. The best demonstration of these reflexes is the observation that a cat can right itself when falling and land on its feet, even though it was dropped feet up (Figure 1-09). It does this by automatically altering the position of its otolith organs. This in turn brings out movements in the rest of its body, which are designed to bring it online with the new position of the head. Down it comes, feet first every time.

Figure 1-09. During free-fall, cats use their righting reflex to rotate their upper body to face downwards and their lower body follows.

So, the otolith organs directly influence the tonus of the muscles of the neck, trunk, and extremities. *The tonic labyrinthine reflex* consists of flexions and extensions of the legs and arms in response changes in head position with respect to the gravity vector. The neck muscles exert a secondary influence on the muscles of the extremities, the so-called *neck reflex*. The neck reflex is elicited by changes in the position of the head relative to the trunk (Figure 1-10). The *labyrinthine righting reflexes* are related to the tonic reflexes, but refer specifically to those reflexes enabling the individual to restore himself to an upright position after his body has been restrained. These reflexes cease to occur when both labyrinths are destroyed (Fukuda 1982).

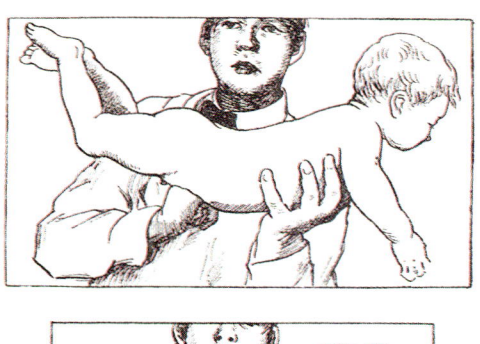
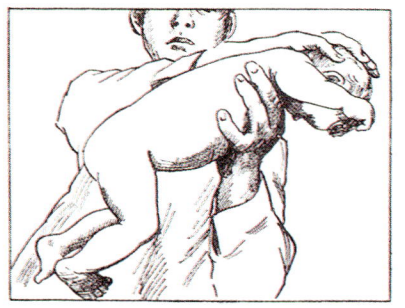
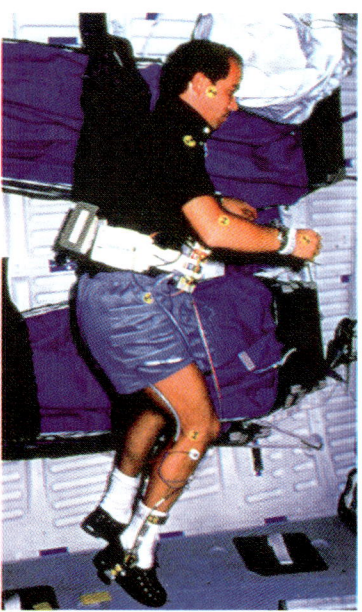

Figure 1-10. When a baby's head is placed horizontally the tonic labyrinthine reflex creates an extensor tone and the baby can extend its arms and legs (upper left panel). As the baby's head is moved forward towards its chest, the extensor tone disappears and the baby curls up into a flexed posture similar to the fetal position (lower left panel). Free-floating astronauts adopt the same flexed posture (right panel). Left panels adapted from Fukuda (1982); photo courtesy of NASA.

The otolith organs also participate in the compensatory eye movements, which serve to maintain a stable image on the retina as the head is moved. For example, the *vestibulo-ocular reflex* (VOR) is a response to the stimulation of the semicircular canals, but the otolith inputs also participate to the VOR during head rotation in pitch or roll. Other compensatory eye movements induced by linear acceleration include *ocular counter-rolling* (OCR) during static roll tilt of the body, *ocular counter-pitching* (OCP) (the doll's eye reflex) during head pitch, and elevations and depressions of the eyes during exposure to decreases and increases in background *gravito-inertial acceleration* (GIA) level in the vertical axis of the upright head, and to linear translations of the body (see Lackner & DiZio 2005 for review).

OCR is the tendency of the normally vertical meridian of the eye to remain vertical as the head is rotated to the right or left about the sagittal axis. As the subject's head rotates about the sagittal axis from upright to 90 deg, his eyes will rotate from their normal position about 6 to 8 deg in the opposite direction about their visual axes, hence a gain of about 7 to 9%. OCR is the only well-defined, entirely involuntary otolith reflex in human. The semicircular canals may contribute to both OCR and OCP during initial movement of the head, but since these reflexes do not decreases significantly with time, their steady-state component can be due only to otolith stimulation.

Compensatory otolith-driven eye movements occur in absence of semicircular canals stimulation, such as during sudden changes in magnitude of gravito-inertial force and during changes in the direction of gravito-inertial force in the sagittal and horizontal (side-to-side) planes, respectively. Therefore, ocular reactions that are dependent on the otoliths and on the processing of otolith information are especially likely to be affected by microgravity. Although OCP has never been tested during space flight, the OCR is absent when astronauts tilt their head in microgravity, and changes in the both vertical VOR and optokinetic reflexes were observed during and after space flight (see Chapter 6).

4.3.2 Posture

The presence of the gravitational forces requires a muscular tone in the extensor muscles that, under the excitatory action of proprioceptive and vestibular signals, allows the maintenance of an erect posture (see Figure 1-10). Consequently, posture is altered to meet the mechanical demands (or the lack of) of weightlessness. Elimination of the need for antigravity postures in microgravity creates a unique context for reinterpretation of sensory inputs and coordination of muscular actions. Studies of postural system adaptation to microgravity are limited, but indicate that there are changes in the interpretation of sensory inputs and in the coordination of muscular actions. Perception of joint and limb position, including the ankle joint, is impaired in microgravity, and naïve subjects assume skewed postures when trying to orient themselves vertically in relation to support surfaces in absence of visual cues. The electrically induced Hoffman reflex, reflecting supraspinal and otolithic control of motoneurons, appears to be depressed in-flight and then enhanced for several days postflight. Returning crewmembers experience difficulties walking and standing with their eyes closed, and in making quick turns. These symptoms occur even after missions of relatively short duration, where changes in muscular strength are limited (Clément 2005). Thus, they cannot be ascribed to skeletal muscle changes, but to an adaptation/re-adaptation of central motor programs to microgravity/Earth gravity.

4.3.3 Locomotion

Considering the strong innate character of human locomotion and the importance of gravity to produce oscillations of the center of mass, a prolonged exposure to weightlessness is not expected to induce large effects on the ability to walk. However, it is known that humans traveling in space develop novel motor strategies inappropriate to the normal Earth environment. Legs are unused except to "perch", either by using foot restraints or by wedging a foot or leg into a convenient spot. Crewmembers learn to push themselves off the floor or walls to get from one place to another. While legs are sometimes used to project the body to another location, the forces are small, brief, and infrequent compared to those generated thousands of times a day in ordinary activity on

Earth. Back and abdominal muscles are unloaded. By contrast, hands and arms are the most-used body segments; they are used for pulling, grasping, and manipulating.

The reductions of force and movement by the major muscles masses directly affect the muscle, tendons, and other supporting tissues, because their normal stress is reduced in microgravity Also, many one-g neuromuscular reflexes (such as the stretch reflex) and coordinated movements are never required or used. The disuse of these strategies is thought to be at the origin of the strong postflight postural disturbances, which are commonly reported after short-duration missions. After six months of weightlessness, returning astronauts are able to walk only two days after their return to Earth. The organization of the gait pattern does not show any major disruption. The coupling between adjacent segments is conserved after the flight, although it tends to decrease when the walk is executed with the eye closed. This indicates that astronauts rely more on visual feedback during postflight walking. Nevertheless, they are unstable and fatigue rapidly for up to several weeks following landing.

4.3.4 Arm Movements

Gravity is an acceleration that not only produces a constant force on a given mass, but also induces dynamic laws of motion of the body with respect to the environment, or of objects with respect to the body. The influence of gravity on the CNS has therefore not only to be looked at in the realm of muscular tone or motor strategies, but also in the full range of the mechanics underlying dynamic behavior and perception.

Astronauts must be able to generate precise movements in weightlessness during space missions. Therefore, they must acquire new internal models in-flight and quickly restore their pre-existing terrestrial models after returning to Earth. How the brain internally represents the dynamics of Earth's gravito-inertial environment has been investigated by examining arm movements. Significant effects have been noted for the limb kinematics (Soechting & Flanders 1989). On Earth, the brain accomplishes arm movements in the vertical plane with different planning processes for movements with or against gravity. The fact that these differences between upward and downward movements persist in the apparent absence of gravity indicates that gravitational forces are anticipated and explicitly used in the feedforward component of the muscle command, rather than being treated as mere disturbances via feedback control.

4.4 Spatial Orientation

4.4.1 Definition

The underpinning of our sense of direction is gravity. As a result of the presence of gravity, we innately perceive six directions along three orthogonal axes. In height, we perceive up and down. We perceive width left and right, and we sense depth in the front and back directions. The anisotropic character of this space is judged by the effort required to move in any given direction. When we are standing on Earth, up and down are distinct and irreversible directions. However, back, front, right and left can be changed in our frame of reference simply by turning around. Hence, when standing in Earth gravity, the number of principle directions is reduced the three: up, down, and horizontal. These correspond or perhaps have given rise to the three basic architectural elements: the ceiling or roof, the floor, and walls.

We are terrestrial creatures. Evolving and living on the Earth has taught us that gravity pulls us downward toward the ground and holds us there. However, when we expose ourselves to alternate environments, such as flying in an aircraft or on a carnival thrill ride, g-forces may be upward or outward and as they are associated with changes in both acceleration magnitude and direction. In these situations, a resultant force is experienced, the effects of which stimulate the vestibular organs and give rise to our recognition of position in space. In the review of orientation, the importance of this will be explained.

Spatial orientation is the perception of attitude and motion in three-dimensional space. It results from the integration of the afferent signals from the vestibular organs with visual inputs from the peripheral retina and proprioceptive and tactile inputs from skin, muscles, and joints. The result of this sensory integration is transmitted to the cerebellum and to higher centers in the thalamus, hippocampus, and cortex. These structures are involved in subjective perception of attitude and motion. The perception involves synthesis and assignment of some meaning to sensory input, taking into account our expectations in the behavioral context, and our prior experience and culture. These areas are also involved in related processes, including learning, adaptation, and habituation.

The threshold of acceleration detectable by the human has been measured by a number of researchers and methods. According to these studies, the subjective perception of motion ranges from 0.002 to 0.01 g. The large variability is due to factors such as differences among subjects, differences in orientation of the subject with respect to the direction of motion, and differences in interpretation of the data, e.g., some subjects determined merely the *presence* or *absence* of motion, whereas others determine the *direction* of motion as well.

Directly related to the otolith organs activity is the perception of the subjective vertical, i.e., how much a subject senses the amplitude of tilt with respect to the gravitational force. This perception can be evaluated by tilting or centrifugating a person and asking him to return to the upright position (subjective postural vertical) or by asking a person to orient a visual line to vertical (subjective visual vertical). Howard (1982) presents an excellent review of the literature on judgment of the postural and subjective vertical. When upright, the subjective visual vertical estimates are quite accurate, but they deteriorate as the subject is tilted away from the vertical. In general, for normal subjects, a vertical reference target will appear inclined in the same direction as the tilt angle for small tilt angles. This illusion is termed the *E-effect*. At a tilt of about 7 deg the E-effect disappears and the target appears to be inclined in a direction opposite to that of the tilt angle. This effect is called the *Aubert-* or *A-effect*. Left-right asymmetries and irregularities in response, especially between successive trials under identical conditions, are the rule rather than the exception.

Whereas, for a given subject, the magnitude of eye movements and rotation motion perception gradually declines with the repetition of angular acceleration, a phenomenon known as *vestibular habituation* (Collins 1973), the same magnitude of eye movements and tilt or translation motion perception is generally observed with the repetition of a simulation of the otolith organs. This difference suggests that habituation or vestibular training does not affect the function of the otolith organs, or if it does, it is not as strongly as for the semicircular canals. Since, theologically speaking, humans are designed to function in a one-g environment, and the otolith organs provide the sensible signals upon which orientation to the vertical is based, logically we would not expect either attenuation in perception or change in dynamics of the otolith organs due to repeated stimulation. By contrast, rotations of the head experienced in uncommon activities would lead to habituation of semicircular canal reflexes. Such habituation has been observed, for example, during activities such as figure skating, gymnastics, and acrobatic flying.

Compared to a "normal" population, astronauts seem to do less error in estimating the subjective vertical (Graybiel *et al.* 1967). Rather than the effects of habituation, the astronaut's superior ability at vertical estimation presumably reflects the high level of training or the highly selective process of choosing astronauts from an already select group of experienced pilots.

4.4.2   Spatial Disorientation

An individual experiencing a false or illusory perception of attitude or motion is said to be *spatially disoriented*. Spatial disorientation may result from the normal reactions of the vestibular system to motion or from visual illusions arising from erroneous interpretation of visual cues. For example, the threshold and dynamic characteristics of the semicircular canals are such that angular accelerations of small magnitude are completely unperceived; the perception of rotations of prolonged duration gradually subsides with time, and at the termination of a prolonged rotation in one direction, an after sensation of rotation in the opposite direction is perceived. The gravity-sensing apparatus, the otolith organs, cannot distinguish between linear acceleration and the acceleration of gravity, and thus their combined vector (the GIA) is interpreted as denoting the vertical.

Spatial disorientation is often referred as *vertigo*. In this context, vertigo has a different meaning than the dictionary definition of a "feeling of dizziness associated with the sensations of rotatory motion of the body or surroundings." Pilots use the term *vertigo* in referring to many types of confusions with respect to attitude and motion during flight that do not correspond to objectively verifiable physical events (Gillingham & Wolfe 1985). The term *disorientation* will be preferably used in this book, because it encompasses the meaning of vertigo as used by the pilots, and avoid the ambiguity in meaning which event causes it. A pilot is disoriented when his sensations of motion and attitude do not correspond to the physical facts, despite his awareness or lack of awareness of the difficulty. This disorientation is also often accompanied by an illusion. For example, a pilot may feel inverted when flying right side up or feel banked to one side when flying straight and level. In these conditions, he is disoriented and suffers the illusions of inversion and "the leans", respectively (Gillingham & Wolfe 1985).

4.4.3   Effects of Space Flight

It is indicated through developmental studies that familiarity with gravity is a learned behavior and is not innate. From infancy, we learn to interact with our environment, including dealing with gravity. By the time an infant has reach four months of age, he or she realizes that it is not possible to roll a ball through an obstacle. In just another month, the infant can discriminate between downward and upward motions. By seven months, sensitivity to gravity is developed. The infant can now sense the appropriate acceleration of a ball rolling up or down an incline. By adulthood, falling objects are judged to move naturally only if they decelerate upward and accelerate downward following a parabolic trajectory. These judgments are based on the visual experience, rather than on mathematical reasoning. Many adults are unable to reason abstractly about such motion (Bower 1992).

A perceptual experience commonly reported by both astronauts and cosmonauts is that of feeling inverted (or experiencing the spacecraft as inverted), even though they are in a familiar, "visually upright" orientation in the cabin. These illusions are relatively persistent at first, even with eyes closed, but usually disappear after a few days in orbit. In addition, the astronauts' perception of the orientation of their bodies is strongly influenced by the presence of familiar objects in their environment such as when catching sight of a fellow astronaut (Figure 1-11). Unlike the inversion illusion, these are easily reversed, under a certain amount of cognitive control, and can occur throughout the mission.

Figure 1-11. Group photo of the Shuttle mission STS-121 crew. It is difficult to recognize some faces when they are upside-down. Photo courtesy of NASA.

In general, at low levels of stimulation of the vestibular system, the conflict between visual and vestibular orientation cues is resolved strongly in favor of visual cues. At increasing levels of vestibular stimulation the resolution in conflict begins to shift toward vestibular cues until at very high levels of stimulation the vestibular cues overwhelm the visual cues. Cues that determine spatial orientation vary during the initial phases of adaptation to weightlessness. Some individuals become strongly dependent on vision as a substitute for the absence of perceived gravity, orienting themselves with respect to familiar vertical references. Others are more "body oriented" and align their sense of the vertical with their longitudinal body axis. The latter group of individuals does not become as disoriented when working in unusual attitudes relative to their external environment or during motion where visual cues for vertical orientation are absent. Postflight alterations of spatial orientation include illusions such as a sense of disorientation when making pitch or roll head movements, changes in linear acceleration thresholds, and unusually strong visual influences on orientation. In addition, during re-entry and shortly thereafter, tilting motions of the head may cause a sense of sudden linear translation in the opposite direction (Young 1984).

Hand-eye coordination, visuo-ocular control, and other bodily movements can be severely disrupted upon astronauts' arrival in Earth orbit. This is because their well-established perceptual-motor programs for use in terrestrial gravity are no longer appropriate for the microgravity environment. A second type of perceptual-motor problems includes mistakes such as allowing hand-held objects, like tools or utensils, to drift away because they have inadvertently imparted motion on the objects. If the astronaut turns his head or other body part quickly to one side, this action may cause the body to rotate uncontrollably in the opposite direction. In the first few minutes in a microgravity environment, an astronaut will reflexively reach to catch an object of which he or she has let go. They are immediately reminded that weightless objects don't fall! An important observation is that with successive trips into space, astronauts report fewer initial problems and more rapid adaptation, evidence of what has been referred to, in general, as *dual adaptation* (Welch *et al.* 1993).

4.5 Cognition

Astronauts and spacecraft in orbital flight are actually continuously free falling. However, the perception is not that of falling (Lackner 1992b). This means that otolith signals *per se* do not determine whether one experiences falling but that additional sensory and cognitive factors must contribute.

Besides its involvement in reflex behavior the vestibular system also plays a role in the internal representation of space and navigation, i.e., the knowledge of directional heading and location in the environment. To determine the position of an object in three-dimensional space the brain must combine the sensory information related to this object, i.e., visual, auditory, olfactory, and tactile, with an internal representation of the body location in space. To form an accurate internal representation of visual space, the brain must also accurately account for movements of the eyes, head, or body. Therefore, both sensory and motor inputs are utilized. Updating of internal representations in response to these movements is especially important when remembering spatial information, such as the location of an object, since the brain must rely on non-visual extra-retinal signals to compensate for self-generated movements.

On Earth, gravity provides a reference that influences how we recognize objects and perceive their shape and orientation. It also influences our expectations of how objects behave when thrown or dropped. The orientation of objects and surfaces in turn influences our own perceived orientation. Gravity also helps the CNS align the various frames of reference used in movement control. An *allocentric* reference frame refers to the orientation of an object external to the observer (e.g., gravity, horizon, buildings). An *egocentric* reference frame refers to locations that are represented with respect to spatial relations of the subject's body (e.g., view-centered, arm-centered, hand-centered). All these processes are fundamentally altered in weightlessness, as evidenced by the visual illusions frequently reported in orbit by astronauts, and also by the surprisingly long-lasting aftereffects seen, particularly after prolonged space flight.

4.5.1 Egocentric Localization

Stimulation of the fovea of the eye produces the sensation that the object causing the stimulation is located in the direction toward which the fovea is pointing. If the fovea is stimulated with the eye in the primary position, i.e., looking straight ahead, the stimulus object will be perceived as being straight ahead of the individual. Similarly, if the fovea is stimulated with the eye turned to the right, the object will appear to be to the right of straight ahead. The same applies if the eye is turned to the left, up, down, or obliquely. In a similar fashion, objects stimulating peripheral areas of the retina are localized in space relative to the foveal direction in accordance with the anatomical positions relative to the fovea of the retinal areas stimulated.

The position in space where one localizes an object is thus determined by the position relative to the fovea of the retinal area stimulated by the object and by information concerning the position of the eye itself. Now consider an individual observing objects in his surroundings. As he directs his gaze from one object to another they appear stationary even though their retinal images move. Clearly, the brain changes some spatial reference to account for the eye movements; otherwise the movements of the retinal images would be interpreted as movements of the objects themselves. The classical experiment of Helmoltz (1925) showed that if the eye is passively moved by physically pushing on the eyeball, objects in the visual field appear to move about. In this situation, movement of the retinal images is interpreted as movement of the objects themselves. Apparently, the brain has not adjusted the spatial reference to account for the passive eye movement. In fact, the space reference changes as a function of the voluntary command to the oculomotor system, and does so regardless of whether or not the command is executed by the oculomotor system (Von Holtz & Mittelstaedt 1950).

A key concept in the field of neuro-motor control is that of defining the *reference frames* used by the CNS to interpret sensory information and to control movements. At the level of individual sensors and actuators, the coordinate systems employed are well defined. It is not in the coordinate system of an individual receptor, but rather, in examining the coordination of sensory and motor activity that the question of reference frames becomes interesting. How does the CNS combine information from a variety of different sensors to produce an appropriate motor response? In what coordinate system is the combined information expressed?

Gravity plays a potential role in calibrating different intrinsic reference frames. The CNS can sense the gravitational force, and thus define a vertical direction, in a variety of ways. The otoliths of the inner ear permit the measurement of gravity in a head-centered reference frame, the force of gravity acting on an outstretched limb permits the perception of vertical in an arm-centered coordinate frame, while visual cues in the environment (walls, falling objects) can be used to define vertical in retino-topic coordinates. The constant direction of gravity can potentially be used to align these different reference frames. To accurately predict the progression of moving targets, the CNS must take into account the effect of gravity on the movement dynamics. Gravity may therefore also be incorporated into the internal models used by the CNS to control movements. The questions to be posed in general concerning the various internal models used by the CNS are, for example: How are these models acquired, and how do they adapt to novel environments? What sensory modalities are involved in the creation of a given model? How accurate are these models, and of what complexity?

4.5.2 Motor Imagery

Another interesting model for studying cognitive representations of gravity is that of motor imagery (covert actions). When one imagines a movement it is presumed that the brain evokes the same motor representation that would be used to actually generate the movement. By studying how overt and covert motor actions are influenced by microgravity one can infer the effectiveness of internal models of gravity on our ability to predict and control our motor actions in one-g and in zero-g. The results show that long exposure to microgravity influences to the same extent the execution of both overt and covert motor actions. This suggests that CNS internally represents the dynamics features of Earth's gravito-inertial force. While the effects of microgravity on the overt actions could be more easily anticipated and attributed either to biomechanical (loss of muscular mass and bone density) or to control factors (re-adaptation of the existing internal models of gravity), its parallel effects on covert actions further emphasize the idea that internal models of Earth's environment dynamics are also fundamental for high level cognitive functions such as motor planning and motor imagery.

4.5.3 Anticipation

Motor activity often needs to be precisely coordinated with motion and the environment. For example, when we duck the head to avoid collision with a tree branch while walking in the woods, or when we catch a ball at the game. According to the psychologist David Lee, animals and humans use first order approximation, based on distance and speed, to evaluate when a moving object will arrive at a given location. The ability to anticipate and predict is one of the CNS basic functions. When we catch a ball, the brain does not wait for it to touch the hand before stimulating arm flexor muscle contraction to compensate for the impact. About one third of a second before impact, the brain elicits just the right amount of contraction to counteract the force exerted, which itself depends on the weight of the object combined with the acceleration of its fall. Lacquaniti & Maioli (1989) have proposed that the CNS uses an *internal model of gravity* to anticipate the ball motion. This hypothesis is supported by a space experiment on board the Space Shuttle where astronauts were catching a ball launched from above their head (Figure 1-12).

In weightlessness they initiated their arm motion a little earlier, as if the CNS calculated the moment of impact still taking into account gravity. The experiment led to the conclusion that the brain works by anticipating the effects of gravity on the ball rather than by making direct measurements of its acceleration (McIntyre *et al.* 2001). This anticipation ability remains even in conditions of weightlessness. It is still unknown why this anticipation of gravity in present although gravity is not perceived by the sensory receptors any more, how long does this anticipation persists during space flight, and how soon it normalizes when returning in a gravitational environment such as the Moon or Mars.

Figure 1-12. In this experiment during the Neurolab STS-90 mission a ball was thrown at sitting crewmembers. On Earth, the ball would accelerate downwards, but in weightlessness the ball was moving at a constant velocity. The trajectory of the subject's arm and the activity of his forearm muscles were recorded as he was trying to catch the ball. Photo courtesy of NASA.

4.5.4 Internal Representation

An internal representation of the real world is necessary for the planning of orienting behavior and the interaction with objects in our environment. This internal representation is used during static perception, motion, and during cognitive processes such as reasoning, learning, and memory. Both sensory input and motor output signals, through a sensorimotor integration, are combined to provide an internal estimate of the state of both the environment and one's own body. For example, reconstruction of the motion of the head in space requires a combination of visual and vestibular measurement of head dynamics parameters. The visual system and the vestibular system alone are not able to provide a completely adequate estimate of head motion parameters, the former measuring head velocity and the latter head acceleration. It has also been recognized that the CNS is equipped with neural *cognitive maps* into which the various senses project themselves. This topographic central representation of visual, tactile, and auditory spaces is now rather well described at anatomical and even neurophysiological levels (Selemon & Goldman-Rakic 1985).

Neural maps are established during development. An interesting question is whether neural space maps can develop in microgravity. Sensory stimulation is implicated in the initial specification of the connections and physiological properties of the neurons that constitutes the maps. The neural maps must have appropriate information regarding the location of the head in the gravitational field, so it follows that the vestibular system must play a key role in the organization of these maps. The absence of otolithic gravitational input in weightlessness may therefore impact the development and maintenance of the components of the spatial maps and neural pathways that play a role in sensing position in three-dimensional space.

4.5.5 Navigation

Most of us are able to navigate the corridors of a familiar house in the dark, recognizing features by touch and texture, by smell and sound. We are doing so because we have an intimate understanding of the topography of the world around us. Our ability to navigate our environment results from the construction of the spatial maps mentioned earlier. Theories vary on the nature of this internal representation. Some suggests that the cognitive map could be egocentric, or centered on the self, that objects in the map would be located relative to the head, and the map representation would change as the self moves through the landscape. Other theories propose an allocentric map, one based on an external frame of reference. In an allocentric view, objects have a fixed location on the map independent of the observer, and self-motion does not affect the map. In this view, the self appears something like a Monopoly token, marking the position and direction of movement of the body relative to its environment.

Both egocentric and allocentric views of the cognitive map allow to test theories about how the brain represents the external world. Scientists recognize that any kind of cognitive map would have certain characteristics that would show up in the neurons that create the map. An allocentric map would be strongly correlated with the physical environment; if you walked into the same room twice, you would expect that your cognitive map of the room would be largely the same both times. If the cognitive map is egocentric, it should change as the position of the self in the environment changes. The testable difference is that an allocentric map would be stable, but an egocentric map would be constantly changing. Other finer points of the cognitive mapping hypothesis are being tested, including the answer to questions such as: Are the maps based on

Cartesian or polar coordinates? Are they measured in units of footsteps, or the time it takes to traverse spaces?

How are the maps built? One hypothesis is that we gain information about our external environment from perceptual experience by navigating through and interacting directly with geographic space, as well as through maps, language, photographs, and other communication media. With all these different experiences, we identify landmarks, such as a word, or a building or a line on a map, that trigger our internal knowledge representation and generate appropriate responses (Piaget 1955).

This spatial memory is robust, and the brain structures responsible for this sense of the world, including the hippocampus in the temporal lobe towards the center of the brain, do more than simply build maps from sensory information. New evidence suggests that our sense of space is an integral part of who we are, and how we relate to the world. In fact, results show that we don't actually use spatial visualization, i.e., visualize the layout of spaces in a mental map as a principal way to help us get around. Rather, it seems that we generate abstract representations of space by paying attention to all of the events that happen, linking them together into short episodes, and then beginning to make connections where the episodes overlap. According to this *memory space theory*, all of this linking and associating is a good way to build an understanding of spatial relationships (Eichenbaum *et al.* 1999). However, although much is known about human terrestrial navigation in two dimensions, little is known about navigation abilities in three dimensions, when body orientation is unconstrained by gravity.

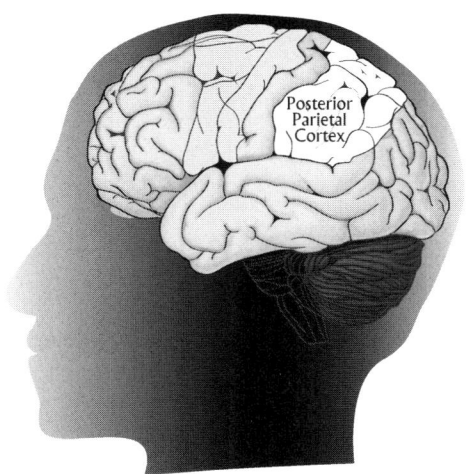

Figure 1-13. Illustration of the human brain showing the location of the posterior parietal cortex. The temporo-parietal cortex and the temporo-insular cortex contain the primary areas through which vestibular and somatosensory sensations are felt. Through their complex network with the visual and auditory areas, these areas play a key role in sensory information processing and spatial orientation.

4.5.6 Vestibular Cortex

During the last 15 years evidence from brain activation studies using *positron emission tomography* (PET) or *functional magnetic resonance imagery* (fMRI) with vestibular stimulation suggested the presence of vestibular cortical areas in humans. Central vestibular projections could be delineated predominantly in the temporo-insular and temporo-parietal cortex in both human hemispheres (Figure 1-13). The same regions also receive visual and somatosensory projections (Bottini *et al.* 1994). Evidence for an involvement of the temporo-parietal cortex in processing vestibular information also derives from electrical stimulation studies carried out directly on the human cortex (Kahane *et al.* 2003).

As mentioned above, neurophysiological findings in monkeys as well as functional imaging results in humans have revealed evidence that our brain uses internal maps of the visual environment, in which the topographical positions of objects reflect their head- and trunk-centered as well as world-centered position in space instead of the retinotopic position of their images. The temporo-insular and temporo-parietal cortex areas are significant sites for the neural integration of multimodal sensory inputs, i.e., vestibular, auditory, neck proprioceptive, visual, and olfactory, into such higher order spatial coordinate systems. The multimodal neurons of this region seem to play an essential role in the spatial encoding of the surrounding space with reference to our body position. They provide us with redundant information about the position and motion of our body relative to external space (Karnath & Dietrich 2006).

Recent functional imaging studies have suggested that a few areas of the human right posterior parietal cortex are important for the processing of head and body orientation in space. Interestingly, these areas seem to correspond to anatomical locations that can provoke spatial neglect in case of their lesion, i.e., lead to a spontaneous bias of eyes and head towards the right, and neglect of information located on the left. Vestibular, auditory, neck proprioceptive, or visual stimulation that reach the multisensory vestibular cortex have compensatory effects on the clinical signs of patients with spatial neglect (Rossetti & Rode 2002).

The vestibular cortex intimately interacts with the visual cortex to match the three-dimensional orientation maps used for the perception of verticality, and mediates self-motion perception by means of a reciprocal inhibitory visual-vestibular interaction. This mechanism of an inhibitory interaction allows a shift of the dominant sensorial weight during self-motion perception from one sensory modality (visual or vestibular) to the other, depending on which mode of stimulation prevails: body acceleration (vestibular input) or constant velocity motion (visual input). These sensory interactions may be at the origin of sensory conflict mechanism or mismatch between expected and actual sensory input (Brandt & Dietrich 1999).

4.6   Integrative Physiology

The weightlessness environment of space flight affects the physiological functioning of major body systems. After insertion into weightlessness there is an immediate redistribution of body fluids due to the changed hydrostatic pressure gradients along the longitudinal body axis. As seen above, the functioning of the vestibular system is disturbed. There is also atrophy in the antigravity muscle and loss of minerals in weight-bearing bone. Other effects of weightlessness include circadian rhythm-related problems involving sleep and performance, and immune-related problems involving infections and immunodeficiency (see Clément 2005 for review). Changes in one of these systems have impact on other systems. For example, the changes in muscle mass could alter proprioceptive signals and the sense of effort. Recent research has demonstrated that information from the vestibular system also influences heart rate, blood pressure, immune responses, circadian rhythms, and arousal. Accordingly, any dysfunction of the vestibular system can potentially induce a number of symptoms including spatial disorientation, postural instability and vertigo, often accompanied by vegetative symptoms such as nausea (Yates 1992). It can also involve psychogenic anxiety or panic attacks (Highstein *et al.* 2004).

Central neural circuits have been identified which enable the integration of vestibular and autonomic information (Balaban & Porter 1998). In particular, stimulation of the vestibular nerve, either through natural stimulation or via selective stimulation of vestibular afferents, is known to produce large effects on sympathetic neural reflexes and blood pressure (Biaggioni *et al.* 1998). Bilateral transection of the vestibular nerves in cats compromises the maintenance of blood pressure during changes in body posture (Doba & Reis 1974). Yates *et al.* (2000) have recently demonstrated that this impaired blood pressure response to changes in posture following vestibular lesions diminishes over weeks, and is less severe when visual and somatosensory cues are available that can provide compensatory information regarding body position in space. While baroreceptors and other cardiovascular orthostatic mechanisms rely on feedback control, the vestibular system likely contributes to the maintenance of stable blood pressure in upright animals via feed forward information regarding changes in body posture (Yates & Miller 1996). The interaction of vestibulo-sympathetic and baroreflexes appears to be additive in humans as well. Natural stimulation of the otoliths during body translations or dynamic tilts results in rapid changes in sympathetic nerve activity and changes in blood pressure (Wood *et al.* 2000, Yates *et al.* 1999).

It has been recently hypothesized that plastic changes in the vestibular otolith processing may be a contributing factor in *orthostatic intolerance*, i.e., the inability to remain in an upright posture due to inadequate blood perfusion to the brain. Approximately two thirds of astronauts experience orthostatic intolerance postflight upon return to normal gravity (Buckey *et al.* 1996). Recent results suggest that orthostatic intolerant astronauts release less of the neurotransmitter norepinephrine when they stand up on landing day than less susceptible astronauts, although it is still being synthesized and available for release (Meck *et al.* 2004). It seems plausible that changes in the central processing of otolith signals following exposure to microgravity result in impaired vestibulo-sympathetic reflexes, and therefore may contribute to postflight orthostatic intolerance.

4.7   Conclusion

Living systems have developed an amazingly refined capability to integrate the inputs from the senses to coordinate global system responses and finally produce a unified whole-body response to these inputs. For example, inputs such as light, sound, touch, attitude, and body motion drive the neck and eye muscles to orient the head toward a fast approaching object and then signal the muscles of the limbs, trunk, and neck to avoid impact. In microgravity, integrating the sensory inputs and coordinating motor responses is challenged. Ambiguities and changes in how the input information is processed can lead to potential errors in perception, which affects spatial orientation. The incorrect impression of self-motion and reflex errors can lead to dysfunctional consequences, such as space motion sickness in orbit and impaired balance after return to Earth.

Because the space around us is represented in the parietal cortex in maps that encode locations and objects of interest in several egocentric reference frames, exposure to microgravity alters the cognitive strategies used in spatially directed tasks, such as navigation and the mental representation of three-dimensional space. Transformations from the coordinates of the receptor surfaces, like the retina or the skin, to the coordinates of the effectors, e.g., the eyes, head, or hands, are executed by both bottom-up and top-down mechanisms. For example, the transformation can be accomplished by dynamic updating of spatial representations in conjunction with voluntary movements.

However, attention, adaptation, training, or experience can also modulate these coordinate transformations (Colby & Goldberg 1999).

Spatial disorientation in microgravity is initially disturbed by the absence of usable graviceptor information from the otolithic organs. In orbit, the reliance on somesthetic and inertial cues shows an apparent decline. Instead, the astronauts seem to use their internal reference frame. A period of reinterpretation of inertial cues to spatial orientation is similarly shown during re-adaptation to one-g. These results are discussed in terms of an internal model representation of body orientation, with time-varying weights applied to extrinsic and intrinsic signals (Young *et al.* 1996).

Other in-flight environmental factors like sleep disturbances and orthostatic intolerance may affect the results of space neuroscience experiments. For example, there is no hydrostatic pressure in a weightless environment. As a result, immediately after insertion into microgravity, a substantial rostral redistribution of body fluids commences. This fluid redistribution causes a decrease in lower-body mass, especially in the legs, and an increase in blood and lymph perfusion in the upper body. Whether this affects muscle viscosity and compliance is not known. In addition, muscle mass and strength both diminish as a result of the decreased load demands with continued exposure to weightlessness. Muscle fiber types may also shift toward fast, fatigable response types. Moreover, approximately 70% of astronauts experience space motion sickness to some degree during the first three or four flight days. The anti-motion sickness drugs that are commonly used to combat sickness, such as scopolamine and promethazine, have soporific effects. These factors, coupled with the many tests involved, can lead to chronic fatigue (Lackner & DiZio 2005).

It is therefore somewhat artificial to separate the in-flight effects of hypogravity from other space flight effects (Aerospace Medicine Advisory Committee 1992). Adaptation to weightlessness involves not just adaptation of vestibularly mediated reflexive and orientation effects, but also accommodation of the entire postural and muscular control system of the body to a radically different force environment. This means that controlling body-relative movements, object manipulation, object use, and body movements relative to the environment requires a remapping of motor commands to the muscles that effect the desired actions (Lackner & DiZio 2005).

Evidently, the human body must be investigated as a whole, and not just as the sum of body parts. All body functions are connected and interact with each other. This is the challenge of *integrative physiology* (Boyd & Noble 1993). In a recent paper in the review *Nature*, Ronald White (2001) presented this challenge as follows: "Body parts will not travel on exploration missions. Instead, the individual space traveler's body must be viewed realistically, with all parts connected and fully interacting. Development and use of such an integrative approach must capitalize on the investments that have been and continue to be made in molecular biology and on the new and emerging capabilities in computing, information storage, modeling, and fast, parallel processing that characterize today's technology. This will not be easy; the problems and challenges that must be faced are many and great." These problems and challenges are described in greater details in the following chapters.

Chapter 2

HISTORY OF SPACE NEUROSCIENCE

This chapter provides a brief history of space flight, with an emphasis on the role of life sciences in the space program. A detailed table including all the neuroscience experiments by all countries from Vostok-3 (August 1962) to the ISS Expedition-15 on board the International Space Station (June-October 2007) completes this overview.

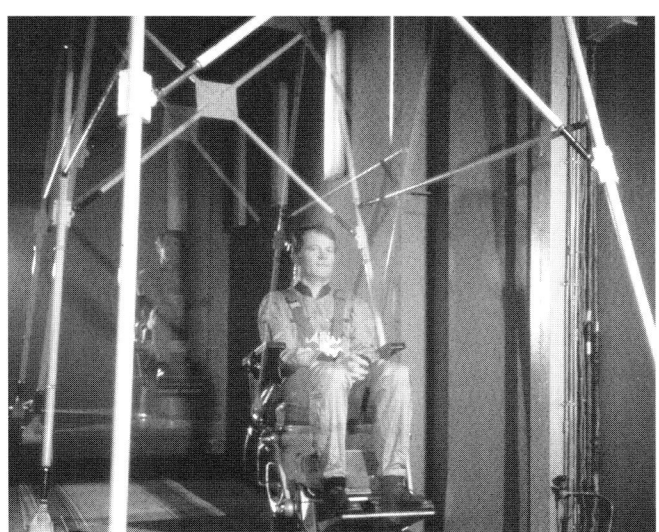

Figure 2-01. The Khilov's swing test was used in Star City (near Moscow, Russia) for selection and training of cosmonauts. Here, a French cosmonaut sat in a chair suspended from the top of a four-post structure. The fore-aft translation of the swing generated linear accelerations that stimulated the otolith organs of his vestibular system. Photo courtesy of CNES.

1   A BRIEF HISTORY OF HUMAN SPACE FLIGHT

To date, astronauts from more than thirty different countries have flown in space and countless more have participated in some capacity with space research. However, only three countries, the United States, Russia, and China, possess the means to launch humans into orbit. The launch of the first living creature on Sputnik-2 on November 3, 1957, marked the beginning of a rich history of unique scientific and technological achievements in space life sciences that has spanned more than fifty years to date.

1.1   The Soviet and Russian Space Program

The Soviet Union initiated the space age with the launch of Sputnik-1 and quickly followed this remarkable achievement by launching a dog, named *Laika,* on board Sputnik-2. The *Sputnik* program (1957-1960) was followed by the *Vostok* program (1961-1963), which after several unmanned sub-orbital and orbital flights launched the first human, Yuri A. Gagarin, into Earth orbit on board Vostok-1 on April 12, 1961. Vostok was followed by the *Voskhod* (1964-1965) flights, an interim program designed to prepare for the more mature *Soyuz* flights (1967-present). The early Soyuz

G. Clément, M.F. Reschke, *Neuroscience in Space.*
DOI: 10.1007/978-0-387-78950-7_2, © Springer Science+Business Media, LLC 2008

flights were designed more with the aspirations of circumlunar and Moon landings, but were quickly adapted to support the Soviet Union's space station programs.

Almaz was the Soviet's first station program scheduled for use in low Earth orbit, and was intended more for military reconnaissance than research. When it became clear that the intended *Proton* launch vehicle could not be man-rated, it was decided to use the *Soyuz* spacecraft as a crew transport vehicle. The modified space station was called *Salyut* (1971-1986). Subsequent Almaz stations were also called Salyut in an attempt to conceal the existence of two separate space station programs. Salyut-1 was launched on April 19, 1971, and became a major step in developing a platform that would help establish a continued human presence in space. Salyut-7 was followed by the *Mir* space station (1986-2001), which was launched on February 19, 1986 (Figure 2-02). Mir was never a static platform, but continued to evolve throughout its lifespan, as a true permanently inhabited space station. Before the Mir station was forced into the Earth's atmosphere, its inhabitants watched the dissolution of the Soviet Union, the formation of the new Russian Republic, and the establishment of cooperative agreements between Russia and the United States, allowing U.S. astronauts to serve as crewmembers along side Russian cosmonauts. The *NASA-Mir* (1994-1995) and *Shuttle-Mir* (1995-1998) programs represented the final scientific endeavors on board Mir, and paved the way for future cooperation on board the *International Space Station* (1998-present).

Figure 2-02. This photograph taken from a Soyuz vehicle shows the Space Shuttle docked to the Russian Mir station. The Shuttle-Mir program consisted of seven Space Shuttle missions to Mir and 1000+ days in space for U.S. astronauts on board Mir between 1994 and 1998. Photo courtesy of NASA.

1.2   The United States Space Program

The *National Space and Aeronautics Administration* (NASA) was created on October 1, 1958 in response to the Soviet Union's launch of Sputnik-1, and charged by the President of the United States, Dwight Eisenhower, with launching a person into space in an environment that would allow effective performance, and to recover that person safely. Project *Mercury* (1958-1963) was the result of that charge initiated by Eisenhower. All together there were two sub-orbital and four orbital missions, the longest lasted for 22 orbits around the Earth (Swenson *et al.* 1966).

Planning for the *Gemini* program (1961-1966) began in May of 1961 even before the first Mercury flight was complete. One of the primary purposes of the Gemini flights was to demonstrate the feasibility of "long duration" space flight. [2] There were a total of twelve Gemini flights, all leading toward the singular idea of putting a man on the Moon and returning him safely home.

With this goal in mind, the *Apollo* program (1967-1972) was singular and straight forward. The previous Mercury and Gemini programs identified no medical or physiological problems that would prevent missions with durations of two weeks or longer (Link 1965). Nevertheless, Apollo was supported by NASA's largest biomedical effort to date, and for the first time a number of significant biomedical findings were identified. These included vestibular disturbances, lower than expected food consumption (most likely attributable to the presence of vestibular disturbances), dehydration and weight loss, decreased postflight orthostatic tolerance, decreased exercise tolerance, recording of postflight cardiac arrhythmias, and a decreased red cell mass and plasma volume (Parker & Jones 1975). Unlike the Soviet program where Titov experienced motion sickness on board Vostok-2, no U.S. astronaut had experienced (or perhaps reported) this malady prior to the Apollo flights.

The *Skylab* program (1973-1974) represented a complete departure in direction. It offered the Unites States the first opportunity to explore the problems of habitability and biology associated with exposure to microgravity over extended periods of time. Skylab was comprised of four separate flights. Skylab-1 placed the orbiting laboratory into space (comprised of the S-IVB stage of a Saturn V booster rocket), and was equipped to house three astronauts for an uninterrupted period of at least three months. Skylab flights 2, 3, and 4 kept crews aloft for 28, 59, and 84 days, respectively. The extended duration of these flights meant that scientists could study and evaluate physiological responses, including long-term adaptation, to microgravity. A secondary feature of Skylab was the volume of the orbital workshop. For the first time, astronauts were free to move about unlike any time before (Figure 2-03). This freedom of movement was instrumental in attaining adaptation levels that were well established (Johnston & Dietlein 1977). Skylab was also the first flight that provided for a complex set of vestibular experiments to be flown (Graybiel *et al.* 1974) (see Figure 7-04).

Figure 2-03. Drawing of the Skylab workshop showing the Orbital Module Laboratory *(with the "transparent" walls, right) and the Apollo crew return vehicle (left). Photo courtesy of NASA.*

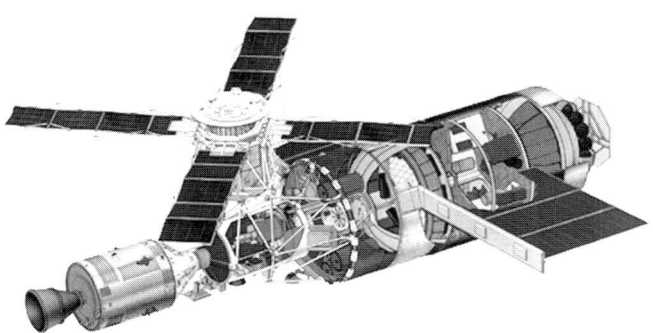

---

[2] "There was concern, even outright fear in the medical community at subjecting the human body to eight days in zero-g. [...] Jim McDivitt and Ed White came back from Gemini 4 visibly tired and drawn, and that one was just four days. [...] [Scientists feared that] the guys might lose the ability to swallow. Air and pressure problems could lead to 'space madness', posed one scientist who feared crew psychosis from oxygen-starved brains. [...] Doctors have always been a pilot's worst enemy." Conrad & Klausner (2005), p. 141.

After a lengthy hiatus NASA participated in the *Apollo-Soyuz Test Project* (ASTP, 1975). Unlike other flights, ASTP was a joint program between the United States and the Soviet Union, whose objectives were primarily political. For the record, ASTP was to test systems for rendezvous and docking that might be useful should the need for an international space rescue ever be needed. Due to an incorrect valve setting during re-entry, most of ASTP postflight science was lost. During descent of the Apollo command module, after nine days in orbit, the United States crew was exposed to toxic gases (nitrogen tetroxide) that entered the command module through a cabin pressure relief valve that had mistakenly been left open in the landing preparation sequence during an inadvertent firing of the reaction control system. This incident is notable only because it was direct evidence of potential effects of space flight on neurological function (Nicogossian 1977).

ASTP was followed by the *Space Shuttle* (or *Space Transportation System*, STS) program (1981-present). The first launch of the Space Shuttle occurred on April 12, 1981, and was uniquely different than previous programs for several reasons:

  a.   It employed a reusable Orbiter.
  b.   Re-entry required the crew to pilot the craft to an un-powered landing.
  c.   The Space Shuttle was the first U.S. spacecraft having a standard sea-level atmospheric pressure and gas mixture (Mercury, Gemini and Apollo operated at 0.33 atmospheres with 100% oxygen. Skylab also operated at 0.33 atmospheres with 70% oxygen and 30% nitrogen).
  d.   The Space Shuttle provided the ability to fly dedicated Spacelab modules where significant science investigations could be conducted in microgravity, opening opportunities for investigators around the world to participate in the United States' space flight program (Nicogossian *et al.* 1994).

The Space Shuttle has been instrumental in NASA's transition to the *International Space Station* (ISS). In its infancy, the ISS is a natural progression from the Russian Mir station to a platform, that once completed, will host the space-faring nations of the world in living and working on board the most complex structure ever assembled in orbit (Figure 2-04).

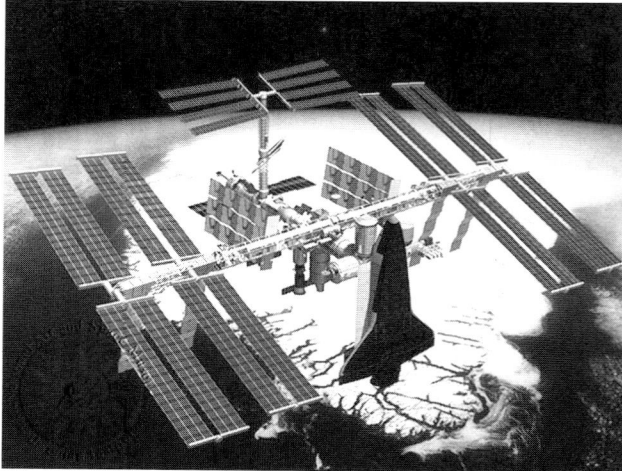

Figure 2-04. In this computer-generated representation, a Space Shuttle is docked to a completed and fully opera-tional International Space Station *(ISS). The ISS will be comprised of scientific modules from the U.S., Europe, Canada, Japan, and Russia. Photo courtesy of NASA.*

1.3   Additional Human Space Programs

While the United States and Russia have dominated space flight, there have been multiple nations from around the world who have participated in various human flight programs primarily through cooperative agreements with either the U.S. or Russia.

Specifically, the *European Space Agency* (ESA), founded in 1975, has been a major contributor to space based research. ESA has participated in multiple flights including *Spacelab* missions 1, 2, and 3, Spacelab D1 and D2 (for *Deutsch*) missions, Spacelab Life Sciences and International Microgravity Laboratory missions, and several missions to the Mir space station. ESA has also developed the *Columbus* research module of the ISS. In addition to those projects sponsored by ESA, individual ESA member states have maintained space flight programs specific to their country. In particular France, Germany, the United Kingdom, Austria, and others have partnered with both the U.S. and Russian flights to fly complex life sciences experiments (see Fitton & Battrick 2001 for review).

The *Japanese Aerospace Exploration Agency* (JAXA), like ESA, has maintained an active flight program, and has undertaken the development of a multipurpose laboratory, *Kibo*, to operate in conjunction with the ISS.

The *Canadian Space Agency* (CSA), established in 1989, has been an active participant in all of the major flight programs, and has developed unique hardware for flight. In addition The CSA, ESA and JAXA have selected and flown astronauts on the Space Shuttle.

China is new to the space age. The Chinese have developed serious launch capabilities and have placed three taikonauts into orbit. They also have plans to develop and build a space station of their own.

2   SPACE FLIGHT: AN ENGINEERING AND SCIENTIFIC MARVEL

Whether the dawn of space flight began with primitive man gazing upon the heavens or with the fatal flight of Icarus, we know that modern man predicted our escape from Earth's atmosphere as early as 1911 when Tsiolkovsky[3] noted in a letter to a friend that "Humanity will not remain on the Earth forever, but in the pursuit of light and space will at first timidly penetrate beyond the limits of the atmosphere, and then will conquer all the space around the Sun." From mythology represented by Daedalus and Icarus, the physics of Archimedes, Newton, Galileo, and Copernicus, the foresight of Leonardo DaVinci, Jules Verne, and H.G. Wells, to the realization of space flight by Tsiolkovsky, Oberth, Von Braun, Korolev, Yuri Gagarin, and Neil Armstrong; the history of modern space travel with its effect on sensory function began in the fifth decade of the twentieth century.

Those familiar with the initial plans to rocket humans into space will recall that flight surgeons expressed concern that the body organs depended on sustained gravity and would not function in a reduced gravity environment. Others worried over the

---

[3] Konstantin Eduardovich Tsiolkovsky (1857-1935) was a Soviet Russian rocket scientist and pioneer of cosmonautics. One of his most famous quotes is usually cited as "Earth is the cradle of humanity, but one cannot live in a cradle forever." However, a more accurate English translation would read "A planet is the cradle of mind, but one cannot live in a cradle forever."

combined effects of acceleration, weightlessness, and the heavy deceleration during atmospheric re-entry. Still other experts were concerned especially about perception and vestibular function. Gauer & Haber (1950) speculated that the brain receives signals on the position, direction, and support of the body from four mechanisms: pressure on the nerves and organs, muscle, posture, and the vestibular organs. Modification of any one of these inputs, they theorized, would disrupt normal functioning of the autonomic nervous system with the ultimate inability to act.

Fortunately the human central nervous system has proven to be enormously plastic. Clearly, humans can adapt to the forces associated with space flight. However, the microgravity environment of space flight does have an impact on the human physiology, and we have recently entered an era where countermeasures must be developed that will not only allow crewmembers to live in space for prolonged periods of time, but also prepare those same crewmembers to encounter the gravitational fields of the Earth and other worlds following flight.

It is interesting to note that soon after NASA began flying humans in space, a series of special symposiums were initiated to address the problems that flight had on astronaut's orientation systems. Addressing the members of the first symposium in 1965 on *The Role of the Vestibular Organs in the Exploration of Space*, Dr. Walton Jones (1968) noted in his opening remarks that "the disturbing symptoms experienced in weightlessness require much detailed study [...] Most experts, I believe, are convinced that we will solve these problems; but, we will not be absolutely sure until we have conducted some experiments in orbit under the weightless condition for considerable time." More than 40 years later, we are still addressing many of the original problems.

While it might appear that what we have learned from the past helps us transition smoothly in the resolution of problems, that progression is at best an illusion. Scientific discovery does not progress linearly, but is born out of revolution. Old paradigms are attacked by the formation of new scientific communities that advance new paradigms. Perhaps we are awaiting a new research community to challenge the old paradigms and initiate a much needed revolution.

The initiation of human space flight and the apparent rational movement from one flight program to the next is perhaps an example of the illusion that science and engineering progress in a linear fashion. When it became clear that space travel would become a reality, most believed that we would leave the Earth for space by progressing on logical building blocks. That is, first we would send animals up in rockets before exposing human beings to the feared rigors of space flight (see Clément & Slenzka 2006 for review). Exactly fifty years ago, in 1957, the Soviets launched the first man-made satellite (Sputnik-1) into low Earth orbit. Later that same year, Sputnik-2 was launched carrying a dog, named Laika, the first living creature to be boosted into space. Sputnik-2 was followed two years later with the sub-orbital launch of one Rhesus and one squirrel monkey in the nose cone of a U.S. ballistic missile. The monkeys survived 38 g and 9 minutes of microgravity (Figure 2-05). Although both monkeys survived the landing, one died later under anesthesia during the removal of implanted electrodes.

Between 1959 and 1961 three other U.S. monkeys made successful sub-orbital flights in Mercury capsules. In 1961, the Chimpanzee, Ham, made the first three-orbit flight in a Mercury-Redstone capsule on January 31, 1961. Prior to human flight, twelve other dogs, many mice, rats, and a variety of plants were sent into space for longer and longer periods of time (Clément & Slenzka 2006).

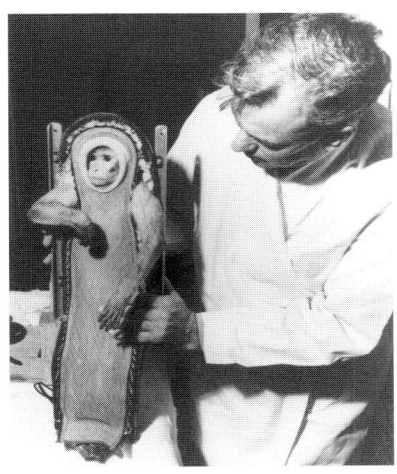

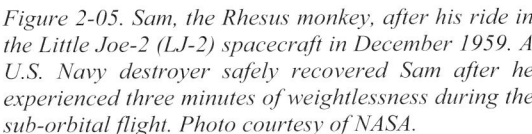

Figure 2-05. Sam, the Rhesus monkey, after his ride in the Little Joe-2 (LJ-2) spacecraft in December 1959. A U.S. Navy destroyer safely recovered Sam after he experienced three minutes of weightlessness during the sub-orbital flight. Photo courtesy of NASA.

Biometric data collected from this menagerie suggested that there were no adverse effects attributable to orbital flight, and on the basis of these results, it was concluded that the physical and mental demands that humans would encounter during space flight would not be a problem. The next steps were obvious. First, a human would be sent into space as a passenger in a capsule (Vostok and Mercury programs). Second, the launch capabilities would increase to include two astronauts, and these crewmembers would be given some control over the capsule (Soyuz and Gemini programs). Third, a reusable space vehicle would be developed to take humans into space and return them (Space Shuttle program). Fourth, a permanent space station would be constructed in low Earth orbit using the reusable vehicle as a transportation system (Mir and ISS programs). Finally, lunar and interplanetary flights could be launched from the station using lower thrust space vehicles. Of course this is not how we have progressed. Scientific, engineering and political revolutions have taken us off course.

By the time of the last Mercury flight in May 1963, the focus of the U.S. space program had already shifted. President John F. Kennedy had announced the goal of reaching the Moon only three weeks after Shepard's relatively simple 15-minute sub-orbital flight, and by 1963, only 500 of the 2,500 people working at NASA's Manned Spacecraft Center were still working on project Mercury. The remainders were already busy on Gemini and Apollo.

It is now acknowledged that the first space flights had little or few impacts on the human sensorimotor systems, and although there may have been hints that spatial orientation was somewhat altered in microgravity, NASA's management had little interest in the life sciences. In a well-written publication on the early history of NASA, Homer Newell (1980) explored the space administration's view of space biology. He wrote that life sciences were something of an enigma to the highest levels of management within NASA. Maybe this was because no one in the upper levels of management had training in the life sciences, but Newell believed that there was more to it than that. His thesis was that you could sense in the life sciences community within the U.S. a fascination with the novelty of space flight, but that there was a real skepticism within the community regarding the application of space flight to the discipline of life sciences. Interestingly, little has really changed over the years.

NASA's philosophy concerning the life sciences was and remains simple: where science was the objective, make the most of space techniques to advance the disciplines; in other areas do only what was essential to meet the need. According to Newell, a natural outcome of this philosophy was to disperse the different life sciences activities throughout the agency, placing each in the organizational entity it served. Even when life sciences administration was concentrated at NASA headquarters in Washington, little was done to modify this underlying practice (Newell 1980).

Throughout the Apollo and Skylab flights, space medicine and the laboratories associated with the clinical aspects made great strides (Parker & Jones 1975, Johnston & Dietlein 1977). Although space medicine, which in the NASA make-up formed a part of manned space flight organization, achieved extensive results, space biology and exobiology produced only modest returns during the 1960s. This is not a complete negative. There are many who view NASA's life sciences as an operational program. "Pure" biological research can be funded by other federal agencies. This philosophy is as appropriate today as it was in 1960. Regardless of its history, the discipline of life sciences within NASA remains a stepchild with little hope of improving in the next several years. It is interesting to note that Newell (1980) entitled his chapter, within his book on the early years of space science, on life sciences as having "No Place In The Sun."

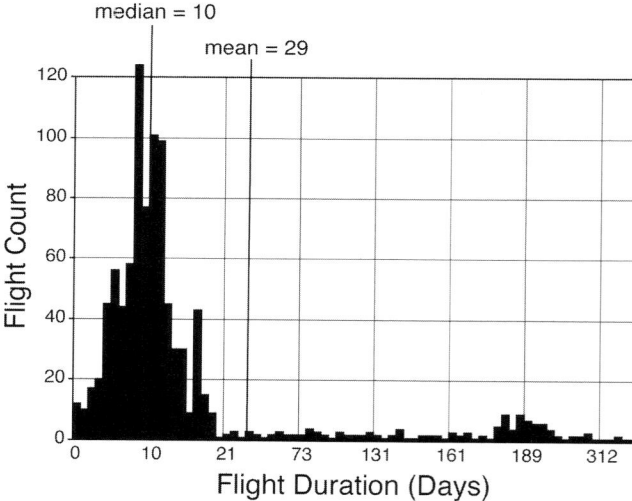

Figure 2-06. Frequency of human space flight as a function of flight duration from 1961 to 2006. Most flights were of short duration with a mean value of 29 days. The median value, however, is in the order of 10 days (Clément 2005).

3 HISTORY OF NEUROSCIENCE RESEARCH DURING SPACE FLIGHT

3.1 Humans in Space

History of manned space flight spans more than forty-five years, beginning with the landmark flight of Yuri Gagarin on April 12,1961 on Vostok 1. Since this first flight there is very little down time when space flight activity did not occur. In fact, activity increased as space flight matured, and since November 2000, there has been a continuous human presence in space on the ISS. It is believed that this trend of continuous human presence in space will progressively persist, as the duration of astronaut time in orbit or in transition between planets and moons becomes common place.

So, how much time have humans spent orbiting the Earth? To get an accurate count, the time must be calculated as man-hours since, at times, there are multiple astronauts flying on the same mission (Clément 2005). Up to ISS Expedition-14, 994 humans (449 not including re-flights) have collectively spent an astounding 707,446 cumulative man-hours, or about 80 man-years in space.

It is interesting to note that, over this period, although the U.S. has launched the most manned vehicles into space (147 for the U.S. as opposed to 103 for Russia) and sent the most humans into space (757 versus 237), the Russians have spent roughly 42% more time in space than the U.S.[4] This is because the majority of the Russian flights were long-duration flights to orbiting laboratories such as the Salyut and Mir stations, while the majority of U.S. flights were short-duration Shuttle missions.

Figure 2-07. Cumulative histogram showing the astronauts count as a function of flight duration. The ordinate axis is truncated at 100 (if not, it would peak at 449). As of today, about 100 human subjects have spent more that three months in space. Among these, less than 50 have spent more than six months in space, and four have flown during continuous missions of one year or more (Clément 2005).

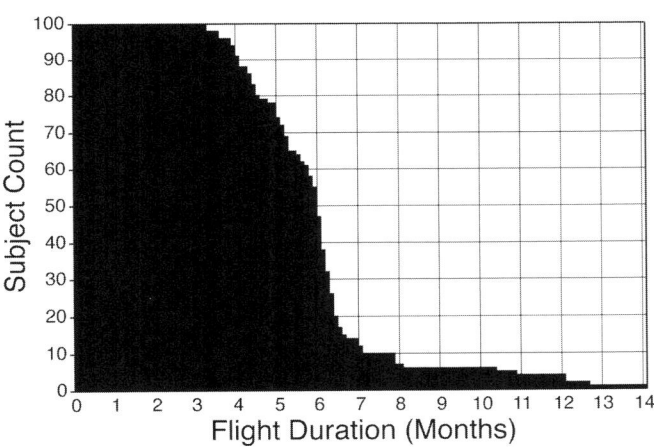

Figure 2-06 is a frequency distribution of flight durations. It shows that the average flight duration is about 29 days and the median flight duration is 10 days, meaning that the majority of life sciences experiments have been performed on crewmembers during short-duration missions. There are only 47 crewmembers with flight durations of six months or greater, and of those, only four have flight durations of one year or greater (Figure 2-07). It is difficult to make conclusions about the effects of long-duration space flight with data from only a handful of subjects, especially when different hardware and protocols were used to collect the data, and the fact that 39 of the 47 long-duration subjects were from Russia, of which we have mostly anecdotal data.

The ISS is currently the only platform available for performing long-duration human physiological experiments. The six-month long ISS flights may not be adequate length for testing the effects of long-duration space flight when, with our current rocket technology, it would take twice this time to reach Mars and return to Earth, plus one and a half year on the surface of the Red Planet (see Figure 1-05).

---

[4] China flew two Shenzhou missions, one in 2003 and one in 2005, which total to about six days in duration.

3.2 Life Sciences Experiments

From our research, we have found that a total of 2,340 human and animal life sciences experiments have been conducted on orbit and pre- and postflight by all countries, excluding Russia, through ISS Expedition-14 (Figure 2-08). We did not include Russia in this total due to our unsuccessful attempts at locating experimental records at the time the metrics were calculated. The only records located were those performed during joint ventures between Russia and other countries, such as the Shuttle-Mir, Euro-Mir, and ISS programs. From these joint ventures, Russia has conducted about 175 life sciences experiments from the periods of 1961-1989 and 2002-2005. We realize that these numbers fall short of an accurate representation of the totals achieved by our Soviet and Russian counterparts. One further point needs to be made. It is important to realize that although many reliable sources were used to compile the life sciences database presented in this chapter, these numbers are not exact since we have no way to verify the information in these sources, but they do give us a good estimate to help illustrate the point.

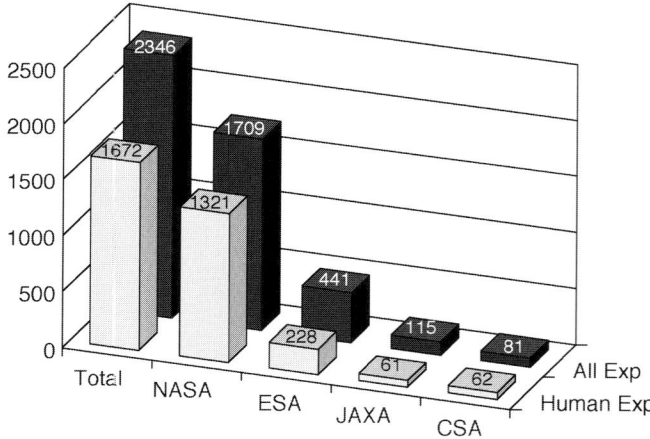

Figure 2-08. Number of space life sciences experiments including both human and animal subjects (All Exp) or just human subjects (Human Exp) for the various space agencies up through ISS Expedition-14.

Although animal studies are a vital aid to understanding space physiology, they are not a perfect analogue to humans. We have detailed these differences in a book published earlier (Clément & Slenzka 2006). Taking only the number of experiments conducted on humans, there is a drop in the number of investigations from 2,346 to 1,672. Even though 1,672 seems like a significant number of human physiological experiments, when it is broken down into the various science disciplines, it is apparent how few experiments have actually been performed (Figure 2-09).

Cardiovascular and neuroscience account for the majority of space life sciences experiments. About 400 space neuroscience experiments have been performed to date. This number is not very encouraging when considering that they were conducted over a 45-year period, i.e., about 8-9 experiments per year on average. In addition, these experiments were performed with different research methods, different hardware, and on mostly short-duration missions. The knowledge attained from these short-duration experiments may not be adequate to predict the physiological changes an astronaut experiences on long-duration missions. Until life sciences and the development of countermeasures become a priority in the space community, astronauts will continue to endure the undesirable neurological and sensorimotor changes brought about by space flight.

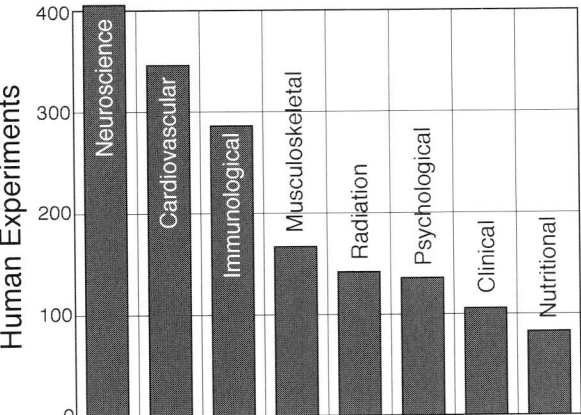

Figure 2-09. Total number of space experiments in human physiology, by disciplines, performed by all countries (except Russia) up through ISS Expedition-14.

4 NEUROSCIENCE EXPERIMENTS CONDUCTED DURING SPACE FLIGHT

To our knowledge, the first documented space neuroscience experiments were performed during the third manned mission on board the Russian Vostok spacecraft. These experiments began after the crew from earlier missions complained from nausea and spatial disorientation in weightlessness. Space neuroscience experiments were typically addressing these operational issues until the Skylab and Salyut space stations were made available for more fundamental research on the effect of gravity (or virtual lack thereof) on central nervous system functions.

The following table lists all the space neuroscience experiments that we have identified between Vostok-3 (1962) and ISS Expedition-15 (2007), sorted by mission launch date.

Mission	Launch Date	Crew-Members	Duration (D:H:M)	Neuroscience Experiments
Vostok-3	11-Aug-62	Nikolayev	3:22:25	ElectroEncephaloGraphy (EEG) ElectroOculoGraphy (EOG) Galvanic Skin Response (GSR) Sensory-Motor Coordination Tests
Vostok-4	12-Aug-62	Popovich	2:22:59	Same as Vostok-3
Vostok-5	14-Jun-63	Bykovsky	4:23:06	Same as Vostok-3
Vostok-6	16-Jun-63	Tereshkova	2:22:50	Same as Vostok-3

Mission	Launch Date	Crew-Members	Duration (D:H:M)	Neuroscience Experiments
Voskhod-1	12-Oct-64	Komarov Feoktistov Yegorov	1:00:17	ElectroOculoGraphy Eyes-Closed Writing Tests with Galvanic Vestibular Stimulation
Voskhod-2	18-Mar-65	Belyaev Leonov	1:02:02	Neurological Investigations including sensory and stereognostic testing
Gemini-5	21-Aug-65	Cooper Conrad	07:22:55	Human Otolith Function (M009) Visual Acuity in the Space Environment (S008)
Gemini-7	4-Dec-65	Borman Lovell	13:18:35	Human Otolith Function (M009) In-Flight Sleep Analysis (M008) Visual Acuity in the Space Environment (S008)
Apollo-7	11-Oct-68	Schirra Eisele Cunningham	10:20:09	Apollo Flight Crew Vestibular Assessment
Soyuz-3	26-Oct-68	Bergovoy	3:22:51	Investigation of Muscle EMG Activity at Rest and After Exercise
Apollo-8	21-Dec-68	Borman Lovell Anders	06:03:01	Apollo Flight Crew Vestibular Assessment
Soyuz-4	14-Jan-69	Shatalov	2:23:23	Same as Soyuz-3
Soyuz-5	15-Jan-69	Volynov Yeliseyev Khrunov	3:00:56	Same as Soyuz-3
Apollo-9	3-Mar-69	McDivitt Scott Schweickart	10:01:01	Same as Apollo-7
Apollo-10	18-May-69	Cernan Stafford Young	08:0:03	Apollo Flight Crew Vestibular Assessment
Biosatellite III	28-Jun-69	Bonnie (Pig-Tailed Monkey)	8:19:00	Digital Computer Analysis of Neurophysiological Data from Biosatellite III Sleep and Wake Activity Patterns of a Pig-Tailed Monkey During Nine Days of Weightlessness Sleep and Wake States in Biosatellite III Monkey: Visual and Computer Analyses of Telemetered Electro Encephalographic Data

Mission	Launch Date	Crew-Members	Duration (D:H:M)	Neuroscience Experiments
Apollo-11	16-Jul-69	Armstrong Aldrin Collins	8:03:09	Apollo Flight Crew Vestibular Assessment
Soyuz-6	11-Oct-69	Shonin Kubasov	4:22:42	Same as Soyuz-3
Soyuz-7	12-Oct-69	Filipchenko Volkov Gorbatko	4:22:41	Same as Soyuz-3
Apollo-12	14-Nov-69	Conrad Gordon Bean	10:4:36	Apollo Flight Crew Vestibular Assessment
Apollo-13	11-Apr-70	Lovell Swigert Haise	5:22:55	Apollo Flight Crew Vestibular Assessment
Soyuz-9	1-Jun-70	Nikolayev Sevastyanov	17:16:59	EEG Monitoring Locomotion Muscle EMG Activity Posture Study Sleep Monitoring
OFO-A (Scout Satellite)	9-Nov-70	Two Bull Frogs	6:00:00	Orbiting Frog Otolith Experiment: Comparison to Control Studies; Preliminary Results; Secondary Spike Analysis
Apollo-14	31-Jan-71	Shepard, Roosa Mitchell	9:00:02	Apollo Flight Crew Vestibular Assessment
Soyuz-11	6-Jun-71	Dobrovolsky Volkov Patsayev	23:18:22	Neurological Testing of Grip Strength, Kinesthetic Sensitivity, Visual Acuity, Color and Contrast Sensitivity, Convergence and Accommodation
Apollo-15	26-Jul-71	Scott Worden Irwin	12:17:12	Apollo Flight Crew Vestibular Assessment
Apollo-16	16-Apr-72	Young Duke Mattingly	11:01:51	Apollo Flight Crew Vestibular Assessment
Apollo-17	7-Dec-72	Cernan Schmitt Evans	12:13:52	Apollo Flight Crew Vestibular Assessment
Skylab-2	25-May-73	Conrad Kerwin Weitz	28:00:50	Human Vestibular Function (M131)
Skylab-3	28-Jul-73	Bean Garriott Lousma	59:12:09	Human Vestibular Function (M131)

Mission	Launch Date	Crew-Members	Duration (D:H:M)	Neuroscience Experiments
Skylab-4	16-Nov-73	Carr Gibson Pogue	84:01:16	Human Vestibular Function (M131) Motor Sensory Performance
Soyuz-17 / Salyut-4	10-Jan-75	Gubarev Grechko	29:13:20	Vestibular Monitoring
Soyuz-19 / ASTP	15-Jul-75	Leonov Kubasov	5:22:31	Achilles Tendon Reflex
Apollo-18 / ASTP	15-Jul-75	Stafford Slayton Brand	9:01:28	Achilles Tendon Reflex Electromyographic Analysis of Skeletal Muscle
Soyuz-21 / Salyut-5	6-Jul-76	Volynov Zholobov	49:06:23	Investigation of Sensitivity Threshold of Vestibular System to Galvanic Stimulation Evaluation of Gustatory Sensations in Weightlessness
Soyuz-24 / Salyut-5	7-Feb-77	Gorbatko Glazkov	17:17:26	Same as Soyuz-21
Soyuz-26 / Salyut-6	10-Dec-77	Romanenko Grechko	96:10:00	Attention and Memory Test Color Sensitivity and Visual Acuity Coordination Tests EEG Monitoring Effect of Plantar Stimulation on Space Motion Sickness First Test of the "Cuban Boot" (to simulate Earth loads on foot proprioceptors) Gustometry Investigation of Tactile Sensation Optokinetic Stimulation Posture Tests Reaction Time Space Motion Sickness (SMS) Questionnaire
Soyuz-27 / Salyut-6	10-Jan-78	Dzhanibekov Makarov	5:22:58	Same as Soyuz-26
Soyuz-28 / Salyut-6	2-Mar-78	Gubarev Remek	7:22:15	Same as Soyuz-26
Soyuz-29/ Salyut-6	15-Jun-78	Kovalyonok Ivanchenkov	139:14:47	Same as Soyuz-26
Soyuz-31 / Salyut-6	27-Jun-78	Klimuk Hermasze-wski	7:22:02	Same as Soyuz-26
Soyuz-32 / Salyut-6	25-Feb-79	Lyakhov Ryumin	175:00:35	Same as Soyuz-26
Soyuz-33 / Salyut-6	10-Apr-79	Rukavishni-kov Ivanov	1:22:23	Same as Soyuz-26

Mission	Launch Date	Crew-Members	Duration (D:H:M)	Neuroscience Experiments
Soyuz-35 / Salyut-6	9-Apr-80	Popov Ryumin	184:20:11	Same as Soyuz-26
Soyuz-36 / Salyut-6	26-May-80	Kubasov Farkas	7:20:45	Same as Soyuz-26
Soyuz-T2 / Salyut-6	5-Jun-80	Malyshev Aksenov	3:22:19	Same as Soyuz-26
Soyuz-37 / Salyut-6	23-Jun-80	Gorbatko Pham	7:20:41	Same as Soyuz-26
Soyuz-38 / Salyut-6	18-Sep-80	Romanenko Tamayo	7:20:43	Same as Soyuz-26
Soyuz-T3 / Salyut-6	27-Nov-80	Kizim Grigoryevich Strekalov	12:19:07	Same as Soyuz-26
Soyuz-T4 / Salyut-6	12-Mar-81	Kovalyonok Savinykh	74:17:37	Same as Soyuz-26
Soyuz-39 / Salyut-6	22-Mar-81	Dzhanibekov Gurragcha	7:20:42	Same as Soyuz-26
STS-1 (Columbia)	12-Apr-81	Young Crippen	2:06:20	Validation of Predictive Tests and Countermeasures for SMS (DSO 401)
Soyuz-40/ Salyut-6	14-May-81	Popov Prunariu	7:20:41	Same as Soyuz-26
STS-2 (Columbia)	12-Nov-81	Engle Truly	2:06:13	Validation of Predictive Tests and Countermeasures for SMS (DSO 401)
STS-3 (Columbia)	22-Mar-82	Lousma Fullerton	8:00:04	Validation of Predictive Tests and Countermeasures for SMS (DSO 401)
Soyuz-T5 / Salyut-7	13-May-82	Berezevoi Lebedev	210:09:04	Attention and Memory Tests Audiometry Color Sensitivity and Visual Acuity Coordination Tests EEG Monitoring Effect of Plantar Stimulation on SMS Gustometry Investigation of Tactile Sensation Optokinetic Stimulation Posture Tests SMS Questionnaire
Soyuz-T6 / Salyut-7 French PVH Mission	24-Jun-82	Dzhanibekov Ivanchenkov Chretien	7:21:50	Same as Soyuz-T5 Posture Experiment: Postural Control during Voluntary Arm and Involuntary Body Movements

Mission	Launch Date	Crew-Members	Duration (D:H:M)	Neuroscience Experiments
STS-4 (Columbia)	27-Jun-82	Mattingly Hartsfield	7:01:09	Validation of Predictive Tests and Countermeasures for SMS (DSO 401)
Soyuz-T7 / Salyut-7	19-Aug-82	Popov Serebrov Savitskaya	7:21:52	Same as Soyuz-T6
STS-5 (Columbia)	11-Nov-82	Brand Overmyer Allen Lenoir	5:02:14	Acceleration Detection Sensitivity (DSO 405) Head and Eye Motion During Shuttle Launch and Entry (DSO 403) Near Vision Acuity and Contrast Sensitivity (DSO 408) Validation of Predictive Tests and Countermeasures for SMS (DSO 401)
STS-6 (Challenger)	4-Apr-83	Weitz Bobko Peterson Musgrave	5:02:14	Acceleration Detection Sensitivity (DSO 405) Extra-Ocular Motion (EOM) Studies, Pre, In and Postflight (DSO 404) Eye Head Motion during Ascent, Entry, and On Orbit (Gyroscopic Head Motion Measurements) (DSO 404) Head and Eye Motion During Shuttle Launch and Entry (DSO 403) Validation of Predictive Tests and Countermeasures for SMS (DSO 401)
Soyuz-T8 / Salyut-7	20-Apr-83	Titov Strekalov Serebrov	2:00:17	Same as Soyuz-T5
STS-7 (Challenger)	18-Jun-83	Crippen Hauck Fabian Ride Thagard	6:02:23	Acceleration Detection Sensitivity (DSO 405) Extra-Ocular Motion (EOM) Studies Pre, In and Postflight (Saccadic Tracking) (DSO 404) Head and Eye Motion During Shuttle Launch and Entry (DSO 403) In-Flight Countermeasures for SMS (DSO 417) Near Vision Acuity and Contrast Sensitivity (DSO 408 On-Orbit Head and Eye Tracking Task (Optokinetic Studies) (DSO 404)

Mission	Launch Date	Crew-Members	Duration (D:H:M)	Neuroscience Experiments
				Validation of Predictive Tests and Countermeasures for SMS (DSO 401)
Soyuz-T9 / Salyut-7	27-Jun-83	Lyakhov Aleksandrov	149:10:45	Same as Soyuz-T5
STS-8 (Challenger)	30-Aug-83	Truly Brandenstein Gardner Bluford Thornton	6:01:08	Acceleration Detection Sensitivity (DSO 405) Extra-Ocular Motion (EOM) Studies Pre, In and Postflight (Saccadic Tracking) (DSO 404) Eye Head Motion during Ascent, Entry and on Orbit (Gyroscopic Head Motion Measurements) (DSO 404) Head and Eye Motion During Shuttle Launch and Entry (DSO 403) In-Flight Countermeasures for SMS (DSO 417) Near Vision Acuity and Contrast Sensitivity (DSO 408) Validation of Predictive Tests and Countermeasures for SMS (DSO 401)
STS-9 (Columbia) First Spacelab Mission (SL-1)	28-Nov-83	Young Shaw Garriott Parker Lichtenberg Merbold	10:07:47	Effects of Rectilinear Acceleration, Optokinetic and Caloric Stimulation on Human Vestibular Reactions and Sensations Eye Movements During Sleep Mass Discrimination During Weightlessness Validation of Predictive Tests and Countermeasures for SMS (DSO 401) Vestibular Experiments Vestibulo-Spinal Reflex Mechanisms using Hoffman Reflex
Soyuz-T10 / Salyut-7	8-Feb-84	Kizim Solovyov Atkov	236:22:49	Same as Soyuz-T5
Soyuz-T11 / Salyut-7	3-Apr-84	Malyshev Strekalov Sharma	7:21:40	Same as Soyuz-T5
STS-41C (Challenger)	6-Apr-84	Crippen Hart Scobee	6:23:40	Near Vision Acuity and Contrast Sensitivity (DSO 408) Validation of Predictive Tests and

Mission	Launch Date	Crew-Members	Duration (D:H:M)	Neuroscience Experiments
		Nelson Van Hoften		Countermeasures for SMS (DSO 401)
Soyuz-T12 / Salyut-7	17-Jul-84	Dzhanibekov Savitskaya Volk	11:19:14	Same as Soyuz-T5
STS-41D (Discovery)	30-Aug-84	Hartsfield Coats Resnik Hawley Mullane Walker	6:00:56	Crew Visual Performance (DSO 440) Near Vision Acuity and Contrast Sensitivity (DSO 408) Validation of Predictive Tests and Countermeasures for SMS (DSO 401)
STS-41G (Challenger)	5-Oct-84	Crippen McBride Sullivan Ride Leestma Garneau Scully-Power	8:05:23	Crew Visual Performance (DSO 440) Near Vision Acuity and Contrast Sensitivity (DSO 408) Validation of Predictive Tests and Countermeasures for SMS (DSO 401)
STS-51A (Discovery)	8-Nov-84	Hauck Walker Fisher Gardner Allen	7:23:44	Validation of Predictive Tests and Countermeasures for SMS (DSO 401)
STS-51C (Discovery)	24-Jan-85	Mattingly Shriver Onizuka Buchli Payton	3:01:33	Crew Visual Performance (DSO 440) Near Vision Acuity and Contrast Sensitivity (DSO 408) Validation of Predictive Tests and Countermeasures for SMS (DSO 401)
STS-51D (Discovery)	12-Apr-85	Bobko Williams Seddon Hoffman Griggs Walker Garn	6:23:55	Extra-Ocular Motion (EOM) Studies, Pre, In and Postflight (DSO 404) Validation of Predictive Tests and Countermeasures for SMS (DSO 401)
STS-51B (Challenger) Spacelab-3 Mission	29-Apr-85	Overmyer Gregory Don Lind Thagard Thornton van den Berg G. Wang	7:00:08	Eye-Hand Coordination During SMS (DSO 451)
Soyuz-T13 / Salyut-7	6-Jun-85	Dzhanibekov Savinykh	112:03:12	Same as Soyuz-T5

Mission	Launch Date	Crew-Members	Duration (D:H:M)	Neuroscience Experiments
STS-51G (Discovery)	17-Jun-85	Brandenstein Creighton Lucid Fabian Nagel Baudry Al-Saud	7:01:38	Clinical Characterization of SMS (DSO 455) Equilibrium and Vertigo: Studies of Postural Control, Vestibulo-ocular Reflex, Optokinetic Nystagmus, and Cognitive Processes Sensory-Motor Adaptation during Visual-Vestibular Interaction
STS-51I (Discovery)	27-Aug-85	Engle Covey van Hoften Lounge Fisher	7:02:17	Clinical Characterization of SMS (DSO 455)
Soyuz-T14 / Salyut-7	17-Sep-85	Vasyutin Grechko Volkov	64:21:52 Vasyutin, Volkov 8:21:13 Grechko	Same as Soyuz-T5
STS-51J (Atlantis)	3-Oct-85	Bobko Grabe Hilmers Stewart Pailes	4:01:44	Eye-Hand Coordination During SMS (DSO 451)
STS-61A (Challenger) German Spacelab Mission (D1)	30-Oct-85	Hartsfiel Nagel Buchli Bluford Dunbar Furrer Messersch-mid Ockels	7:00:44	European Experiments on the Vestibular System: Tonometry, Spatial Disorientation, Cognitive Adaptation, Causation of Inversion Illusions, Space Motion Sickness, Mass Discrimination Vestibular Adaptation Using the European Vestibular Sled
STS-61B (Atlantis)	26-Nov-85	Shaw O'Connor Cleave Spring Ross Vela Walker	6:21:04	Clinical Characterization of SMS (DSO 455)
STS-61C (Columbia)	12-Jan-86	Gibson Bolden Chang-Diaz Hawley Nelson Cenker Nelson	6:02:03	Clinical Characterization of SMS (DSO 455) Eye-Hand Coordination During SMS (DSO 451) Otolith Tilt-Translation Reinterpretation (DSO 459) Visual Observations from Space (DSO 204)

Mission	Launch Date	Crew-Members	Duration (D:H:M)	Neuroscience Experiments
Kosmos-1887	29-Sep-87	Rhesus Monkeys; Various bugs; Rats	13:00:00	Effect of Microgravity on Metabolic Enzymes of Hippocampus and Spinal Cord in Rat
Soyuz-TM5 / Mir-3	7-Jun-88	Solovyev Savinykh Alexandrov	9:20:10	Psychomotor Studies
STS-26 (Discovery)	29-Sep-88	Hauck Covey Lounge Nelson Hilmers	4:01:00	Otolith Tilt-Translation Reinterpretation (DSO 459) Visual Observations from Space (DSO 204)
Soyuz-TM7 / Mir-4 French Aragatz Mission	26-Nov-88	Volkov Krikalev Chretien	24:18:07	Physalie Experiment: Study of Postural, Oculomotor, and Cognitive Systems
STS-28 (Columbia)	8-Aug-89	Shaw Richards Adamson Leestma Brown	5:01:00	Otolith Tilt-Translation Reinterpretation (DSO 459) Postural Equilibrium Control During Landing/Egress (DSO 605)
Kosmos-2044	15-Sep-89	Rhesus Monkeys; Male specific pathogen free Wistar Rats	14:00:00	Adaptation of Optokinetic Nystagmus to Microgravity Effect of Microgravity on: Metabolic Enzymes, Neurotransmitter Amino Acids, and Neurotransmitter Associated Enzymes in Selected Regions of the Central Nervous System Functional Neuromuscular Adaptation to Space Flight Metabolic and Morphologic Properties of Muscle Fibers and Motor Neurons after Space Flight: II. Ventral Horn Cell Responses to Space Flight and Suspension Studies of Vestibular Primary Afferents in Normal, Hyper- and Hypogravity
STS-33 (Discovery)	22-Nov-89	Gregory Blaha Musgrave Carter Thornton	5:00:06	Preflight Adaptation Training (PAT) (DSO 468)
STS-36 (Atlantis)	28-Feb-90	Creighton Casper Mullane	4:10:18	Postural Equilibrium Control During Landing/Egress (DSO 605)

Mission	Launch Date	Crew-Members	Duration (D:H:M)	Neuroscience Experiments
		Hilmers Thout		Preflight Adaptation Training (PAT) (DSO 468)
STS-41 (Discovery)	6-Oct-90	Richards Cabana Shepherd Melnick Akers	4:02:10	Postural Equilibrium Control During Landing/Egress (DSO 605) Visual-Vestibular Integration as a Function of Adaptation (Eye and Head Movements) (DSO 604OI-3)
STS-35 (Columbia)	2-Dec-90	Brand Gardner Hoffman Lounge Parker Durrance Parise	8:23:05	Postural Equilibrium Control During Landing/Egress (DSO 605) Preflight Adaptation Training (PAT) (DSO 468)
STS-39 (Discovery)	28-Apr-91	Coats Veach McMonagle Hieb Harbaugh Bluford Hammond	8:07:22	Postural Equilibrium Control During Landing/Egress (DSO 605) Visual-Vestibular Integration as a Function of Adaptation (Eye and Head Movements) (DSO 604OI-3)
STS-40 (Columbia) Spacelab Space Life Sciences 1 Mission (SLS-1)	5-Jun-91	O'Connor Gutierrez Bagian Jernigan Seddon Gaffney Hughes Fulford	9:02:14	Postural Equilibrium Control During Landing/Egress (DSO 605) Vestibular Experiments in Spacelab: Smooth Pursuit, Optokinetic Nystagmus, Vestibulo-Ocular Reflex
STS-43 (Atlantis)	2-Aug-91	Blaha Baker Lucid Adamson Low	8:21:21	Head and Gaze Stability During Locomotion (DSO 614) Postural Equilibrium Control During Landing/Egress (DSO 605) Visual-Vestibular Integration as a Function of Adaptation (Eye and Head Movements) (DSO 604 OI-3B)
STS-48 (Discovery)	12-Sep-91	Creighton Reightler, Brown Gemar Buchli	5:08:27	Head and Gaze Stability During Locomotion (DSO 614) Visual-Vestibular Integration as a Function of Adaptation (Perceptual Reporting) (DSO 604OI-1)
Soyuz-TM13 / Mir-10	2-Oct-91	Viehboeck	7:22:12	Directional Hearing in Microgravity Eye-Head-Arm Coordination and

Mission	Launch Date	Crew-Members	Duration (D:H:M)	Neuroscience Experiments
AustroMir Mission				Spinal Reflexes in Weightlessness Orientation Effects from Optokinetic Stimulations Sleep Experiment
STS-44 (Atlantis)	24-Nov-91	Gregory Henricks Runco Voss Musgrave Hennen	6:22:50	Head and Gaze Stability During Locomotion (DSO 614) Postural Equilibrium Control During Landing/Egress (DSO 605) Visual-Vestibular Integration as a Function of Adaptation (Eye and Head Movements & Perceptual Reporting) (DSO 604)
STS-42 (Discovery) IML-1 Mission	22-Jan-92	Oswald Thagard Hilmers Readdy Bondar Merbold	8:01:14	Microgravity Vestibular Investigations: Studies of Visual-Vestibular Interactions during Passive Body Rotation in Yaw, Pitch and Roll
Soyuz-TM14 / Mir-11 German '92 Mir Mission	17-Mar-92	Flade	7:21:52	Illusions of Verticality Sleep and Circadian Rhythm
STS-45 (Atlantis)	24-Mar-92	Bolden Duffy Sullivan Leestma Foale Lichtenberg Frimout	8:22:09	Head and Gaze Stability During Locomotion (DSO 614) Visual-Vestibular Integration as a Function of Adaptation (Perceptual Reporting) (DSO 604OI-1)
STS-49 (Endeavour)	7-May-92	Brandenstein Chilton Thout Thornton Hieb Akers Melnick	8:21:17	Head and Gaze Stability During Locomotion (DSO 614) Postural Equilibrium Control During Landing/Egress (DSO 605) Visual-Vestibular Integration as a Function of Adaptation (Eye and Head Movements & Perceptual Reporting) (DSO 604)
STS-50 (Columbia) USML-1 Mission	25-Jun-92	Richards Bowersox Dunbar Baker Meade DeLucas Trinh	13:19:30	Head and Gaze Stability During Locomotion (DSO 614) Physiological Evaluation of Astronaut Seat Egress Ability at Wheels Stop (DSO 620) Postural Equilibrium Control

Mission	Launch Date	Crew-Members	Duration (D:H:M)	Neuroscience Experiments
				During Landing/Egress (DSO 605)
Soyuz-TM15 / Mir-12 French Antares Mission	27-Jul-92	Tognini	13:18:56	Haptic Perception in Weightlessness Study of Adaptive Process in Human Proprioceptive Functions at Cognitive and Sensory-Motor Levels in Weightlessness Symmetry Detection
STS-46 (Atlantis)	31-Jul-92	Shriver Allen Hoffman Chang-Diaz Nicollier Ivins Malerba	7:23:15	Head and Gaze Stability During Locomotion (DSO 614) Visual-Vestibular Integration as a Function of Adaptation (Perceptual Reporting) (DSO 604OI-1)
STS-47 (Endeavour) Japanese Spacelab-J Mission	12-Sep-92	Gibson Brown Lee Jan Davis Apt Jemison Mohri	7:22:30	Autogenic Feedback Training Exercise (AFTE) as a Preventative Method for Space Adaptation Syndrome Head and Gaze Stability During Locomotion (DSO 614)
STS-52 (Columbia)	22-Oct-92	Wetherbee Baker Veach Shepherd Jernigan MacLean	9:20:56	Postural Equilibrium Control During Landing/Egress (DSO 605) Visual-Vestibular Integration as a Function of Adaptation (Eye and Head Movements & Perceptual Reporting) (DSO 604)
STS-53 (Discovery)	2-Dec-92	Walker Cabana Bluford Voss Clifford	7:07:19	Head and Gaze Stability During Locomotion (DSO 614) Postural Equilibrium Control During Landing/Egress (DSO 605) Visual-Vestibular Integration as a Function of Adaptation (Eye and Head Movements & Perceptual Reporting) (DSO 604)
Kosmos 2229	29-Dec-92	Two Rhesus Monkeys	12:00:00	Adaptation to Microgravity of Oculomotor Reflexes Functional Neuromuscular Adaptation to Space Flight Studies of Vestibular Neurons in Normal, Hyper-, and Hypogravity

Mission	Launch Date	Crew-Members	Duration (D:H:M)	Neuroscience Experiments
STS-54 (Endeavour)	13-Jan-93	Casper McMonagle Runco Harbaugh Helms	5:23:38	Postural Equilibrium Control During Landing/Egress (DSO 605) Visual-Vestibular Integration as a Function of Adaptation (Eye and Head Movements & Perceptual Reporting) (DSO 604)
STS-56 (Discovery) German Spacelab Mission (D2)	8-Apr-93	Cameron Stephen Oswald Michael Foale Cockrell Ochoa	9:06:08	Postural Equilibrium Control During Landing/Egress (DSO 605)
STS-57 (Endeavour)	21-Jun-93	Grabe Duffy Low Sherlock Wisoff Voss	9:23:44	Head and Gaze Stability During Locomotion (DSO 614) Visual-Vestibular Integration as a Function of Adaptation (Eye and Head Movements & Perceptual Reporting) (DSO 604)
Soyuz-TM17 / Mir-14 French Altair Mission	1-Jul-93	Haignere	21:16:08	Haptic Perception in Weightlessness Mental Rotation Study of Limb/Body Movement in Microgravity Study of Visual-Motor Interactions during Operational Activities Symmetry Detection
STS-51 (Discovery)	12-Sep-93	Culbertson Readdy Newman Bursch Walz	9:20:11	Postural Equilibrium Control During Landing/Egress (DSO 605) Visual-Vestibular Integration as a Function of Adaptation (Eye and Head Movements & Perceptual Reporting) (DSO 604)
STS-58 (Columbia) Space Life Sciences 2 Mission (SLS-2)	18-Oct-93	Blaha Searfoss Seddon McArthur Wolf Lucid Fettman	14:00:12	Head and Gaze Stability During Locomotion (DSO 614) Physiological Evaluation of Astronaut Seat Egress Ability at Wheels Stop (DSO 620) Postural Equilibrium Control During Landing/Egress (DSO 605) Vestibular Experiments in Spacelab

Mission	Launch Date	Crew-Members	Duration (D:H:M)	Neuroscience Experiments
				Visual-Vestibular Integration as a Function of Adaptation (Eye and Head Movements) (DSO 604OI-3B)
STS-61 (Endeavour)	2-Dec-93	Covey Bowersox Musgrave Thornton Nicollier Hoffman Akers	10:19:58	Visual-Vestibular Integration as a Function of Adaptation (Eye and Head Movements) (DSO 604OI-3B)
STS-60 (Discovery)	3-Feb-94	Bolden Reightler Davis Sega Chang-Diaz Krikalev	8:07:09	Alterations in Postural Equilibrium Control Associated with Long Duration Space Flight Autonomic and Gastric Function Associated with SMS Biomechanics of Movement During Locomotion Eye-Head Coordination During Target Acquisition The Effects of Long-Duration Space Flight on Eye, Head, and Trunk Coordination During Locomotion
STS-62 (Columbia)	4-Mar-94	Casper Allen Thuot Gema Ivins	13:23:16	Head and Gaze Stability During Locomotion (DSO 614) Postural Equilibrium Control During Landing/Egress (DSO 605) Visual-Vestibular Integration as a Function of Adaptation (Eye and Head Movements) (DSO 604 OI-3B)
STS-59 (Endeavour)	9-Apr-94	Gutierrez Chilton Godwin Apt Clifford Jones	11:05:49	Visual-Vestibular Integration as a Function of Adaptation (Eye and Head Movements) (DSO 604OI-3B)
STS-65 (Columbia) IML-2 Mission	8-Jul-94	Cabana Halsell Hieb Walz Chiao Thomas Mukai	14:17:55	Head and Gaze Stability During Locomotion (DSO 614)
STS-64 (Discovery)	9-Sep-94	Richards Hammond,	10:22:49	Head and Gaze Stability During Locomotion (DSO 614)

Mission	Launch Date	Crew-Members	Duration (D:H:M)	Neuroscience Experiments
		Linenger Helms Meade Lee		Postural Equilibrium Control During Landing/Egress (DSO 605) Visual-Vestibular Integration as a Function of Adaptation (Eye and Head Movements) (DSO 604 OI-3B)
STS-68 (Endeavour)	30-Sep-94	Baker Wilcutt Jones Smith Bursch Wisoff	11:05:46	Head and Gaze Stability During Locomotion (Path Integration) (DSO 614B) Postural Equilibrium Control During Landing/Egress (DSO 605) Visual-Vestibular Integration as a Function of Adaptation (Eye and Head Movements) (DSO 604OI-3C)
Soyuz-TM20 / Mir-17 ESA EuroMir '94 Mission	4-Oct-94	Merbold	31:12:35	Adaptation of Basic Vestibulo-Oculomotor Mechanism to Altered Gravity Conditions Circadian Rhythms and Sleep During a 30-Day Space Mission Otolith Adaptation to Different Levels of Microgravity Perception of Figure Symmetry by the Two Cerebral Hemispheres (STAMP) Posture and Movement in Microgravity Spatial Orientation and SMS
STS-66 (Atlantis)	3-Nov-94	McMonagle Brown Ochoa Parazynski Tanner Clervoy	10:22:34	Head and Gaze Stability During Locomotion (Path Integration) (DSO 614B) Postural Equilibrium Control During Landing/Egress (DSO 605) Visual-Vestibular Integration as a Function of Adaptation (Eye and Head Movements) (DSO 604OI-3C)
STS-63 (Discovery)	3-Feb-95	Whetherbee Collins Foale Voss Harris Titov	8:06:28	Anticipatory Postural Activity (POSA) Autonomic and Gastric Function Associated with SMS Biomechanics of Movement During Locomotion Eye-Head Coordination During Target Acquisition

Mission	Launch Date	Crew-Members	Duration (D:H:M)	Neuroscience Experiments
STS-67 (Endeavour)	2-Mar-95	Oswald Gregory Jernigan Grunsfeld Lawrence Parise Durrance	16:15:08	Head and Gaze Stability During Locomotion (Path Integration) (DSO 614B) Postural Equilibrium Control During Landing/Egress (DSO 605) Visual-Vestibular Integration as a Function of Adaptation (Eye and Head Movements) (DSO 604OI-3C)
Mir-18 NASA-Mir Mission 1	14-Mar-95	Dezhurov Strekalov Thagard	115:08:43	Alterations in Postural Equilibrium Control Associated with Long Duration Space Flight Anticipatory Postural Activity (POSA) Eye-Head Coordination During Target Acquisition The Effectiveness of Manual Control During Simulation of Flight Tasks (PILOT) The Effects of Long-Duration Space Flight on Eye, Head, and Trunk Coordination During Locomotion
STS-71 (Atlantis) Spacelab-Mir Mission (Mir-19)	27-Jun-95	Gibson Precourt Baker Dunbar Harbaugh Solovyev Budarin	9:19:22 STS-71 crew 75:11:20 Mir 19 crew: Solovyev, Budarin	Alterations in Postural Equilibrium Control Associated with Long Duration Space Flight Anticipatory Postural Activity (POSA) Autonomic and Gastric Function Associated with Space Motion Sickness Biomechanics of Movement During Locomotion Eye-Head Coordination During Target Acquisition The Effects of Long-Duration Space Flight on Eye, Head, and Trunk Coordination During Locomotion
STS-70 (Discovery)	13-Jul-95	Henricks Kregel Currie Thomas Weber	8:22:20	Visual-Vestibular Integration as a Function of Adaptation (Perceptual Reporting) (DSO 604OI-1)
Soyuz-TM22 / Mir-20	3-Sep-95	Reiter	179:01:41	Differential Effects of Otolith Input on Ocular Lateropulsion, Cyclorotation,

Mission	Launch Date	Crew-Members	Duration (D:H:M)	Neuroscience Experiments
ESA EuroMir '95 Mission				Perceived Visual Vertical, Straight Ahead, and Tonic Neck Reflexes in Man Influence of Gravity on the Preparation and Execution of Voluntary Movements Postural Modifications in Microgravity
STS-69 (Endeavour)	7-Sep-95	Walker Cockrell Voss Newman Gernhardt	10:20:28	Postural Equilibrium Control During Landing/Egress (DSO 605) Visual-Vestibular Integration as a Function of Adaptation (Eye and Head Movements) (DSO 604OI-3B)
STS-73 (Columbia)	20-Oct-95	Bowersox Rominger Thornton Coleman Lopez-Alegria Leslie Sacco	15:21:53	Postural Equilibrium Control During Landing/Egress (DSO 605) Visual-Vestibular Integration as a Function of Adaptation (Eye and Head Movements & Perceptual Reporting) (DSO 604)
STS-74 (Atlantis)	12-Nov-95	Cameron Halsell Ross McArthur Hadfield	8:04:31	Visual-Vestibular Integration as a Function of Adaptation (Perceptual Reporting) (DSO 604OI-1)
STS-72 (Endeavour)	11-Jan-96	Duffy Jett Chiao Barry Scott Wakata	8:22:01	Visual-Vestibular Integration as a Function of Adaptation (Eye and Head Movements & Perceptual Reporting) (DSO 604)
Soyuz-TM23 / Mir-21	21-Feb-96	Onufrienko Usachev	193:19:07	Anticipatory Postural Activity During Long Duration Space Flight Sensory and Motor Mechanisms in Vertical Posture Control After Long Duration Exposure to Microgravity (Ravnovesie) The Effects of Long Duration Space Flight on Gaze Control The Effects of Long-Duration Space Flight on Eye, Head, and Trunk Coordination During Locomotion

Mission	Launch Date	Crew-Members	Duration (D:H:M)	Neuroscience Experiments
STS-76 (Atlantis) / Mir-21 Shuttle-Mir Mission 2	22-Mar-96	Lucid	188:04:01	Anticipatory Postural Activity During Long Duration Space Flight The Effects of Long Duration Space Flight on Gaze Control
STS-78 (Columbia) Spacelab LMS Mission	20-Jun-96	Henricks Kregel Helms Linnehan Brady Favier Thirsk	16:21:48	Canal Otolith Integration Studies (COIS) Torso Rotation Experiment (TRE)
Soyuz-TM24 / Mir-22	17-Aug-96	Korzun Kaleri	196:17:26	Sensory and Motor Mechanisms in Vertical Posture Control After Long Duration Exposure to Microgravity (Ravnovesie) The Effects of Long-Duration Space Flight on Eye, Head, and Trunk Coordination During Locomotion
STS-79 (Atlantis) / Mir-22 Shuttle-Mir Mission 3	16-Sep-96	Blaha	128:05:29	Recovery of Neurological Function in Long Duration Crewmembers
STS-81 (Atlantis) / Mir-23 Shuttle-Mir Mission 4	12-Jan-97	Linenger	132:04:01	Countermeasures and Correction of Adaptation to Space Syndrome and SMS (Sensory Adaptation) Kinematic and Dynamic Locomotion Characteristics Prior and After Space Flight (Lokomotsi) Microgravity Impact on Induced Muscular Contraction (Tendometria) Recovery of Neurological Function in Long Duration Crewmembers Sleep Investigations Study of Hypo-Gravitational Ataxia Syndrome (Motor Control) Sensory and Motor Mechanisms in Vertical Posture Control After Long Duration Exposure to Microgravity

Mission	Launch Date	Crew-Members	Duration (D:H:M)	Neuroscience Experiments
				(Ravnovesie)
				The Effects of Long-Duration Space Flight on Eye, Head, and Trunk Coordination During Locomotion
Soyuz-TM25 / Mir-23 ESA-Mir Mission	10-Feb-97	Lazutkin Tsibliev Ewald	184:22:07 Lazutkin Tsibliev 19:16:34 Ewald	Sensory and Motor Mechanisms in Vertical Posture Control After Long Duration Exposure to Microgravity (Ravnovesie) Sleep and Vestibular Adaptation Sleep Investigations Study of Hypo-Gravitational Ataxia Syndrome (Motor Control)
STS-84 (Atlantis) / Mir-23 & 24 Shuttle-Mir Mission 5	15-May-97	Foale	144:13:48	Sleep Investigations
Soyuz-TM26 / Mir-24	5-Aug-97	Solovyov Vinogradov	197:17:34	Countermeasures and Correction of Adaptation to Space Syndrome and SMS (Sensory Adaptation) Kinematic and Dynamic Locomotion Characteristics Prior and After Space Flight (Lokomotsi) Microgravity Impact on Induced Muscular Contraction (Tendometria) Sensory and Motor Mechanisms in Vertical Posture Control After Long Duration Exposure to Microgravity (Ravnovesie) Sleep Investigations Study of Hypo-Gravitational Ataxia Syndrome (Motor Control)
STS-86 (Atlantis)	25-Sep-97	Wetherbee Bloomfield Titov Parazynski Chretien Lawrence	10:19:22	Adaptation to Linear Acceleration After Space Flight (DSO 207)
STS-86 / Mir-24	25-Sep-97	Wolf	127:20:02	Sleep Investigations

Mission	Launch Date	Crew-Members	Duration (D:H:M)	Neuroscience Experiments
Shuttle-Mir Mission 6				
STS-89 (Endeavour) / Mir-25	22-Jan-98	Thomas	140:15:13	Recovery of Neurological Function in Long Duration Crewmembers
Shuttle-Mir Mission 7				
Soyuz-TM27 / Mir-25	29-Jan-98	Musabayev Budarin	207:12:51	Sensory and Motor Mechanisms in Vertical Posture Control After Long Duration Exposure to Microgravity (Ravnovesie) Study of Hypo-Gravitational Ataxia Syndrome (Motor Control)
STS-90 (Columbia) Spacelab Neurolab Mission	17-Apr-98	Searfoss Altman Linnehan Williams Hire Buckey Pawelczyk	15:21:50	Artificial Neural Networks and Cardiovascular Regulation Autonomic Neurophysiology in Microgravity Autonomic Neuroplasticity in Weightlessness Frames of Reference and Internal Models Integration of Neural Cardiovascular Control in Space Role of Visual Cues in Spatial Orientation Spatial Orientation of the Vestibulo-Ocular Reflex Visual-Otolithic Interaction in Microgravity Visuo-Motor Coordination during Space Flight Sleep and Respiration in Microgravity
Soyuz-TM28 / Mir-26	13-Aug-98	Padalka Avdeyev Baturin	379:14:51 Avdeyev 198:16:31 Padalka 11:19:41 Baturin	Sensory and Motor Mechanisms in Vertical Posture Control After Long Duration Exposure to Microgravity (Ravnovesie) Study of Hypo-Gravitational Ataxia Syndrome (Motor Control)
STS-95 (Discovery)	29-Oct-98	Brown Lindsey Parazynski Robinson	8:21:44	Postflight Recovery of Postural Equilibrium (DSO 605)

Mission	Launch Date	Crew-Members	Duration (D:H:M)	Neuroscience Experiments
		Duque Mukai Glenn		
Soyuz-TM29 / Mir-27 ESA Mir Mission	20-Feb-99	Afanasyev Haignere Bella	188:20:16 Afanasye Haignere 7:21:56 Bella	Sensory and Motor Mechanisms in Vertical Posture Control After Long Duration Exposure to Microgravity (Ravnovesie) Study of Hypo-Gravitational Ataxia Syndrome (Motor Control)
Soyuz-TM30 / Mir-28	4-Apr-00	Zalyotin Kaleri	72:19:42	Sensory and Motor Mechanisms in Vertical Posture Control After Long Duration Exposure to Microgravity (Ravnovesie) Study of Hypo-Gravitational Ataxia Syndrome (Motor Control)
STS-106 (Atlantis)	8-Sep-00	Wilcutt Altman Burban Lu Mastracchio Malenchen- ko Morukov	11:19:12	Eye Movements and Motion Perception Induced by Off-Vertical Axis Rotation (OVAR) at Small Angles of Tilt After Space Flight (DSO 499)
Soyuz-TM31 / ISS Expedition-1	31-Oct-00	Shepherd Gidzendko Krikalev	140:23:28	Countermeasures and Correction of Adaptation to Space Syndrome and SMS (Sensory Adaptation) Functional Neurological Assessment (Posture) Kinematic and Dynamic Locomotion Characteristics Prior and After Space Flight (Lokomotsi) Microgravity Impact on Induced Muscular Contraction (Tendometria) Sensory and Motor Mechanisms in Vertical Posture Control After Long Duration Exposure to Microgravity (Ravnovesie) Study of Hypo-Gravitational Ataxia Syndrome (Motor Control)

Mission	Launch Date	Crew-Members	Duration (D:H:M)	Neuroscience Experiments
STS-102 (Discovery)	8-Mar-01	Wetherbee McNeal Thomas Richards	12:19:50	Effects of Altered Gravity on Spinal Cord Excitability (H-Reflex)
STS-102 / ISS Expedition-2	8-Mar-01	Usachev Helms Voss	167:06:41	Same as Expedition-1
STS-100 (Discovery)	19-Apr-01	Rominger Ashby Hadfield Parazynski Phillips Guidoni Lonchakov	10:19:58	Eye Movements and Motion Perception Induced by Off-Vertical Axis Rotation (OVAR) at Small Angles of Tilt After Space Flight (DSO 499)
STS-104 (Atlantis)	12-Jul-01	Lindsey Hobaugh Gernhardt Reilly Kavandi	12:18:36	Spatial Reorientation Following Space Flight (DSO 635)
STS-105 (Discovery)	10-Aug-01	Horowitz. Sturckow Barry Forrester	11:21:13	Spatial Reorientation Following Space Flight (DSO 635)
STS-105 / ISS Expedition-3	10-Aug-01	Culbertson Dezhurov Tyurin	128:20:45	Same as Expedition-1
Soyuz-TM33 / ISS ESA Andromede Mission	21-Oct-01	Afansyev Andre-Deshays Kozeyev	8:59:35	Cognitive Process for 3D Orientation Perception and Navigation in Weightlessness (COGNI)
STS-108 (Endeavour) / ISS Expedition-4	5-Dec-01	Onufrienko Bursch Walz	195:19:39	Same as Expedition-1
STS-109 (Columbia)	1-Mar-02	Altman Carey Grunsfeld Currie Newman Linnehan Massimino	10:22:11	Spatial Reorientation Following Space Flight (DSO 635)
Soyuz-TM34 / ISS Italian Marco Polo Mission	25-Apr-02	Gidzenko Vittori Shuttleworth	9:21:25	An Investigation of Space Radiation Effects on the Functional State of the Central Nervous System and an Operator's Working Capacity (ALTEINO)

Mission	Launch Date	Crew-Members	Duration (D:H:M)	Neuroscience Experiments
STS-111 (Endeavour)	5-Jun-02	Cockrell Lockhart Chang-Diaz Perrin	13:20:36	Eye Movements and Motion Perception Induced by Off-Vertical Axis Rotation (OVAR) at Small Angles of Tilt After Space Flight (DSO 499) Spatial Reorientation Following Space flight (DSO 635)
STS-111 (Endeavour) / ISS Expedition-5	5-Jun-02	Korzun Whitson Treschev	184:22:14	Same as Expedition-1 Promoting Sensory-Motor Response Generalizability: A Countermeasure to Mitigate Locomotor Dysfunction After Long-duration Space Flight (Mobility)
STS-112 (Atlantis)	7-Oct-02	Ashby Melroy Wolf Sellers Magnus Yurchikhin	10:19:58	Eye Movements and Motion Perception Induced by Off-Vertical Axis Rotation (OVAR) at Small Angles of Tilt After Space Flight (DSO 499) Spatial Reorientation Following Space flight (DSO 635)
Soyuz-TMA1 / ISS Belgian Odissea Mission	30-Oct-02	Zalyotin De Winne Lonchakov	10:20:53	Directed Attention Brain Potentials in Virtual 3D Space in Weightlessness (NEUROCOG) Sleep-Wake Actigraphy and Light Exposure During Space Flight (SLEEP) (DSO 634) Stress, Cognition and Physiological Response During Space Flight (COGNISPACE) Sympathoadrenal Activity in Humans During Space Flight (SYMPATHO)
STS-113 (Endeavour)	24-Nov-02	Wetherbee Lockhart Lopez-Alegria Herrington Bowersox	13:18:48	Eye Movements and Motion Perception Induced by Off-Vertical Axis Rotation (OVAR) at Small Angles of Tilt After Space Flight (DSO 499)
STS-113 / ISS Expedition-6	24-Nov-02	Bowersox Pettit Budarin	161:01:17	Same as Expedition-1 Promoting Sensory-Motor Response Generalizability: A Countermeasure to Mitigate Locomotor Dysfunction After Long-duration Space

Mission	Launch Date	Crew-Members	Duration (D:H:M)	Neuroscience Experiments
				Flight (Mobility) Study of the Action Mechanism and Efficacy of Various Countermeasures Aimed at Preventing Locomotor System Disorders in Weightlessness (Profilaktika)
STS-107 (Columbia)	16-Jan-03	Husband McCool Clark Chawla Brown Anderson Ramon	15:22:20	Sleep-Wake Actigraphy and Light Exposure During Space Flight
Soyuz-TMA2 / ISS Expedition-7	26-Apr-03	Lu Malenchen-ko	13:18:48	Same as Expedition-6
Soyuz-TMA3 / ISS Expedition-8	18-Oct-03	Foale Kaleri Duque	194:18:35 Foale, Kaleri 09:21:02 Duque	Same as Expedition-6 Directed Attention Brain Potentials in Virtual 3D Space in Weightlessness (NEUROCOG)
Soyuz-TMA4 / ISS Expedition-9	19-Apr-04	Padalka Fincke Kuipers	187:21:17 Padalka, Fincke 10:20:52 Kuipers	Same as Expedition-8 Effects of Weightlessness on Eye Movements, Body Coordination, and Posture Effects of Weightlessness on Motion Perception and Susceptibility to Space Sickness (MOP)
Soyuz-TMA5 / ISS Expedition-10	14-Oct-04	Chiao Sharipov Shargin	192:19:2 Chiao, Salizhan 9:21:29 Shargin	Same as Expedition-9
Soyuz-TMA6 / ISS Expedition-11	14-Apr-05	Krikaliev Phillips	179:00:23	Bioavailability and Performance Effects of Promethazine During Space Flight Foot/Ground Reaction Forces During Space Flight Hand Posture Analyzer Promoting Sensorimotor Response Generalizability: A Countermeasure to Mitigate Locomotor Dysfunction After Long-Duration Space Flight (Mobility)

Mission	Launch Date	Crew-Members	Duration (D:H:M)	Neuroscience Experiments
STS-114 (Discovery)	26-Jul-05	Collins Kelly Camarda Lawrence Noguchi Robinson Thomas	13:21:32	Eye Movements and Motion Perception Induced by Off-Vertical Axis Rotation (OVAR) at Small Angles of Tilt After Space Flight (DSO 499) Spatial Reorientation Following Space flight (DSO 635)
Soyuz-TMA7 / ISS Expedition-12	3-Oct-05	McArthur Tokarev Olsen	189:19:53 McArthur Tokarev 9:21:21 Olsen	Bioavailability and Performance Effects of Promethazine During Space Flight (PMZ) Cognitive Cardiovascular Experiment (CARDIOCOG-2) Sleep-Wake Actigraphy and Light Exposure During Space Flight
Soyuz-TMA8 / ISS Expedition-13	29-Mar-06	Vinogradov Williams Reiter	182:23:44 Vinogrod Williams	Same as Expedition-12 Cultural Determinations of Co-working, Performance and Error Management in Space Operations (CULT)
STS-121 (Discovery)	4-Jul-06	Fossum. Kelly Wilson Lindsey Sellers Reiter Nowak	12:18:37	Eye Movements and Motion Perception Induced by Off-Vertical Axis Rotation (OVAR) at Small Angles of Tilt After Space Flight (DSO 499) Spatial Reorientation Following Space flight (DSO 635)
STS-115 (Atlantis)	9-Sep-06	Jett Ferguson Stefanyshyn-Piper Tanner Burbank MacLean	12:02:34	Perceptual Motor Deficits in Space (PMDIS) Spatial Reorientation Following Space Flight (DSO 635)
Soyuz-TMA9 / ISS Expedition-14	18-Sep-06	Lopez-Alegria Tyurin Reiter Ansari	215:08:23 Lopez-Alegria, Tyurin 10:21:04 Ansari	Same as Expedition-13 Anomalous Long Term Effects in Astronauts' Central Nervous System (ALTEA) Countermeasures for Space Adaptation Syndrome and Space Motion Sickness Functional Neurological Assessment (Posture) (MR042L) Kinematic and Dynamic Locomotion Characteristics Prior and After Space Flight

Mission	Launch Date	Crew-Members	Duration (D:H:M)	Neuroscience Experiments
				(Lokomotsi)
				Microgravity Impact on Induced Muscular Contraction (Tendometria)
				Researching for Individual Features of State Psychophysiological Regulation and Crewmembers Professional Activities during Long Space Flights (Pilot)
				Sensory and Motor Mechanisms in Vertical Posture Control after Long Duration Exposure to Microgravity (Ravnovesie)
				Studies of Listing's Plane under Different Gravity Conditions (ETD)
				Study of Hypo-Gravitational Ataxia Syndrome (Motor Control)
				Study of the Action Mechanism and Efficacy of Various Countermeasures Aimed at Preventing Locomotor System Disorders in Weightlessness (Profilaktika)
				Test of Reaction and Adaptation Capabilities (TRAC)
STS-116 (Discovery) / ISS Expedition-14	9-Dec-06	Polansky Oefelein Curbeam Higginbo-tham Patrick Fuglesang Williams	12:20:45 (STS-116) 194:18:03 Williams	Functional Neurological Assessment (Posture) (MR042L) Perceptual Motor Deficits in Space (PMDIS) Spatial Reorientation of Sensorimotor Balance Control in Altered Gravity (DSO 635)
Soyuz-TMA10 / ISS Expedition-15	7-Apr-07	Yurchikhin Kotov Simonyi	173:06:28 Yurchikin, Kotov 13:19:00 Simonyi	Same as Expedition-14 Elaboratore Imagini Televisive - Space 2 (ELITE-S2)
STS-117 (Atlantis) / ISS Expedition-15	8-Jun-07	Sturckow Archam-bault Forrester Swanson Olivas Reilly Anderson	13:20:12 (STS-117) 111:00:21 Anderson	Perceptual Motor Deficits in Space (PMDIS) Functional Neurological Assessment (Posture) (MR042L)

Chapter 3

OPERATIONAL ASPECTS

Space motion sickness is the most clinically significant phenomenon that occurs during the first few days of space flight, and immediately following flight. It has significant impact on human spacecraft operations, including delays in the performance of *extra-vehicular activities* (EVA), and accomplishing critical activities on a specified timeline. Other, more chronic problems include sleep disorders, decreased head-eye coordination and precision of movements, increased reaction time, memory problems, and fatigue.

Some problems also occur after the mission, during the re-adaptation to the one-g environment. The longer the mission, the longer the after-effects. For example, crewmembers have major difficulty walking after landing, and running, while possible, is difficult until five days after landing. Impairment in standing and walking is partly due to head movements, which cause persistent illusory spinning and pitching sensations as long as seven days after the flight. After a three-month mission, subjective heaviness and spatial disorientation episodes can last up to one month after landing, and in some cases have lasted for as long as a year or more following the flight.

In addition to disrupting the well being of crewmembers, these disturbances have the potential to decrease their operational efficiency. Several crewmembers have felt that unassisted emergency egress was not possible immediately after landing, and that motion illusions impair the ability to function normally in everyday activities. After long-duration space flight, full recovery of balance, as measured by a posture platform, takes up to four weeks. However, some crewmembers felt like they did not return to baseline until between ten weeks and five months later.

Figure 3-01. This humorous picture was taken during the IML-1 Spacelab mission (Shuttle mission STS-42). The astronauts were illustrating their participation in "another puking experiment." This experiment utilized a rotating chair for the investigation of eye movements and perception during controlled stimulation of the vestibular system during space flight. Photo courtesy of NASA.

Neurovestibular problems may also have a significant impact on during sub-orbital flights participants envisioned for space tourism. This is particularly relevant as participants are likely to have an active role in aspects of vehicle operations related to

G. Clément, M.F. Reschke, *Neuroscience in Space*.
DOI: 10.1007/978-0-387-78950-7_3, © Springer Science+Business Media, LLC 2008

health and safety. For example, these individuals may be expected to return to their designated seats after a period of weightlessness and reattach their own harness for re-entry. In addition, like the astronauts during off-nominal operations, they may be expected to egress the vehicle without assistance upon landing, or take particular actions in an emergency situation. To avoid these problems, research is directed toward neurophysiological, behavioral, and psychological investigations, and the development and evaluation of effective countermeasures.

1   SPACE MOTION SICKNESS

The anomalous perceptual, sensory, sensorimotor, and autonomic reactions that develop during the initial period of adaptation to weightlessness are reminiscent of the clinical form of standard, terrestrial motion sickness. This similarity led many U.S. and Russian scientists researching the physiological effects of weightlessness to refer to this phenomenon as *space motion sickness* (SMS). Accompanying SMS are reports of visual, orientational, proprioceptive, and self-motion illusions. Other disturbances include sleep disorders, a decreased ability to perform high-precision tasks, as well as cognitive and performance impairments. It is important to differentiate between what some call the *Space Adaptation Syndrome* (SAS) and SMS. SAS may include SMS, but it also refers to the tendency of the physiological systems to assume a zero-g normal state, or flight homeostasis. SMS, on the other hand, is specific to the motion sickness developed during space flight, and by definition its appearance requires motion of either the self or the surrounding visual environment.

1.1   Signs and Symptoms

On Earth, exposure to provocative real or apparent motion leads to the progressive cardinal symptoms of terrestrial motion sickness. These symptoms typically include pallor, increased body warmth, cold sweating, dizziness, drowsiness, nausea, and vomiting (Figure 3-01). The constellation of signs and symptoms of SMS when taken together with the time course of symptom development and movements encountered upon exposure to microgravity, suggests that sickness experienced during space flight is similar to terrestrial motion sickness.

There are, however, several important features associated with SMS that may not be part of motion sickness experienced on Earth. The symptoms experienced as part of SMS may differ slightly from those exhibited during acute provocation on the ground. In particular, there is virtually no occurrence of sweating in SMS, and flushing is more common than pallor (Homick & Miller 1975, Oman *et al.* 1990) (Figure 3-02). Nearly universal are malaise, anorexia or loss of appetite, lack of initiative, and irritability. In microgravity there are more reports of stomach awareness, vomiting, headache (due perhaps to headward fluid shifts), impaired concentration, lack of motivation, and drowsiness than are typically seen during acute motion sickness on the ground (Homick & Vanderploeg 1989, Thornton *et al.* 1987, Davis *et al.* 1988). Some of these symptoms, particularly the lack of motivation to work or interact with others, drowsiness, fatigue, and the inability to concentrate, could be due to *sopite syndrome*, which may be a byproduct of a dizziness experienced by astronauts during space travel.

During SMS, vomiting is usually sudden and often without prodromal nausea. A well-known Shuttle astronaut recalled, "I checked and double-checked and then triple-checked that my numerous barf bags were ready for a quick draw. The veterans had warned us the sickness would come on very suddenly. They were right. The curse hit." (Mullane 2006, p. 171).

The bouts of SMS are not frequent, usually separated by one to three hours, with no dry heaves. Bowel sounds, obtained by auscultation, have been found to be decreased or absent in astronauts suffering from SMS (Thornton *et al.* 1987). Typically, the gastrointestinal symptoms have their onset from minutes to hours after orbital insertion.

To our knowledge, symptoms (at least none as catastrophic as vomiting) have not been experienced during EVA. However, there have been one, and possibly two vomiting episodes while donning the space suit. To prevent SMS during an EVA, NASA mission rules restrict crews from performing EVA before the third day of flight since SMS usually abates by then, and a minimum flight duration has been set at three days to ensure that no astronauts are sick prior to re-entry and landing (Davis *et al.* 1988).

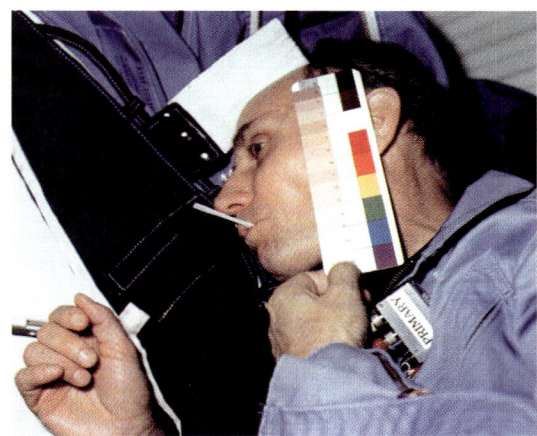

Figure 3-02. A Space Shuttle astronaut is measuring his skin pallor for an experiment on space motion sickness. Pallor in the face often occurs during early stomach symptoms of epigastric discomfort. Pallor is the result of vasoconstriction, due to increased sympathetic activity controlled by hypothalamic centers. Vasoconstriction of the skin and facial pallor may also be due, in part, to the increased levels of vasopressin during motion sickness. Photo courtesy of NASA.

1.2   Incidence

The Russian Cosmonaut Titov was the first to experience (and report) symptoms of SMS. Symptoms have been reported by 48% of the cosmonauts in the Russian space program (Gorgiladze & Bryanov 1989). There were no reports of SMS in the American Mercury and Gemini programs, while 35% of the Apollo astronauts developed symptoms and the incidence in the Skylab missions was approximately 60% (Davis *et al.* 1988). These incidence numbers are, however, probably underestimated. According to Mike Mullane, a retired Shuttle astronaut, "Astronauts didn't want to admit to an episode of vomiting out of fear that it would eliminate them from consideration for future spacewalk missions. As a result, many astronauts were less than truthful about their symptoms. Some blatantly lied. We would hear stories of crewmembers who were seriously sick, yet the data would never appear on the flight surgeon's bar charts." (Mullane 2006, p. 107).

Rightly or wrongly, the U.S. space program categorizes the severity of SMS depending on its impact upon crew performance (Table 3-01). For example, 'mild' SMS has no operational impact, because the crewmember can still perform all the required activities. 'Moderate' or 'severe' SMS are operational concerns since the workload must be redistributed among the remaining, unaffected crew.

None	No signs or symptoms reported
Mild	One to several transient symptoms No operational impact All symptoms resolved in 36-48 hours
Moderate	Several symptoms of a persistent nature Minimal operational impact All symptoms resolved in 72 hours
Severe	Several symptoms of a persistent nature Significant performance decrement Symptoms may persist beyond 72 hours

Table 3-01. NASA categorization of Space Motion Sickness according to the severity of symptoms. It is important to note that even with a classification of 'Mild', vomiting can still occur, making this classification system less than adequate. The 'Mild', 'Moderate' and 'Severe' categories are also referred to as "One bag", "Two bags" and "Three bags", respectively[5] (Oman 2007).

In the first 36 missions of the Space Shuttle program, about 71% of the 109 crewmembers making their first flight reported symptoms of SMS. Of these astronauts, about 33% reported 'mild' symptoms, 27% 'moderate' symptoms, and 11% 'severe' symptoms (Davis *et al.* 1988). In a more recent postflight survey on 112 crewmembers during Shuttle missions flown between 1996 to 2000, about 37% of crewmembers reported 'mild' symptoms, 9% 'moderate' symptoms, and 2% 'severe' symptoms (Locke 2003).

In all these flights, there were no statistically significant differences in symptom occurrence between career vs. non-career astronauts, commanders and pilots vs. mission specialists, males vs. females, different age groups, or first-time vs. repeat flyers. Also, aerobic fitness was not related to SMS symptoms or severity (Jennings *et al.* 1988). Those who are susceptible on their first flight usually have SMS on subsequent flights. The severity of SMS among those making a second flight remained unchanged in 56% of crewmembers, whereas a slight improvement was observed in 35%, but even more symptoms were noted in 9% (Davis *et al.* 1988). Thus, previous SMS is the best predictor for future SMS.

Space motion sickness is self-limiting. Most crewmembers recover by the end of the third day in space (Thornton *et al.* 1987). In a few cases in the Russian and U.S. missions, however, crewmembers were ill for 7-14 days. After complete adaptation occurs, crewmembers appear to be immune to the development of further symptoms. This development of immunity to further SMS symptoms was eloquently demonstrated by rotating chair tests conducted in-flight during the Skylab missions, which were specifically designed to provoke an SMS response[6] (Graybiel *et al.* 1975, 1977) (see Figure 7-04). Preflight, most of the nine participating astronauts were highly susceptible to the Coriolis and cross-coupled stimulations when they made head movements while spinning on the rotating chair. In-flight, five of the nine astronauts experienced SMS

---

[5] Typically astronauts may also refer to their level of sickness in the Garn unit, with one Garn being as sick as a human can possibly be, named after the infamous Senator Jake Garn (NASA's first space tourist) who experienced extreme SMS during his entire mission on the Space Shuttle in 1985.

[6] Perhaps since that time, every time an investigator mentions the use of a rotating chair during and after space flight, astronauts and flight managers alike frown at the thought of "another puking experiment."

during their first several days in orbit, delaying the initial rotation tests. Crewmembers were encouraged to restrain themselves against a "wall" of the orbiting laboratory and make pitching head movements with their eyes open and then closed. This adaptation protocol may have helped develop some immunity to SMS. When initially exposed to the Coriolis and cross-coupled angular acceleration of the rotating chair on or after the sixth day in-flight, the astronauts were insusceptible to motion sickness. They did not become even slightly sick when they made 150 head movements while rotating at 30 rpm. This resistance to motion sickness during the chair rides continued on the ground for several weeks after the flight.

However, adaptation to provocative motion during flight does not always convey immunity to motion sickness immediately afterward. Approximately 30% of the Skylab astronauts experienced seasickness in the *Command Module* and on board the recovery ship. Since the seas were rough, the contribution of space flight to these symptoms is unclear. Within two to three days, these same crewmembers (along with those not susceptible to seasickness) were immune to the programmed Coriolis and cross-coupled accelerations experienced during passive rotation, and this immunity lasted for several weeks (Graybiel *et al.* 1977). In the Russian space program, about 27% of cosmonauts following short-duration flights (4 to 14 days), and 92% returning from longer missions present symptoms of motion sickness after landing similar to the 'mal de débarquement' in seafarers (Gorgiladze & Bryanov 1989).

No reports of this *postflight motion sickness* (PFMS) were noted in the U.S. Shuttle program through the mid 1980's (Thornton *et al.* 1987). However, it now appears that this syndrome affects a similar percentage of both U.S. and Russian crews. The Russian reports indicate that PFMS symptoms generally occur in cosmonauts who have SMS in-flight. However, 11% of those who experience little or no SMS on orbit do experience mal de débarquement (Bryanov *et al.* 1986). Postflight medical debriefs were examined for Shuttle missions from the beginning of the program, in April 1981, through January 1999, which involved 241 crewmembers having flown between one and six missions. Postflight, 32% of crewmembers reported vertigo, 14.7% reported nausea, and 8% vomiting (Bacal *et al.* 2003).

PFMS onset occurs in a time pattern similar to that of SMS. Within minutes of g-force onset during re-entry symptoms may already be developing. Crewmembers who have no symptoms during re-entry and landing may develop symptoms as soon as they stand up to exit the vehicle. The severity of the symptoms and the functional recovery seem to be directly proportional to the time on orbit. There have been reports of a "relapse" phenomenon in the post-landing recovery course. Astronauts who are exposed to certain types of inertial environments, like turning a corner in a car or lying in bed in the dark, can bring on a sudden return to an early postflight state of maladaptation, which may elicit 'mild' to 'severe' PFMS symptoms several days up to a week after return to Earth. Recovery from this "relapse" generally occurs more rapidly than the recovery immediately after returning from orbit (Ortega & Harm 2007).

Like SMS, PFMS does not appear to correlate with gender, age, crew position, or number of previous flights. Past experience with postflight re-adaptation does not seem to affect incidence (Bacal *et al.* 2003). PFMS is likely complicated by the relative dehydration upon return and orthostatic intolerance following flight.

Figure 3-03. Astronauts often complained that going through the tunnel connecting the Space Shuttle middeck and the Spacelab module was particularly disorienting and provocative of space motion sickness symptoms. Photo courtesy of NASA.

1.3   Provocative Stimuli

Factors that may initiate or worsen SMS include distasteful, unpleasant, or uncomfortable, sights noxious odors, certain foods, excessive warmth, loss of one-g orientation (Figure 3-03), and head movements (Jennings 1998). Similarly, postflight symptoms may be induced and/or exacerbated by warmth and head movements during re-entry and immediately after landing.

1.3.1   Head and Body Movements

Microgravity by itself does not induce SMS. There were no reports of motion sickness during the Mercury and Gemini space flights. However, as the volume of spacecraft has increased, the incidence of SMS has increased as well. Astronauts and cosmonauts quickly observed that excessive movement early on-orbit commonly increased symptoms (Graybiel 1980). Ground-based studies also showed that motion sickness arises when movements are made during exposure to inertial backgrounds higher in magnitude than the one-g Earth gravity (Lackner & Graybiel 1986, 1987).

In fact, the reduced overall incidence of SMS in cosmonauts (48%) compared to about 71% in astronauts during the first Shuttle missions may be explained by the smaller cabin sizes in Russian space vehicles, which equate to less freedom of movement. Shuttle crews are suddenly released into a large volume after an eight-minute rocket ride. They must doff the launch and re-entry suit (known as the LES, which is designed to protect crews from the sudden loss of cabin pressure), an activity that involves significant head movements. Russian Soyuz crews spend one to two days on board a much smaller spacecraft before entering in the larger Mir or ISS volumes. The large volume of the Shuttle combined with the high activity level immediately upon orbital insertion, compared to the small volume of the Soyuz and low activity levels for the first few days on orbit, may account for the differences in SMS incidence rates in astronauts and cosmonauts.

It has been well documented that movements that produce changes in orientation, particularly whole body or head movements, are the most provocative. Among those, pitch head movements are initially more provocative than head movements made in other planes (Thornton *et al.* 1987, Oman *et al.* 1990). However, once sickness has been established, head movements in any plane are generally minimized by the affected crewperson. Indeed, movement of any kind is frequently restricted until the astronaut is on the road to recovery. Some crewmembers who have suffered repeated, persistent bouts of vomiting have restrained themselves to the structure of the spacecraft until SMS has been resolved. Movement, on the other hand, is necessary to overcome SMS and adapt.

Head or body movements made upon transitioning from microgravity to a gravitational field less than the Earth's, and vice versa, may not be as provocative. It is interesting to note that of the twelve Apollo astronauts who walked on the Moon, only three reported 'mild' symptoms, such as stomach awareness or loss of appetite, prior to their moon walk. None reported symptoms while in the one-sixth gravity of the lunar surface and no symptoms were noted upon return to weightlessness (Homick & Miller 1975, Schmitt & Reid 1985).

1.3.2 Orientation Cues

Visual cues are known to play a role in spatial orientation and SMS symptoms can be elicited in orbit during episodes of reorientation, i.e., when the visual scene does not correspond to that expected by the astronaut, requiring an adjustment in perceived "up" and "down" (Figure 3-04).

Individual astronauts differ with respect to the adaptation strategies they use to compensate for the sensory disturbances encountered following exposure to microgravity. Some astronauts (Type VS for 'visuo-spatial') report that they become predominantly "visual creatures" in weightlessness. Apparently these astronauts develop a spatial orientation framework by increasing the weighting of visual spatial orientation signals. Type VS astronauts may experience discomfort or symptoms when visual objects that exhibit a consistent polarization with respect to gravity on Earth are seen in unusual orientations in space, i.e., when their feet are not oriented to the deck. For example, seeing a fellow crewmember floating "upside down" (Oman *et al.* 1990) or viewing the Earth in unexpected orientations through the Shuttle windows may be reported as disturbing by a Type VS astronaut. Other astronauts (Type IZ for 'internal z-axis') appear to compensate for the absence of gravity primarily by increasing the weighting assigned to internally-generated orientation vectors. These astronauts appear to ignore visual polarity information and use the direction of their feet to define "down."

However, although about 80% to 93% of Russian cosmonauts reported orientation illusions associated with SMS (Gorgiladze & Bryanov 1989), these illusions are seldom reported by U.S. crews.

1.4 Time Course

There are considerable individual differences in susceptibility to SMS, and currently it is not possible to predict with any accuracy those who will have some difficulty with sickness while aloft. Symptom resolution usually occurs between 30 and 48 hours. However, the recovery rate, degree of adaptation, and specific symptoms vary widely between individual astronauts.

Although anti-motion sickness drugs offer some protection against SMS (see Chapter 8, Section 2.3), some drugs (i.e., scopolamine) may interfere with the adaptation process and symptoms controlled by these drugs are experienced once treatment ceases. This "state dependency" has been reported by several astronauts who experienced episodes of nausea and vomiting three to four hours following the last dose of a three-day prophylactic regimen of scopolamine (0.4 mg) and dexedrine (5.0 mg).

Unless all movement is inhibited, most susceptible astronauts begin to experience space sickness within the first hour of orbital flight. Again, there are wide variations in the latency and intensity of SMS. Several crewmembers after leaving their seats have vomited only minutes following cut-off of the Shuttle's main engines (known as MECO). However, both the frequency and intensity of symptoms appear to decrease as the flight progresses. By the sixth day of the Skylab flights, all astronauts were immune to SMS, and, as described above, could not be made sick with rotation and programmed head movements (Graybiel *et al.* 1977).

Symptoms (at least none as catastrophic as vomiting) have never occurred during *extra-vehicular activity* (EVA). However, since most SMS has abated by the third day of flight, mission rules now restrict EVA until the third mission day, and the minimum flight duration is three days to ensure that no astronauts are sick prior to re-entry and landing (Davis *et al.* 1988).

Figure 3-04. Inside the ISS modules, up and down are a matter of personal perspective. Note, however, than the instrument keyboard and the booklet are in the "upright" (e.g., ground) configuration. It is therefore likely that the astronaut playing the keyboard took an inverted pose just for the photo shot. Photo courtesy of NASA.

1.5   Theories for Space Motion Sickness

Two major theories advanced to account for SMS are the *fluid shift* theory and the *sensory conflict* (also known as the neural mismatch, sensory mismatch, or sensory rearrangement) theory (Crampton 1990). Although both theoretical positions have some merit and neither is ideal, the fluid shift theory may have the most difficulty in dealing with the development of motion sickness during space flight.

1.5.1 Fluid Shift Theory

According to this theory, the headward fluid shifts accompanying weightlessness would produce concomitant changes in intracranial pressure, the cerebrospinal fluid column, or the inner ear, thereby altering the response properties of vestibular receptors. However, Graybiel & Lackner (1979) did not find an increased susceptibility to provocative motion stimulation during head-down tilt, even when the subjects were rotated around their longitudinal body axis in the tilted position. Also, anecdotal reports from Space Shuttle astronauts and limited in-flight measurements of responses presumably sensitive to increased intracranial fluid pressure (e.g., auditory evoked potentials and intraocular pressure) do not favor the fluid shift hypothesis (Thornton *et al.* 1985, Lackner & DiZio 2006).

1.5.2 Sensory Conflict Theory

On the other hand, the theory of sensory conflict advanced by Reason & Brand (1975) best explains SMS, and is parsimonious. Briefly, the sensory conflict theory of motion sickness assumes that human orientation in three-dimensional space, under normal gravitational conditions, is based on at least four sensory inputs to the central nervous system. The otolith organs provide information about linear accelerations and tilt relative to the gravity vector; angular acceleration information is provided by the semicircular canals; the visual system provides information concerning body orientation with respect to the visual scene or surround; and touch, pressure, and somatosensory (or kinesthetic) systems supply information about limb and body position. In normal environments, information from these systems is compatible and complementary, and matches that expected on the basis of previous experience. When the environment is altered in such a way that information from the sensory systems is not compatible and does not match previously stored neural patterns, motion sickness may result.

The sensory conflict theory postulates that motion sickness occurs when patterns of sensory inputs to the brain are markedly rearranged, at variance with each other, or differ substantially from expectations of the stimulus relationships in a given environment. In microgravity, sensory conflict can occur in several ways. First, there can be conflicting information (i.e., regarding tilt) transmitted by the otoliths and the semicircular canals. Sensory conflict may also exist between the visual and vestibular systems during motion in space; the eyes transmit information to the brain indicating body movement, but no corroborating impulses are received from the otoliths (such as during car sickness). A third type of conflict may exist in space because of differences in perceptual habits and expectations. On Earth, we develop a neural store of information regarding the appearance of the environment and certain expectations about functional relationships (e.g., the concepts of "up" and "down"). In space, these perceptual expectations are at variance, especially during the inversion illusion described above.

It is important to note that no single course of sensory conflict appears to entirely account for the symptoms of space sickness. Rather, it is the combination of these conflicts that somehow produces sickness, although the exact physiological mechanisms remain unknown. Thus, sensory conflict explains everything in general, but little in the specific. Shortcomings of the sensory conflict theory include: (a) its lack of predictive power; (b) the inability to explain those situations where there is conflict but no sickness; (c) the inability to explain specific mechanisms by which conflict actually gives rise to vomiting; and (d) the failure to address the observation that without conflict, there can be no adaptation. The hypotheses outlined below may be helpful in overcoming some of the weaknesses associated with the construct of this theory.

1.5.3 Treisman's Theory

Treisman (1977) suggested that the purpose of mechanisms underlying motion sickness, from an evolutionary perspective, was not to produce vomiting in response to motion, but to remove poisons from the stomach. He believed that motion was simply an artificial stimulus that activated these mechanisms or, more specifically, that provocative motions act upon mechanisms designed/developed to respond to minimal physiological disturbances produced by absorbed toxins. According to Treisman, neural activity to coordinate inputs from all the sensory systems in order to control limb and eye movements would be disrupted by the central effects of neurotoxins. Therefore, disruption of this activity by unnatural motions is interpreted as an early indication of the absorption of toxins, which then activates a mechanism to produce vomiting.

Money & Cheung (1983) hypothesized that if a vestibular mechanism exists to facilitate vomiting in response to poisons, then surgical removal of the vestibular apparatus in animals should result in a defective vomiting response to poisons. This hypothesis was tested by administering four emetic poisons intramuscularly to seven dogs and a fifth poison to four of the animals; each poison was given a total of six times over a two week period prior to surgical removal of the vestibular apparatus. Seven weeks following the surgical procedure, the dogs were again tested with each emetic poison. The vomiting response to pilocarpine and apomorphine was not influenced by removal of the vestibular apparatus, whereas the response to lobeline, levodopa, and nicotine was delayed and failed to occur in 56 of the 108 postoperative tests. The investigators concluded that the mechanism to facilitate vomiting in response to toxins is partly vestibular.

1.5.4 Otolith Mass Asymmetry Hypothesis

Von Baumgarten *et al.* (1982) have proposed a mechanism complementary to the sensory conflict theory to explain individual differences in SMS susceptibility. They suggest that some individuals possess slight functional imbalances, for example, mass differences, between the right and left otolith receptors that are compensated for by the central nervous system in one-g. A mass imbalance between the left and right otoconia is reasonable since there is a continual turnover of otoconia, and it is unlikely that the two otolith membranes would ever weigh exactly the same. This compensation is inappropriate in zero-g, however, since the mass differential is nullified and the compensatory response (either central or peripheral) is no longer correct for the new inertial environment. The result would be a temporary asymmetry producing vertigo, inappropriate eye movements, and postural changes until the imbalance is compensated or adjusted to the new situation. A similar imbalance would be produced upon return to one-g, resulting in postflight vestibular disturbances. Individuals with a greater degree of asymmetry in otolith morphology would thus be more susceptible to SMS.

Because the otoliths govern *ocular counter-rolling* (OCR), a reflexive rotation of the eyes in the direction opposite to that of a head tilt, it has been proposed that otolith asymmetry should be reflected in the stability of ocular torsion (Diamond & Markham 1991). Experiments conducted in parabolic flight found that two significantly different groups could be distinguished on the basis of variability of disconjugate eye torsion (the standard deviation of the differences in amplitude of torsion between the two eyes) and an index of ocular torsion instability (an indicator of the two eyes torquing in opposite directions while in an upright position) exhibited in hypo- and hypergravity. Two former astronauts with opposing SMS histories also participated in the experiment; one, who had experienced SMS symptoms, was a member of the high disconjugate torsion/high instability group while the other, who had no SMS symptoms, was in the low disconjugate torsion/low instability group. The two astronauts did not exhibit differences in the mean amplitude of OCR and that same measure did not divide the remaining subjects into the same groups as the other two measures. The significant group differences observed in the hypo- and hypergravity force environments were not evident in measurements taken in one-g, perhaps indicating why susceptibility to terrestrial-gravity motion sickness tests does not correlate well with SMS incidence and, in particular, why eye torsion studies on the Earth are not predictive (Diamond & Markham 1998).

In another study, the random ocular torsion that occurs normally while in an upright position was measured in nine former astronauts during parabolic flight. Prior to the flight, the subjects completed SMS history questionnaires. Disconjugate torsional eye movements were measured during both hypo- and hypergravity phases. When ranked by increasing asymmetry during 0 g and 1.8 g, the four subjects with the least asymmetry had not experienced SMS symptoms while the five subjects with higher asymmetries had suffered from SMS. These two groups were significantly different on the basis of the asymmetry measures. For the latter group, the latency, severity and duration of SMS were positively correlated with torsional asymmetry scores (Diamond & Markham 1991).

Other tests of otolith and canal function conducted under terrestrial gravity conditions by the Russians have not been adequate predictors of SMS. Yakovleva *et al.* (1982) reported that cosmonauts with heightened canal threshold sensitivity, otolithic asymmetry, or non-standard canal-otolith interaction were more prone to SMS. However, some cosmonauts exhibiting responses within the normal range also experienced SMS, implicating the involvement of other factors. The model developed by Diamond & Markham (1991) may also fail as a predictor, once an adequate number of subjects have been examined, unless additional predictors are included in the analysis.

In addition, there are at least three problems with a hypothesis based on otolith mass asymmetry:

a.   If an Earth-based compensation for otolith mass asymmetry was released in microgravity, and this compensation was enough to produce SMS, then crewmembers should develop symptoms *without* head and body movements. However, the majority of astronauts who have experienced SMS have said that movement was required in the development of symptoms.

b.   This hypothesis does not explain the role of visual orientation in the development of SMS symptoms. For example, the illusion of a self-motion when there is a movement in the visual field could be perceived without head or body movements.

c.   There is no correlation of SMS in orbital flight with SMS during parabolic flight. According to the otolith asymmetry hypothesis, SMS should occur without head movements during the weightless phase of the flight, and also during the hypergravity phase of flight. Only about half the subjects tested during parabolic flight develop symptoms in the absence of head movements and both susceptible and insusceptible subjects demonstrate earlier onset of symptoms and increased symptom severity when head movements are performed in either hypo- or hypergravity force environments vs. head stationary conditions (Lackner & Graybiel 1985). Parabolic flight is not a good predictor of who will have difficulty on-orbit.

1.5.5   Sensory Compensation Hypothesis

Sensory compensation occurs when the input from one sensory system is attenuated and signals from others are augmented. In the absence of an appropriate graviceptor signal (or perhaps the presence of atypical signals) in microgravity, information from other spatial orientation receptors such as the eyes, the semicircular canals, and the neck position receptors would be used to maintain spatial orientation and movement control (Parker & Parker 1990). In fact, astronauts frequently report that they increase their reliance on visual cues for spatial orientation and motion control (Young *et al.* 1984, Clément *et al.* 1984). The increase in reliance on visual cues for spatial orientation could be explained by this mechanism. Closely related to this sensory compensation hypothesis is the OTTR hypothesis described below.

1.5.6   Otolith Tilt-Translation Reinterpretation Hypothesis

The *otolith tilt-translation reinterpretation* (OTTR) hypothesis was proposed on the basis of data from experiments performed as early as the eighth Shuttle flight (Parker *et al.* 1985), and represents a refinement of an otolith reinterpretation hypothesis developed by Young and his colleagues (Young *et al.* 1984). The rationale for this hypothesis is as follows:

a.   Weightlessness is a form of stimulus rearrangement to which people adapt.

b.   Because of the fundamental equivalence between linear acceleration and gravity, graviceptors signal both the head orientation with respect to gravity (tilt) and a linear acceleration of the head that is perceived as translation.

c.   As a consequence of the absence of sensed gravity during orbital flight, graviceptors do not respond to static pitch or roll in weightlessness; however, they do respond to linear acceleration. Because stimulation from gravity is absent during space flight, interpretation of the graviceptor signals as tilt is meaningless. Therefore, during adaptation to weightlessness, the brain reinterprets all graviceptor output to indicate translation.

As presented above, the OTTR hypothesis would predict that following space flight a forward pitch head movement would be perceived as a rearward translation, and a counter-clockwise roll head movement would be perceived as linear translation to the right (see Figure 4-10). However, several reports from astronauts suggest that the OTTR hypothesis is incorrect as currently formulated. Specifically, the direction of self-motion associated with roll or pitch head movement during re-entry and immediately following landing may be opposite to that predicted by OTTR. Also, the Neurolab experiment with a human-rated centrifuge failed to demonstrate that crewmembers experienced translation rather than tilt when exposed to a steady-state linear acceleration in space (see Chapter 4, Section 5.4.2). There are several possibilities that may account for these differences.

Graviceptor Ambiguity

As for any linear accelerometer, signals from graviceptors, including the otolith organs are fundamentally ambiguous. Graviceptor signals may indicate either tilt (pitch or roll) with respect to gravity or accelerated motion along a straight path. Consequently, veridical perception based on graviceptor signals must rely on other factors. These may include graviceptor signal temporal dynamics and/or signals from other systems.

Motion Attribution Ambiguity

The vestibular nuclei and the vestibulo-cerebellum receive inputs from visual, somatic, proprioceptive, and vestibular receptors. These brain structures may be associated with a hypothesized spatial orientation and motion perceptual system (Gibson 1966). Signals from the vestibular nuclei and the vestibulo-cerebellum sent to oculomotor and spinal-motor control systems contribute to gaze and postural stabilization. A third output from the vestibular nuclei is to the cerebral cortex. Activity in this pathway may be associated with perception of orientation and motion.

Neurons in the vestibular nuclei respond to motion of either the observer or the visual surround (Waespe & Henn 1979). The time constant for the neural responses generated by observer motion (vestibular signals) is shorter than that for neural responses generated by surround-motion (visual signals). Otherwise, the neural signals generated by surround-motion and observer motion are indistinguishable. This leads to surround-motion/observer-motion ambiguity: slow-onset signal changes from the visual or vestibular systems can be perceived by the observer either as self-motion or surround-motion.

Resolution of Graviceptor Ambiguity: Signal Temporal Dynamics

During natural, observer-initiated motion, linear accelerations can be sustained only briefly, whereas head tilts can be sustained for prolonged periods. This has led Benson (1974) to suggest that the temporal dynamics of graviceptor signal changes may contribute to resolution of graviceptor ambiguity. Specifically, short-duration signal changes would tend to be perceived as linear self-motion or linear surround-motion. Long-duration graviceptor signal changes would tend to be perceived as altered orientation of the receptor with respect to gravity. During the Neurolab experiment, crewmembers continued to experience tilt during centrifugation in space. However, centrifugation can be characterized as "0 Hz." Tilt-translation interaction may be a low-frequency phenomenon, with a tilt "singularity" at DC.

Resolution of Graviceptor Ambiguity: Semicircular Canal Contributions

Observer roll and pitch motion from an upright position on Earth elicits both graviceptor and angular motion receptor signal changes. When vision is absent, the observer's angular motion is detected primarily by the semicircular canals. The brain may process neural information so that graviceptor signals that are concurrent with semicircular canal signals are interpreted as angular self-motion, whereas graviceptor signals alone are interpreted as linear motion.

Resolution of Graviceptor Ambiguity: "Top Down" Processes

It is well known that a person's expectations and/or actions contribute significantly to perception (Schachter 1975, Gibson 1966). One approach to this "top down" processing uses the reference model developed by von Holst & Mittelsteadt (1950). This model suggests that perception is the consequence of a comparison between reaffent signals (afferent signal changes associated with observer-initiated motion) and efference copy signals (a neural signal representing the command to move). The brain may process neural information so that graviceptor signal ambiguity is resolved by taking into account the efferent copy generated during voluntary motion.

Implications for OTTR Hypothesis

The above considerations may help to account for the discrepancy between astronaut reports of self- or surround-motion during head tilts and predictions from the OTTR hypothesis. First, there may be an increased probability that a graviceptor signal change results in perceived translation. Secondly, the direction of the perceived translation may be determined by concurrent semicircular canal signals and/or signals associated with top-down processes.

Most of the findings reported later in this chapter concerning locomotion, postural control, and sensory reports associated with space flight can be explained by either the OTTR hypothesis, a concept of sensory compensation, or a combination of the two. Consideration of OTTR and sensory compensation are important in understanding the mechanisms underlying sensory and sensorimotor adaptation to space flight.

1.6   Prediction

The prediction of susceptibility to motion sickness has long been of interest. Since motion sickness treatments are more effective when administered prior to the development of symptoms, the identification of those individuals susceptible to SMS would allow preventive measures to be taken only by those requiring them, and free insusceptible persons from the undesirable side effects of anti-motion sickness medications and/or the scheduling requirements of pre-training (Diamond & Markham 1991).

A number of predictors for motion sickness have been investigated and can be grouped into the following categories:

a.   Exposure history – typically obtained by the use of questionnaires.

b.   Physiological predisposition – measurement of autonomic nervous system tendency (sympathetic or parasympathetic dominance) and sensitivity of sensory end organs.

c.   Psychological predisposition – personality type and perceptual style.

d.   Plasticity – ability to adapt, which may be determined by the physio- logical or psychological predisposition of the individual.

e.   Provocative tests – attempts to recreate in the laboratory the nauseogenic force environment of interest.

f.   Operational measures – prediction of susceptibility to motion sickness in one environment based on susceptibility in another.

1.6.1 Preflight Terrestrial Motion Sickness Susceptibility

In an early attempt to predict SMS, 29 Space Shuttle crewmembers performed a preflight *Coriolis sickness susceptibility index* (CSSI) test to 'severe' malaise endpoint and completed questionnaires related to previous motion sickness experience. Other crewmembers were exposed to *off-vertical axis rotation* (OVAR). Of the approximately one-third of the Shuttle astronauts who performed either a CSSI or an OVAR test prior to space flight, no significant correlation was found between the number of symptoms provoked by the ground-based test and the severity of SMS symptoms, suggesting that susceptibility to ground-based motion stressors is not predictive of susceptibility to SMS.

No correlations were found between any of the biochemical parameters measured and the number of symptom points scored on the OVAR test. However, the number of symptom points accumulated on the CSSI test was found to be significantly and positively correlated with preflight levels of serum sodium, serum chloride and urinary ADH. Significant negative correlations between ground-based motion sickness susceptibility and preflight serum uric acid, serum calcium, plasma angiotensin I, and plasma insulin were also noted (Reschke *et al.* 1994).

1.6.2 Preflight and Postflight Susceptibility

In Space Shuttle crewmembers, the severity of SMS symptoms was significantly and positively correlated with preflight levels of serum chloride and with postflight levels of urinary sodium, chloride, calcium, and uric acid.

A significant negative correlation was found between SMS severity and preflight serum phosphate and uric acid and both pre- and postflight plasma TSH. Non-susceptible crewmembers were found to have significantly higher preflight serum chloride, serum uric acid, plasma cortisol, plasma TSH, and urine volume than those who reported space motion sickness symptoms while also exhibiting lower preflight urine specific gravity, osmolality and phosphate than their more susceptible crewmates (Reschke *et al.* 1994).

1.6.3 In-flight Susceptibility

Available in-flight data for Shuttle crewmembers was divided into two time periods: 24 to 42 hours after launch, during the time SMS symptoms normally peak, and 175 to 190 hours into the mission, after symptom resolution would be expected to occur. During the first time period, a positive correlation was found between serum chloride and magnesium levels and severity of SMS symptoms. Also during this period, a negative correlation was found between symptom severity and plasma ACTH and ADH levels. However, none of these correlations reached statistical significance. A positive correlation between serum creatinine and symptom severity was observed for the second time period, as was a negative correlation between symptomatology and plasma cortisol, aldosterone and ACTH. Only the correlation with cortisol was statistically significant during the second time period (Leach 1987).

The real test of any predictive method is in the use of astronaut data. Ground-based measures on normative subjects, while useful, are not true measures of the criterion of interest, SMS. Until enough flight data become available, along with ground-based tests for flight personnel, the relationships between various predictors and SMS susceptibility will remain unclear.

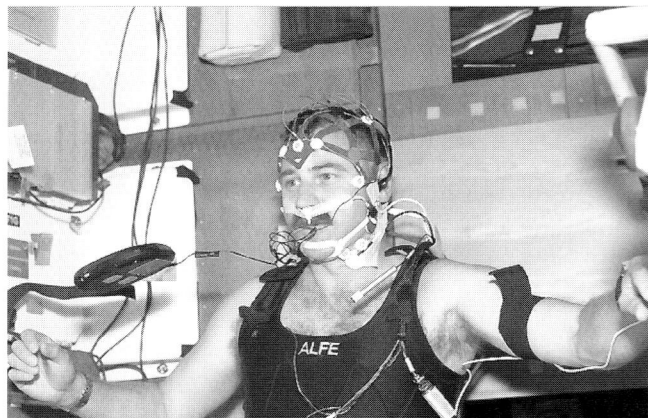

Figure 3-05. Sleep was monitored during the Neurolab mission to evaluate the altered sleep patterns of crewmembers during space flight and to validate clinical trial of melatonin as hypnotic for the crew. Photo courtesy of NASA.

2   SLEEP DISORDERS

While sleep requirements vary from individual to individual, an individual's sleep requirements remain relatively constant over the short term. When the sleep-wake cycle is disrupted, however, it can result in decreased motivation to perform one's job. This is particularly true when the job requires a great deal of concentration. Such a situation could pose a serious risk during a mission to Mars to the crewmembers whose safety often depends upon job performance (Mallis & DeRoshia 2005).

Sleep loss, fatigue, and poor quality sleep have been reported on numerous space missions (Stampi 1994, Frost *et al.* 1997). Qualitative and quantitative studies indicate that crewmembers on Shuttle missions have an average daily sleep period time of five to six hours, compared to their typical period of seven to eight hours on Earth. Daily sleep period time is reduced even more when critical operations occur, such as nighttime Shuttle dockings on ISS, or during an emergency (e.g., equipment failure). On some nights, total sleep time was reduced to as short as 3.8 hours (Dijk *et al.* 2003).

One survey shows that more than 50% of crewmembers use sleeping medication at some point during a mission (Santy *et al.* 1988). Sleep medications are reportedly used by astronauts throughout the mission, in contrast to the space motion sickness medications that are used primarily during the first few days. Along with the inconvenience of disturbed sleep, crewmembers also face the cumulative effects of sleep loss or the carry-over effects of a sleeping pill. These effects manifest themselves by a deterioration of alertness and cognitive performance during the active hours of the day.

2.1   Causes for Sleep Disturbances during Space Flight

Sleep disruption may be attributable to the absence of *Zeitgebers*, i.e., the cues from the environment, which give information about time such as light, sound, and temperature. Ground-based studies have shown that the absence of Zeitgebers tends to disrupt the subjects' circadian rhythms, i.e., the physiological patterns in the body that recur every 24 hours and regulate body temperature, hormone levels, sleep-wake episodes, and many other periodic physiological processes as well as certain behavioral, cognitive and personality variables. In low Earth orbit, a sunrise or a sunset occurs every 45 minutes, sending potentially disruptive signals to the circadian pacemaker in those subjects exposed to the 90-min "days." The timing of sleep might then be disrupted because astronauts' biological clocks are no longer exposed to the Earth's 24-hour day-night cycle.

Changes in respiration may also be one of the reasons why sleep in space is disturbed. During space flight, the respiratory pattern and motions of the chest and abdominal wall are altered. Irregular breathing patterns, high carbon dioxide levels in the blood, or low blood oxygen levels cause sleep problems on Earth. However, recent data suggest that the sleep disruption in microgravity is not caused by an increase in respiratory events. On the contrary, apneas and hypnoas are reduced in microgravity (Eliott *et al.* 2001, Prisk *et al.* 2003).

Finally, there are also factors that are not specific to space flight. Jetlag, anxiety, excitement, and an uncomfortable sleeping environment are well-know factors that on Earth may lead to sleep disturbances. On board space vehicles, there is little privacy, quarters are confined, and noises or other interruptions may occur. Astronauts often complain of cold or warm sleeping environment. Informal reports also suggest that cosmonauts on long-duration flight have experienced disrupted sleep and wake cycles that may be attributable to the psychological adaptation to isolation, confinement, risk, anxiety, and stress (Lebedev 1988).

Work-sleep cycle, shift work, and high workloads can also be an issue. To comply with mission rules based on minimal awake time before landing, astronauts are often scheduled to rise earlier every day, progressively advancing their bed and wake up time by five hours or more during the course of a two-week space mission. In fact, this circadian shifting begins during the seven-day pre-launch quarantine to align the astronauts' work-sleep cycle with launch operations. This circadian shifting is comparable to jetlag.

On some flights, some astronauts may have to work while the others sleep to handle the numerous tasks related to the mission. In general, scheduling difficulties have been problematic for many space crews. A well-known example is those astronauts from the second Skylab mission who went on strike by interrupting communications with ground control because they were not allowed enough uninterrupted pre-sleep time to "wind down." Additionally, many crewmembers have had to forego sleep in lieu of pressing operational or scientific tasks (Frost *et al.* 1977).

During long-duration Russian missions, cosmonauts often reported periodic sleep disturbances during expeditions lasting over two months. Complaints by cosmonauts of increased susceptibility to fatigue were linked with either the prolonged and stressful work of the expedition or with changes in the crew's work and rest regimen. Pronounced sleep disorders were observed in all crewmembers over the course of one of the expeditions, during which an extraordinary number of emergencies (Figure 3-06), off-nominal situations, and intensive repair operations occurred both inside and outside the Mir station. As a result, cosmonauts periodically received soporific and sedative medications and tranquilizers (Gontcharov *et al.* 2005).

2.2   Sleep Monitoring

Only in a few occasions was traditional polygraphic-type monitoring of sleep, i.e., using *electro-encephalography* (EEG), *electro-cardiography* (ECG), *electro-myography* (EMG), and *electro-oculography* (EOG), performed during space missions. Such studies provide valuable information regarding changes in the quality of sleep experienced by crewmembers on a daily basis. Radioimmunoassay analyses of salivary cortisol levels provide information regarding the longer-term changes in circadian rhythms. "Daytime" activity of the crew has also been monitored using a wrist activity recorder. Information gathered from this device provides an indicator of the impact of the disturbance of sleep on normal daily activity. When complemented by stress and mood measurements, these studies provide information regarding the particular variables associated with any given sleep disturbances.

The most recent comprehensive sleep studies during space flight were performed during the sixteen-day LMS and Neurolab missions on board Space Shuttle Columbia in 1996 and 1998, respectively (Monk *et al.* 1998). Results showed that the average daily sleep period time was on average 30-40 min less per night during the flight than prior to and after the flight. However, the overall structure and temporal organization of sleep, based on polygraphic recording, was not markedly altered during and after space flight, nor were neurobehavioral performance measured from vigilance and memory tests (Dijk *et al.* 2003). Since body temperature and urinary cortisol have strong circadian variations, they were used to follow the changes in circadian rhythms. However, body studies showed conflicting results. The circadian system was unable to keep pace with the advancing sleep-wake schedule during Neurolab (Dijk *et al.* 2003), whereas this was not the case during the LMS mission (Monk *et al.* 1998), although the rest-activity schedules were very similar for both missions. The authors attribute this apparent discrepancy to differences in methods of analysis.

A better understanding of the neurophysiological mechanisms of circadian rhythms and the effects of space flight can come from studies on animal models. In these biological studies, animal models can be monitored continuously over long period of time, and are less subject to changes in work-rest schedule or stress than humans. An experiment is currently being performed with scorpions on board the ISS. Changes in heart rate, respiration, and eye and motor activity are recorded over long periods. The advantage of the scorpion model is supported by the fact that data can be recorded preflight, in-flight and postflight from the same animal. With this animal model, it is hoped that basic insights will be obtained about the decoupling of circadian rhythms of multiple oscillators and their adaptation to the entraining Zeitgeber periodicity during exposure to microgravity for several biological parameters recorded simultaneously (Riewe & Horn 2000).

Loss of circadian entrainment to Earth-based light-dark cycles, and chronic reduction of sleep duration in space, result in fatigue and jeopardize astronaut performance. Fatigue is a common symptom in prolonged space flight. Additionally, the most frequent medications taken in-flight by astronauts are hypnotics for sleep disturbances. Extensive ground-based scientific evidence documents that circadian disruptions and restriction of sleep at levels commonly experienced by astronauts can severely diminish cognitive performance capability, posing risks to individual astronaut safety and mission success. Consequently, new methodologies are needed for continuous acquisition of data on sleep, circadian rhythms, sleeping environment, and performance of astronauts before, during, and after short- or long-duration space missions. The determination of sleep, work, and recreation schedules, which could optimize human performance and adaptation to space, is also essential for long-duration space missions (Robbins 1988).

3 BEHAVIOR AND PERFORMANCE

Astronauts execute a wide variety of tasks in space. They do things like operating complex technical systems, e.g., docking or undocking a spacecraft, controlling robot arms, conducting scientific experiments, or performing specific tasks during extra-vehicular activities. Tasks like these place demands on different cognitive and psychomotor functions and skills. As a consequence, astronauts must maintain a high level of performance efficiency during their stay in space. This represents an important pre-condition of success and safety in the perspective of long-duration missions. During physiological adaptation to the space environment and during specific stress states induced by disruption of the daylight-darkness cycle, sleep deprivation, altered carbon dioxide levels, confinement, heavy workload, lack of privacy, isolation from friends and family, fatigue, stress, and SMS, performance impairments might be expected (Kanas *et al.* 2006).

From time-to-time, when confronted with even well practiced procedures and simple instructions, some astronauts have anecdotally reported initial befuddlement. This condition is casually referred to as "space stupids", "space fog", "space dementia", or "mental viscosity." This problem is typically first reported soon after arriving in orbit and typically abates after just a few days. This is perhaps in direct relation to the astronaut's course of adaptation. Because this reduction in cognitive function occurs just at a time when quick and accurate cognitive performance is necessary, especially in the event of an emergency, it clearly poses a potentially serious hazard (Figure 3-06). It has been suggested that this problem occurs because astronauts are trying to perform more than one cognitive task at a time, or by distraction from the concurrent demands of adapting to the altered gravitational environment (Clark 2007).

Reports indicate that Shuttle astronauts are sometimes unable to locate objects, especially small ones, even when they are in plain view. This deficit may occur because of microgravity *per se*, given the fact that these objects are oriented in atypical ways. It may also simply be that there are many more places apart from the floor that the objects can be located. During a Shuttle mission, there was one instance during which a crewmember was "lost." Several of his crewmates looked for all over for him, but couldn't find him. Yet, all the while they were searching, he was right in front of them! The lost crewmember was actually upside-down relative to those looking for him.

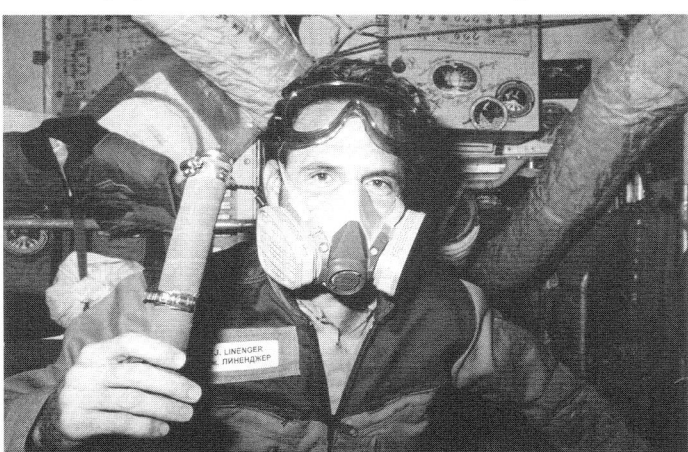

Figure 3-06. Astronaut Jerry Linenger wearing an oxygen mask following the fire accident on board the Russian Mir station in June 1997. Photo courtesy of NASA.

Astronauts have also reported to experience time as passing unusually rapidly in orbit (Lebedev 1988, Linenger 2001, Jones 2006). Time perception tests were executed on four crewmembers during a Shuttle mission. The data obtained indicate that short-duration task time (2 sec) is progressively overestimated as the mission progresses, as compared to preflight values. The tasks that required longer periods of time to estimate (8, 12, and 16 sec) were less affected (Ratino *et al.* 1988). It remains unclear as to whether this time compression is due directly to microgravity or any of the other factors that characterize space flight, as listed above.

3.1   Factors for Impairment in Behavior and Performance

The congruence between vestibular signals and other receptors, like visual and tactile, is disrupted due to the lack of usual gravitational force. This is also the case for the signals between the vestibular otolith and semicircular receptors caused by the altered outputs from the gravity-sensitive otoliths. Another direct effect of microgravity is related to mechanical and proprioceptive changes during the execution of movements, leading to a disruption of the usual relationships among efferent and afferent signals (Bock 1998). Complex adaptive processes, such a re-weighting of afferent information, can be expected because of both of these effects. Thus, the efficiency of cognitive and perceptual-motor skills that have been established on the ground is affected.

The working and living conditions of space can induce stress states in astronauts in addition to these specific effects of microgravity. These stress states can lead to degradations of cognitive and psychomotor performance. Examples of such stress states include decreased alertness and fatigue, high workload, and emotional stress due to interpersonal tension or the long-term effects of confinement and isolation. A shift in the pattern of physiological activation, including arousal, into a region that is non-optimal for efficiency may result in performance impairments. Applying performance strategies that require less effort may lead to another performance impairment that results from trying to actively cope with stress and high workload (Hockey 1997).

In much the same way that physiologists have been concerned with prevention of bone loss or muscle atrophy in microgravity, behavior and performance training has traditionally been concerned with the prevention of performance decrements. Prevention of performance decrement is inextricably linked with performance enhancement provided by human factors engineering. The role of humans in complex systems, the design of equipment and facilities for human use, and the development of environments for comfort and safety is the focus of human factors engineering. Accordingly, psychological, cultural, and human factors could affect the behavior and performance of crewmembers during space missions.

Crew composition and selection, ground-crew interaction, and isolation are psychosocial factors that are relevant to the space environment. Workload and its association with personal autonomy and control and individual mood is also a contributing factor. The lack of autonomy or control over workload by a crewmember could negatively influence health, well-being and performance. Other factors include the ability to cope with both mundane and unexpected events and circumstances, the relationship between crewmembers and their organizations both pre- and in-flight, and the differential access to support and resources among crewmembers (Kanas & Manzey 2003).

Cultural factors include the organizational cultures involved in these special settings; the culturally-influenced set of expectations and norms that regulate social interaction, including social support and rules of conflict; and language and cultural differences that influence the ability of crew members to communicate with one another both formally and informally.

Human factors include the influence of workload on individual and group performance, time allocation, training, and opportunity for exercise and leisure activities; and the influence on environmental conditions such as noise, crowding, storage, and light, on health, sleep patterns, mood, and task performance.

3.2   Effects of Space Flight on Operational Performance

Operational performance may be impaired by spatial disorientation, perceptual illusions, and disequilibrium that may occur during and after transitions between gravity levels. If some form of artificial gravity is used, the transitions between gravitational and dynamic acceleration environments will be associated with sensorimotor adaptation mechanisms and potential adverse sensory conflict reactions may occur. In addition, there are mechanical and proprioceptive changes, as well as a reduced efficiency of central motor programs, that have been established under normal gravity conditions. Performance impairments resulting from these changes include a loss of precision of voluntary movements, or a slowing of movement times, and changed kinematics during execution of these movements, as compared to one-g conditions (Bock *et al.* 2001). Crew work capacity, vigilance, and motivation may also be impaired by motion sickness symptoms occurring during and after g-transitions. These may be problematic during periods requiring crew control of vehicles or other complex systems. Crew performance of routine and critical actions during launch, landing, and the periods immediately following these events may be compromised (Figure 3-07). This risk may be exacerbated by vehicle and habitat designs that do not maintain consistent architectural frames of reference or those presenting ambiguous visual orientation cues. It may also be exacerbated by low visibility situations (smoke, landing weather, poor lighting), environmental vibration, or unstable support surfaces (floors, seats).

Capability to egress the vehicle in an emergency or to perform post landing tasks may be compromised by impaired movement and coordination caused by long-term exposure to microgravity. Crewmembers may be unable to accomplish certain postflight physical activities involving upright posture, locomotion and handling loads. This risk may be exacerbated by duration of microgravity exposure, cardiovascular deconditioning, muscle atrophy, orthostatic intolerance, relative hypovolemia, diminished aerobic capacity, and/or poor task, equipment or vehicle/habitat design.

Performance of mission-related physical activities may also be impaired due to loss of muscle mass, strength and endurance associated with prolonged exposure to reduced gravity. There is a growing database demonstrating that skeletal muscles, particularly postural muscles of the lower limb, undergo atrophy and undergo structural and metabolic alterations during space flight. Decreased loading of skeletal muscle during space flight is associated with decreased muscle size, reduced muscle endurance, and loss of muscle strength. These alterations, if unabated, may affect performance of mission tasks. Exercise countermeasures have to-date not fully protected muscle integrity. The risk may also be influenced by sensorimotor deficits, contractile protein loss, changes in contractile phenotype, reduced oxidative capacity, bone loss, poor nutrition, or insufficient exercise.

Although infrequent, serious neurobehavioral problems involving stress and depression have occurred in space flight, especially during long-duration missions. In some of these instances, the distress has contributed to performance errors (Lebedev 1988, Linenger 2001). In other instances, emotional problems led to changes in motivation, diet, sleep and exercise, i.e., all critical factors to behavioral and physical health in-flight. No matter how prepared crews are for long-duration flights, the U.S. and Russian experiences reveal that at least some subset of astronauts will experience problems with their behavioral health, which will negatively affect their performance and reliability, posing risks both to individual crewmembers and to the mission (Kanas & Manzey 2003). Earth-analog studies show that there is a significant likelihood of psychiatric problems emerging during long-duration missions. A recent report notes that the incidence rate ranges from 3-13 percent per person per year. The report transposes these figures to 6-7 crewmembers on a three-year mission (Ball 2001, p. 106).

Radiation exposure during space flight, especially within the Van Allen radiation belt and in deep space with a long travel time, may affect neural tissues, which in turn may lead to changes in motor function and behavior, or neurological disorders function or behavior during the mission or after return. Acute and late radiation damages to the CNS may be caused by occupational radiation exposure or the combined effects of radiation and other space flight factors such as microgravity or physiological changes (Planel 2004).

Finally, human performance failure in accomplishing certain tasks may occur due to human factors inadequacies in the physical work environments, such as habitats, workplaces, equipment, protective clothing, and tools and tasks. Additionally, tasks not designed to accommodate human physical limitations, including changes in crew capabilities resulting from mission and task duration factors, may lead to crew injury or illness or reduced effectiveness or efficiency in nominal or predictable emergency situations. Performance may be further affected by state of fitness, effectiveness of exercise countermeasures, and training.

Figure 3-07. Attired in orange launch and re-entry suits, Space Shuttle astronauts are practicing an emergency egress of the launch tower. In case of an emergency until T-minus 30 sec in the countdown, the crew should exit the vehicle, walk onto the platform and board baskets that quickly skid along a 400-m long wire down to the emergency shelter. Photo courtesy of NASA.

3.3   Effects of Space Flight on Cognitive and Psychomotor Functions

Most of the empirical studies addressing the possible effects of space flight-related stressors on human cognitive and psychomotor performance have been conducted during short-duration (< 30 days) space flights (Manzey & Lorenz 1998, Manzey 2000). The majority of these have been focused primarily on elementary cognitive and psychomotor tasks. These studies have revealed a fairly consistent pattern of effects in spite of their different methodological approaches. Manzey *et al.* (1993) have found no impairments of the speed and accuracy of elementary cognitive functions, grammatical reasoning, mental arithmetic, or memory-search in space. However, significant disturbances of perceptual-motor tasks have been found, especially while simultaneously engaged in a secondary cognitive task *(*Manzey *et al.* 1995). Newman & Lathan (1999) reported similar results. It is appropriate to consider these results a measure of cognitive capacity because these tasks did not involve the kinds of hand-eye coordination or neurovestibular control that would be directly influenced by microgravity (Sangals *et al.* 1999). It appears, therefore, that astronauts are most likely to experience cognitive deficits when forced to engage in two or more tasks at the same time, tasks performed in a rapid sequence, or are distracted by other demands on their attention (Manzey *et al.* 2000).

Cognitive performance was also evaluated during the *Cognilab* experiment flown on space missions ranging from two weeks to one month on board the Mir space station. This experiment studied the mechanisms used by CNS in the perception of visual stimuli, force or duration, and tested the ability of the crewmembers to manipulate objects through a force-actuated joystick (Figure 3-08). Motors attached to each axis of the joystick were programmed by a control computer to resist certain movements attempted by the subject, or to apply forces or torque to the hand of the subjects. Results showed that weightlessness affected both the haptic perception of applied forces and the control of arm movements (Lipshits *et al.* 2001, McIntyre *et al.* 2005).

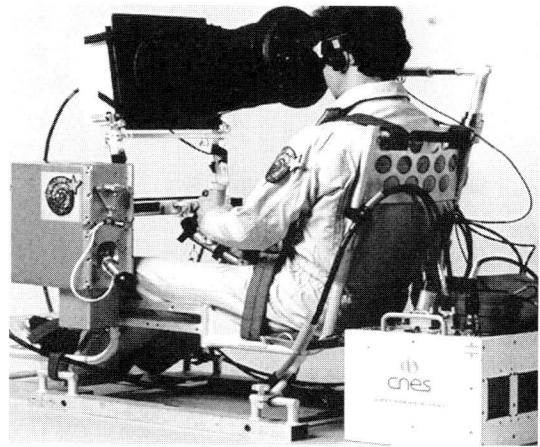

Figure 3-08. The Cognilab hardware presented 2D stimuli to the subject on a small flat-screen video monitor. A force feedback joystick recorded tracking responses from the subject via push button inputs and control knob turns. Note that the subject was restrained in a seat that was itself fixed to the spacecraft at the same location in both the ground mock-up and the Mir station. The resulting tactile cues and the knowledge of the subject's own orientation relative to the spacecraft were playing a significant role in this cognitive performance task. Photo courtesy of CNES.

Other research studies have used EEG recordings to monitor and measure working memory and other indicators of cognitive ability. A recent experiment conducted over the course of three space flights quantified the EEG oscillations at 10

Hz (alpha and mu rhythms), which are the most prominent rhythms observed in awake subjects with the eyes closed. This activity increased in five cosmonauts in-flight compared to preflight. The authors of this study attribute this increase to a reduction in graviceptive inputs to cortical networks participating in the mental representation of space (Chéron *et al.* 2006).

During very long-duration space flights, only a few studies were performed on perceptual, cognitive, and psychomotor performance. A comprehensive performance monitoring study was conducted during one long-term mission involving one Russian cosmonaut during a 437-day stay on board Mir. The results of this study suggest remarkably little impact of space flight-related stressors on elementary cognitive and perceptual-motor functions (Manzey *et al.* 1998). Only during the first two weeks of the flight were impairments of performance, alertness, and subjective well being found. These impairments were noted immediately after the mission when the cosmonaut was back on Earth. These disturbances occurred during the critical phases of transition between gravitational force levels, which were also associated with comparatively high workload. Subjective mood and performance functions returned to preflight levels after successful adaptation to the space environment. These functions remained stable throughout the remaining 400+ days in space.

Decrements in operational task performance have been observed during isolation studies and space missions. Therefore, maintenance of operational efficiency and complex skills may become a serious problem. As an example, Salnitski *et al.* (1999) investigated cosmonaut performance levels in a simulated manual-docking maneuver during their stay in space. A considerable loss of skill was found after a period of three months in space, which was mainly attributed to a lack of on-orbit refresher training under the altered gravity conditions. Other degradations of operational performance in cosmonauts, i.e., errors made by crewmembers in conducting mission tasks, have been related to non-specific stress effects arising from disturbances of the usual sleep-wake cycle, high workload, or psychosomatic discomfort (Nechaev 2001).

Systematic research addressing acquisition and retention of complex cognitive and perceptual-motor skills during space missions is lacking. Research efforts have been limited to just a few studies performed during ground-based simulations of space flight, apart from a few early studies of operational efficiency executed during Skylab missions and the Russian studies mentioned above (Sauer *et al.* 1999). The results of these studies did not provide any consistent effect pattern. Most of the investigations suffered from methodological constraints, such as persistent learning effects throughout the experiment. Other areas of potential concern, including the effects of the space environment on higher-order cognitive processes relevant for space operations like decision-making, have not been studied to date.

The different stressors to which astronauts are exposed can cause detrimental effects on basic cognitive, attention, and psychomotor processes, as suggested by the aforementioned research. Even less is known about how or if these effects are moderated by culture. It does not seem likely that the effects of microgravity on performance are different for members of different cultures. However, performance may depend on the cultural background of astronauts in other ways. One example is the effect of lack of privacy and crowding in a confined environment, like a space habitat. The cultural background of individuals significantly influences whether the lack of privacy is perceived as a stressor, and thus entails detrimental effects on mood and performance (Raybeck 1991). Similar effects may be assumed for other psychosocial stressors as well. These include monotony and boredom, time pressure, and workload. In addition, culture might be a factor influencing higher-order cognitive processes, like decision-making or the use of schemes in information processing.

3.4   Research Program

The studies described above shed no light on whether the observed impairments of cognitive function and perceptual-motor performance are related to direct effects of microgravity or to non-specific stress effects. Such non-specific stress effects may arise from physiological changes, inadequate work-rest schedules, or sleep disturbances during adaptation to microgravity, not to mention the extreme living conditions in a space habitat. Clearly there must be additional research conducted to ascertain the adverse effects of space flight on behavior and performance. The results of this research could be used to provide appropriate support mechanisms and countermeasures. In particular, tools are needed which can discern subtle performance decrements before they escalate (Kanas *et al.* 2006).

Specific screening tests can be used to assess astronaut performance. These tests would be administered repeatedly with the results compared to a self-referenced baseline established during preflight training. The *Windows-spaceflight cognitive assessment tool* (WinSCAT) is one example of such a tool. It is, in fact, currently being used on the ISS. WinSCAT is a clinical tool for tracking performance through simple tests comprising code substitution, running memory, mathematics, and match-to-sample. These tests were selected from among a group of tests, called the *automated neuropsychological assessment metrics*. The United States Navy and Army developed these tests to assess the cognitive impairment of military personnel who have been subjected to medication or are suspected to have sustained brain injuries. Now in the public domain, these tests have been validated in a variety of clinical settings. The tests are presented in a Microsoft Windows shell that facilitates administration and enables immediate reporting of test scores in numerical and graphical forms (Kane *et al.* 2005).

MiniCog is another cognitive monitoring tool in use on the ISS. Nine cognitive functions can be tested using hand-held PDA. These functions include attention (vigilance, divided attention, filtering), memory (verbal, spatial working), problem solving (verbal, spatial), cognitive set switching, and motor control. MiniCog performs automatic calculations of mean response time and error rate. If an astronaut is suffering from stress-related deficits such as fatigue that may affect performance, then MiniCog can potentially provide an "early warning alert" to the situation. The information can be used by the astronauts themselves: they can be warned to pay additional attention and take extra care, or better yet, take a break, consume food or caffeine, or even take a nap (Shepard *et al.* 2006).

The application of more integrated test batteries is required to obtain a complete description and monitor of the nature of cognitive and psychomotor deficits during space missions. These tests should assess not only the behavioral aspects of information processing, such as the speed and accuracy of performance, but also subjective and psychophysiological measures. In particular, information about the astronauts' emotional state should be provided. The objectives and methods of these studies are listed in Table 3-02.

Research specialists are developing new assessment tools. For example, because eye blink activity is highly correlated with performance on a compensatory tracking task, video recording of eye movement coupled with software that automatically extracts blink activity is interesting. Plans are on the drawing board to develop and test optically based computer algorithms to effectively detect emotional distress, neurocognitive degradation, and neuroendocrine responses to behavioral stressors. At NASA, Schlegel and his colleagues have taken this monitoring effort one step further in their program to develop and validate a methodology for self-assessment of cognitive and sensorimotor state. This methodology could be integrated with prescriptions for in-flight training and countermeasures. This new research effort builds upon their previous work in the development of a *performance assessment workstation* (PAWS) for use in monitoring the cognitive performance of crewmembers during the Spacelab IML-2 and LMS missions (Eddy *et al.* 1998).

- *Objectives*
  - *Effects of the space environment on sensory systems, motor systems and the cognitive process that enable them*
  - *Cognitive processing and effects of emotion, age and gender with respect to space missions*
  - *Identification of neural mechanisms, neurotransmitters, neural substrates that impact on optimal psychological performance: especially in areas of mood, well-being, motivation, and reward*
  - *Research on the behavioral side effects of pharmacological countermeasures*
  - *Study of circadian physiology and biochemical indices during space flight and their implications on the quality of rest and performance*
  - *Effect of noise, light, and crowding on the health, sleep patterns, and mood of space mission and related crews*
- *Methodologies*
  - *Quantify task errors, such eye-hand coordination, visual memory of scenes, orientation of self and objects*
  - *Measure performance of meaningful tasks essential to mission safety and mission completion that is transparent to crew daily work*
  - *Validation of monitoring methods that improve objective assessment of essential behavior and performance factors in-flight*
  - *Development of proven methods for defining circadian cycles of humans that requires less intrusive implementation*

Table 3-02. Research program on behavior and performance during space flight.

4   HUMAN FACTORS

The primary drivers on crew accommodation design are volume and mass. The affects of habitats, work environments, workplaces, equipment, protective clothing, tools and tasks on human performance in a space context can be assessed from anecdotal information from crew reports and extrapolations from physiological studies. However, there is insufficient data on physical changes in strength, stamina, and motor skill as functions of time in the space flight environment. Returning crewmembers generally manifest substantial motor and physical deficits. Incorporating appropriate human factors into vehicle, task, and equipment design will no doubt enhance overall performance.

Several systems to supplement human cognitive information are being developed in the area of human factors. As an example, noise reduction headsets that employ feedforward disturbance cancellation techniques are being tested for application to objective hearing assessment in noisy environments. Haptic (tactile) displays arranged on the user's body (Figure 3-09) have been developed to provide orientation and spatial situation awareness (Rupert 2000). These tactile displays could be particularly useful during complex extra-vehicular activities (Figure 3-10). Spatial orientation cues can be reinforced by virtual reality displays. Kinesthetic cues (the sense of effort), stereoscopic tracking as well as supervisory control teleoperation system can assist with demanding teleoperation tasks.

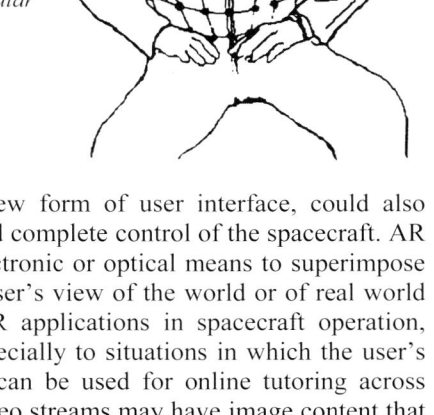

Figure 3-09. Researchers at the Naval Aerospace Medical Research Laboratory *(NAMRL) in Pensacola (Florida) have developed* the Tactile Situation-Awareness System *(TSAS), which uses the sense of touch to provide spatial orientation and situational awareness information to aviators. The system could be applied to astronauts during teleoperation tasks or extra-vehicular activity.*

Augmented reality (AR), which is a new form of user interface, could also provide astronauts with increasingly flexible and complete control of the spacecraft. AR systems are displays that incorporate either electronic or optical means to superimpose synthetic, spatially conformal imagery onto a user's view of the world or of real world imagery. There are a plethora of possible AR applications in spacecraft operation, simulation training, and ground operations, especially to situations in which the user's hands are occupied. Relatively low-level AR can be used for online tutoring across distributed environments. Camera-generated video streams may have image content that can be quickly annotated with text or symbols using feature tracking and uplinked to users. AR has already been implemented for neurosurgery and can be adapted for use in remote manipulator system type tasks.

The internal appearance of a spacecraft could be dramatically transformed by replacing the many large and fixed location workstations with smaller, portable, wearable personalized AR displays, dramatically empowering the crew while lowering power consumption and mass. However, implementation of an untethered, portable display appropriate for augmented reality will require the development of wireless technology. Actually, such a technology exists – *Ultra Wide Band* (UWB). It's new, but works very well in small confined places and the bandwidth is more than sufficient to support this application.

Expert systems can be used for assessing and improving individual cognitive performance in-flight. Dr. Larry Young and his colleagues at MIT have developed an expert system called the *Principal-Investigator-in-a-Box*. It provides an excellent model of how an expert system might be used in space to improve cognitive performance, not merely in the completion of a specific task, but in training and reviewing performance effectiveness. The PI-in-a-Box was found to be significant enabler to individuals monitoring biomedical instrumentation in a sleep experiment. Its potential, however, extends beyond single experiments.

Figure 3-10. During extra-vehicular activity astronauts are relying primarily on visual inputs and prior experience during training for their orientation. Sensory enhancement or virtual reality training are being evaluated to improve performance. Photo courtesy of NASA.

5   SUMMARY

Sensory and sensorimotor disturbances during the initial adaptation to microgravity are well documented, the most well known of these being *space motion sickness*. Individual susceptibilities, spacecraft size, and room available for movement influence specific SMS symptoms. Typically lasting the first several days of weightlessness, symptoms include everything from headaches and fatigue to nausea and vomiting. The consequences range from simple discomfort to possible incapacitation, creating potential problems during extra-vehicular activity, re-entry, and emergency egress from the spacecraft.

The body receives a variety of conflicting neurosensory inputs from the visual, tactile, proprioceptive, and vestibular organs in weightlessness. The conflicting inputs are thought to be the primary cause of SMS. Testing equipment such as rotating chairs and accelerating sleds have been used both in orbit and on the ground in attempts to understand the complex mix of sensory inputs responsible for SMS. The precise mechanisms of the sensory conflict are not well understood. Evidence exists that indicates that neurosensory changes take place continuously during space flight, and even after landing, long after the acute symptoms of SMS have subsided.

Medications currently used in-flight to alleviate the SMS symptoms produce undesirable side effects. Various medications, combinations, and doses are tested in laboratories for their effectiveness in countering motion sickness in volunteers usually seated in rotating chairs. Different ways of administering of the medication are also being tested. Low-dosage intra-muscular injection of anti-SMS drug offers an alternative to oral administration, given that orally administrated drugs seem to be absorbed differently in weightlessness than in normal gravity. Finally, simulators designed to adapt space crewmembers to the sensory disorientation of space flight before they fly have produced encouraging results.

The success of human space flight depends on astronauts remaining alert and vigilant while operating complex, state-of-the-art equipment. Therefore, getting enough sleep is a crucial factor of mission success. Weightlessness, a confined environment, and work demands coupled with the loss of the 24-hour day-night cycle make sleep difficult in space. Astronauts typically average only about six hours of sleep each night. Cumulative sleep loss and sleep disruption could lead to performance errors and accidents that pose significant risk to mission success.

Sleep and circadian rhythmicity also temporally modulate a broad range of physiological functions, including body temperature, cardiovascular activity, respiration, and immune responses; hormonal functions, including growth hormone, melatonin, cortisol, and thyroid hormones; behavioral functions like movement, posture, and reaction time; and cognitive functions like fatigue, alertness, vigilance, memory, and cognitive throughput. No individual, even an astronaut, regardless of the amount of training, preparation, nutrition, psychosocial support, or environmental protection is provided, is immune from the daily control of physiology and performance by the homeostatic drive for sleep and the endogenous circadian timing system. Failure to take these two interactive neurobiological imperatives into account when planning human activities in space will have catastrophic consequences (Prisk & Fuller 2001).

Many biomedical systems essential for maintaining astronaut physical condition, mental health, and performance capability are continually influenced by the need for sleep and the circadian pacemaker. Dysfunction of sleep and circadian systems can adversely affect the ability to respond to environmental challenges. This condition has been linked to physiological and psychological disorders. Therefore, this subject has a high degree of relevance to a number of other space life science technical areas including research on muscle and bone loss, cardiovascular and immune changes, neurovestibular alterations and nutritional needs, and behavioral and psychological health in space flight.

Current investigations aim at developing methods to prevent sleep loss, promote wakefulness, reduce human error, and optimize mental and physical performance during long-duration space flight. Particular concerns for long-duration missions include the impact of the space environment on higher-order cognitive processes like decision-making, the impact of culture on performance, and the impact of transient exposure to artificial gravity on mental functions, which will be important if artificial gravity is considered as a countermeasure for future interplanetary space missions.

It is also necessary to develop human-response measurement technologies to assess the crew's ability to perform flight-management tasks effectively. Behavioral and psychophysiological response measurement systems are needed to assess mental loading, stress, task engagement, and situation awareness. Measurement capabilities could include monitoring of pulse, heart and muscle electrical activity (ECG and EMG), skin temperature and conductance, respiration, and tracking of the eye look point (oculometry) and overt behavior (video analysis), coupled with topographic brain mapping (EEG and evoked responses). A real-time mobile physiological monitoring and behavioral response capture-stations are required to refine these measurements for flight-management research.

Chapter 4

SENSORY FUNCTIONS IN SPACE

To be aware of the environment, one must *sense* or *perceive* that environment.[7] All living organisms on Earth have the ability to sense and respond appropriately to changes in their internal and external environment. Organisms, including humans, must sense accurately before they can react, thus ensuring survival. If our senses are not providing us with reliable information, we may take an action that is inappropriate for the circumstances, and this could lead to injury or death.

The body senses the environment by the interaction of specialized sensory organs with some aspect or another of the environment. The central nervous system utilizes these sensations in order to coordinate and organize muscular movements, shift from uncomfortable positions, and adjust properly. In common speech, five different senses are usually recognized: vision, auditory (hearing), olfaction (smell), gustation (taste), and somatosensation. Of these, the first four use special organs (the eye, ear, nose, and tongue, respectively), whereas the last use nerve endings that are scattered everywhere on the surface of the body (the muscles, joints, tendons, skin), as well as inside the body (visceral sensations).

One other sense, the sense of self-motion, is the ability to sense body movement combined with our ability to maintain balance. Maintaining postural equilibrium, sensing movement, and maintaining an awareness of the relative location of our body parts requires the precise integration of several of the body's sensory systems including visual (peripheral retina), vestibular (inner ear), somatosensory (touch, pressure, stretch receptors in our skin, muscles, and joints), somaesthetic (viscerae), and auditory inputs. Acting together, these systems constantly gather and interpret sensory information from all over the body and usually allow us to act on that information in an appropriate and helpful way.

Figure 4-01. Human beings have the ability to walk a tightrope, skate on sliding surfaces, combine twists and turns when diving, or stand upside-down… all without losing balance and while keeping track of the relative position of their arms and legs with respect to the rest of the body.

---

[7] The words '*sense*' and '*perceive*' are from Latin words: "sense" means "to feel", whereas "perceive" means "to take in through", i.e., to receive an impression of the outside world through some sensory organs.

G. Clément, M.F. Reschke, *Neuroscience in Space*.
DOI: 10.1007/978-0-387-78950-7_4, © Springer Science+Business Media, LLC 2008

1   VISION

Because accurate perception of the environment is critical for orientation and adaptation to the space environment, vision has received particular attention in space neuroscience studies. Initially, it was expected that exposure to the space environment would induce changes in visual performance. These changes could result from several factors including altered light conditions, modified receptor physiology, and disturbances of the eye movement control system as a secondary effect of vestibular changes.

The visual environment in space is altered in several ways. First, objects are brighter under solar illumination. Earth's atmosphere absorbs at least 15% of the incoming solar radiation. Water vapor, smog, and clouds can increase this absorption considerably. In general, this means that the level of illumination in which astronauts work during daylight is about one-fourth higher than on Earth. Second, there is no atmospheric scattering of light. This causes areas not under direct solar illumination to appear much darker and results in a transformation of normal visual intensity relationships.

An interesting visual phenomenon that evidently was related to space radiation was first noted in the Apollo program. During the time of trans-Earth coast, crewmembers of Apollo-11 reported seeing faint spots or flashes of light when the cabin was dark and they had become dark-adapted. From these reports, it was assumed that the light flashes resulted from high-energy cosmic rays penetrating the spacecraft structure and the crewmembers' eyes (Hoffman *et al.* 1977). Later experiments performed on board the Skylab and Mir stations showed that the number of flashes greatly increases in the South Atlantic Anomaly, probably as a result from trapped protons or particles with heavy nuclei in this region (Parker *et al.* 1989). The fact that prior dark adaptation is necessary suggests that the phenomenon is connected with the retina rather than with a direct stimulation of the optic nerve or visual cortex (Fuglesang *et al.* 2006). An experiment is currently being conducted on board the ISS to verify this hypothesis. Astronauts wear a helmet-shaped device holding silicon particle detectors designed to measure the trajectory, energy and species of individual ionizing particles. At the same time an EEG is measuring the brain activity of the crewmember to determine if radiation strikes cause changes in the electrophysiology of the brain in real time (Narici *et al.* 2004) (Figure 4-02).

Figure 4-02. An ISS crewmember is wearing the ALTEA helmet holding six silicon particle detectors designed to measure cosmic radiation passing through the brain. Data collected will help quantify risks to astronauts on future long-distance space missions and propose optimized countermeasures. Photo courtesy of NASA.

   Observations taken on board early Soviet spacecraft found cosmonauts' ability of the visual system to estimate the direction and number of certain focusing patterns of dashed lines under standard conditions to be sub-normal. Decrements for all crewmembers were reported, with losses averaging 20%. Reductions in color perception during space flight have also been noted. Measurements made on Vostok 2 and Soyuz 9 indicated a 25% diminution of color intensity, with losses particularly marked for purple, light blue, and green (Popov & Boyko 1967). Contrast sensitivity decrements of as much as 40% have also been reported. In spite of these changes, however, it is unclear if these sensory deficits were direct effects of microgravity or due to other factors in the spacecraft such as high contrast effects and the requirement to adapt to rapidly changing brightness levels in low Earth orbit.

   Reports during early U.S. space missions suggested that astronauts were able to see vehicles moving on the highway, an airplane moving on an airfield while in orbit (Duntley *et al.* 1966). To investigate a possible change in visual acuity, Gemini astronauts were tested using a small, self-contained binocular optical device containing an array of high- and low contrast rectangles. Astronauts judged the orientation of each rectangle and indicated their response by punching holes in a record card (Figure 4-03).

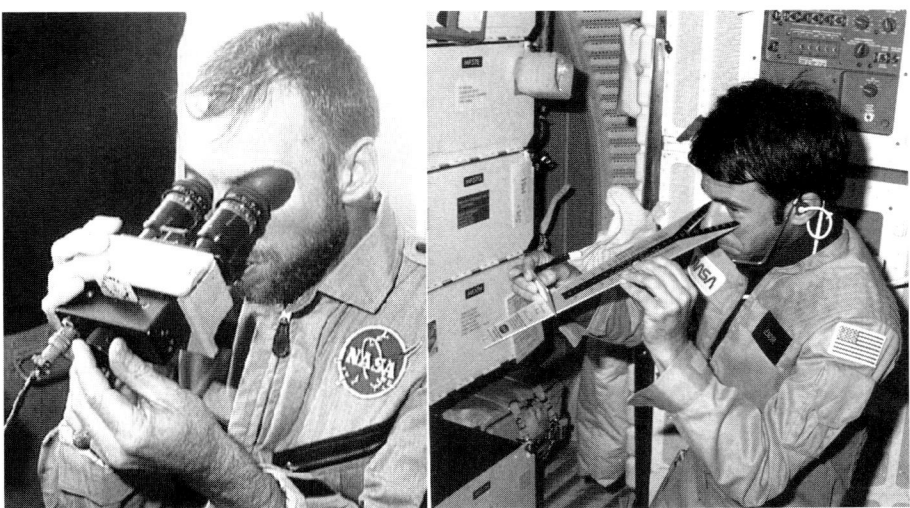

Figure 4-03. Left. The visual acuity of the Gemini-5 and 7 crew was tested each day using an in-flight vision tester, which was a binocular optical device containing a transilluminated array of 36 high-contrast and low-contrast rectangles. Half of the rectangles were oriented vertically in the field of view, and half were oriented horizontally. The flight crew made forced-choice judgments of the orientation of each rectangle and indicated their responses by punching holes in a record card. No significant difference was noted between preflight and in-flight measurements. Right. No significant differences in diopter measurements were measured neither in the near point of accommodation (near-vision acuity) among the pre-, in-, and postflight conditions in 23 crewmembers of the Space Shuttle. Photos courtesy of NASA.

   Another method, taking into account the particularity of the visual environment of space described above, also used large rectangular patterns displayed at ground sites in Texas and Australia. Astronauts had to report the orientation of the rectangles.

Display were changed in orientation between passes and adjustments for size were made in accordance with slant range, solar elevation, and the visual performance of astronauts on preceding passes. Results with both measurement methods indicated that visual performance was neither significantly degraded nor improved during space flight. The astronauts' reported ability to detect trucks or airplanes was probably due to dust clouds or shadows associated with traffic, which could significantly enlarge visual images (Parker *et al.* 1989).

Other visual performance testing on Mercury, Gemini and Apollo missions revealed few significant changes in visual function with the exception of constriction of visual field, changes in intraocular tension, and changes in the caliber of retinal vasculature Some constriction of visual field was noted postflight as well as a decrease in unaided seven-meter visual acuity, although the latter was not statistically significant. Postflight decrease in intraocular tension was significant and returned to preflight levels more slowly than expected. Retinal photography revealed no lesions, but showed some signs of decrease in the size of retinal vessels. Some of these effects could be due to the 100% oxygen atmosphere on board the earlier space vehicles. However, the degree of constriction of retinal vasculature was greater and persisted for a longer time than could be accounted for by the vasoconstrictive effect of atmosphere oxygen alone (Hawkins & Zieglschmid 1975).

Anecdotal reports from Shuttle crewmembers describing decreases in visual performance, such as difficulty in reading checklists and changing focus in the cabin, led to additional visual performance testing. Near visual acuity was examined during the NASA's *Extended Duration Orbiter Medical Project* (EDOMP) to measure changes in accommodation and other parameters affecting vision, such as contrast sensitivity, phoria, eye dominance, flicker fusion frequency, and stereopsis. Contrast sensitivity was examined to provide a more accurate measure of general vision loss as well as insight into the possible physiological location and nature of vision changes. Except for contrast sensitivity, in-flight experiments on a group of astronauts revealed no statistically significant changes in any of the parameters measured compared to preflight baseline (Task & Genco 1987). Contrast sensitivity exhibited a 10% loss immediately after orbital insertion and continued to decline to a 40% loss after five days. Even at these levels of change, it was concluded that the effect of space flight on visual function was relatively small (Nicogossian & Parker 1982).

Photographic studies continued to show a significant decrease in the size of the retinal vasculature after flight. Intraocular pressure rose during flight and dropped below preflight levels after landing, as was shown during the Spacelab D-1 mission. Using a hand-applanation tonometer, three crewmembers measured a mean rise in intraocular pressure of 20 to 25% (Draeger *et al.* 1986). These intraocular tension changes are probably associated with the fluid shift toward the head following insertion into weightlessness and the subsequent pooling of fluid in the lower extremities after return to Earth, which can in turn result in changes in the physical characteristics of the eye (Ginsburg & Vanderploeg 1987). The effects of long-duration flights on these visual function mechanisms are unknown and remain be determined. In addition to causing a possible decrement in visual performance, increase in intraocular pressure could lead to permanent loss of vision secondary to ocular pathologies, e.g., glaucoma and retinal vascular disease, if left untreated (Mader 1991). It will be necessary to continue performing tonometry, fundoscopy, and visual performance testing to monitor visual dysfunction on long-duration missions.

Possible changes during weightlessness in the ability to maintain visual fixation on targets while moving the head and track moving targets have not been fully examined. However, results of several experiments on eye movement control suggest that performance of these tasks might be diminished during the initial period of exposure to weightlessness and immediately after landing (Bloomberg *et al.* 1999). The vestibular nuclei located in the brain stem are part of a system that allows one to fix the gaze on a stationary target during voluntary head motions as well as to track moving targets (see Chapter 6, Section 4). This system appears to be disturbed during flight, perhaps as a consequence of altered vestibular receptor function due to the absence of gravity. Viéville *et al.* (1986) reported that the amplitudes (gains) of vertical eye movements were diminished during the first three days of weightlessness, both when astronauts moved their heads voluntarily and when they attempted to track a moving visual target. After four days in orbit, the gains returned to the preflight level, perhaps as a consequence of substituting neck receptor cues for vestibular receptor cues. Grigoryan *et al.* (1986) reported that after 237 days of space flight, the pattern of movements was significantly altered when cosmonauts moved their heads and eyes to fixate a laterally displaced target. The authors reported that similar disturbances have been observed in patients suffering from cerebellar disorders.

A diminished dynamic visual acuity, the ability to read and make other fine visual discriminations during active head and/or body movement, poses a potentially serious perceptual problem during space flight. If this capacity is impaired during weightlessness because of instability of the head and body, it is difficult or impossible to compensate for because the movement is involuntary and random. Furthermore, since inadvertent bodily movements are likely to continue even after perceptual-motor adaptation is otherwise complete, reduced dynamic visual acuity may remain a problem as long as the flight continues.

Being weightless means having to depend more on visual cues for orientation than is the case on Earth, where the otoliths share this task. Evidence for this increased "visual dominance" includes the observations of enhanced self-motion illusions of circular and linear vection (Young 1993), visual orientation effects (Young & Shelhamer 1990), visually induced tilt (Clément & Lestienne 1988), and the anecdotal reports of astronauts becoming more reliant on the spacecraft walls, ceiling, and floors as spatial references (see Chapter 7, Section 2).

2   HEARING

Several aspects of space flight can have an impact on hearing capability: (a) life support equipment is continuously running (ranging from 64 dBA for the air conditioning to 100 dBA for some vent relief valves) and the noise reverberates through the spacecraft's structure; (b) astronauts spend 24-hour a day in the office, always close to noise sources; and (c) there is no privacy, with a constant interaction with other crewmembers. Thus quietness periods such as on Earth do not exist: earplugs can reduce noise but not vibrations.

Space flight raises the spectrum of noise questions: its effect on perception and performance, adaptation effects, the fatiguing and annoying aspects of noise, and individual sensitivity differences. Because certain minimum noise levels are always present, space flight potentially constitutes a more stressful noise environment than a simple consideration of decibel levels would imply.

The investigation of hearing in astronauts is difficult to conduct during space flight because classical hearing assessment techniques do not work in the noisy environments often found in spacecraft (no soundproof laboratory). Since crews are at risk for hearing loss due to noise levels often encountered during space flight, techniques and investigation to track this loss are needed during and after the mission.

Nevertheless, auditory brain stem response recordings were investigated during Shuttle flights (Thornton *et al.* 1985). Auditory evoked potentials are particularly useful in the study of increased pressure or decreased vascular perfusion in both the labyrinth and lower portions of the CNS. This investigation was aimed at verifying if fluid shift toward the upper body in microgravity could induced such increased pressure, which could play a role in space motion sickness. Auditory evoked potentials were recorded on seven astronauts in-flight at various times during the space missions, beginning at 12 hours. No significant difference was observed between mean latency values for any potential on the ground or during flight (Thornton *et al.* 1985). Slight changes in the morphology of the responses were attributed to in-flight noise and electrical interference on board the Shuttle. Since the utility of these auditory potentials has primarily been associated with auditory threshold determination in clinic (Galambos & Hecox 1978), the absence of significant decrement in-flight objectively suggests that the auditory function is not altered in microgravity.

Persons with normal hearing are capable of localizing sound sources in the environment. This faculty depends on *binaural hearing*, i.e., the central processing involved in the comparison of the sounds received by one ear with the sounds received by the other ear. Directional hearing is utilized for determining the direction of a sound source and is a cue to spatial orientation (Barfield *et al.* 1997). One could assume that, in the apparent absence of gravity, orientation by acoustic cues would play a much more important role for spatial orientation. To address this hypothesis, an experiment flown on board Mir during the Austrian-Russian mission in 1991 was designed to determine whether directional hearing becomes a more important sense in microgravity than in normal gravity (Persterer *et al.* 1992). The test stimuli included white noise and a few bars of a Viennese waltz and simulated a sound source in a fixed location or moving around the subject's head. Both eye movements (EOG) and subjective comments of the subjects were recorded on two cosmonauts. The first phase of the experiment attempted to determine how precisely a subject could locate the direction of fixed sound sources presented binaurally over headphones. Results showed that the localization error in the horizontal plane in microgravity was within the same range as on Earth, i.e., between 1 and 2 deg. However, a significant downward shift in elevation judgments by approximately 10 deg downwards was observed. In other words, the test subjects thought that sounds came from 10 deg below their actual location or that the center of the auditory field had shifted up by 10 deg.

The second phase of this experiment tried to trigger an illusion of movement by auditory cues. It was hypothesized that if sound localization is more important in microgravity, then it would be easier to "trick" a cosmonaut into feeling a sense of motion due to hearing a moving sound source. In agreement with this hypothesis, the presentation of a sound moving counterclockwise induced an illusion of a clockwise body rotation, which was never observed on ground in the two subjects tested. This sensation was stronger with the waltz than with the white noise. Corresponding eye movements were observed, with slow phase directed counterclockwise. Reversing the direction of rotation of the sound source provoked a reversal of the direction of eye movements but not of the illusion of rotation (Persterer *et al.* 1992), suggesting that spatial orientation was disturbed, but not hearing.

Figure 4-04. Crewmembers eating their meal on board the ISS. Food is individually packaged and stowed for easy handling in weightlessness. Most of the food is frozen (e.g., entrees, vegetables, dessert items), refrigerated (fruits, vegetables, dairy products) or thermostabilized for (heat-processed and canned food). Photo courtesy of NASA.

3   TASTE AND OLFACTION

Both U.S. and Russian crews have reported changes in the taste and smell of food during space flight (Figure 4-04). During the Skylab flights, astronauts asked for more spices and condiments to add taste to the prepared food. Astronauts frequently complain of the foods on the Space Shuttle being bland, and of a dislike for coffee while on orbit. Cosmonauts on long-duration space missions have reported a reduced appetite for sweets and a desire for pungent food flavors (Lebedev 1988).

During the 84-day flight of the Skylab-4 mission, taste and odor thresholds were measured using slips of paper impregnated with different flavors. The three crewmembers tasted orange and onion flavors in addition to the four basic tastes. Five concentrations were used with each flavor. The crewmembers were told in the beginning of the test about the different flavors and were asked to report the flavor when first detected and when definitely confirmed. Odor identification thresholds were measured similarly for lemon, orange, onion, pepper, chicken, wintergreen, chocolate, cherry, spearmint, and cinnamon, in addition to a blank. No evidence of change in the ability to identify odors was found, but there was an increased threshold in the taste tests for certain sensations, although these results were highly individualized (Heidelbaugh *et al.* 1975).

In a later experiment on board STS-41G, olfactory recognition was tested by having two astronauts identifying the smell of solutions of lemon, mint, vanilla, and distilled water. In addition, taste recognition and detection thresholds were investigated with solutions of sucrose, urea, sodium chloride, and citric acid. Preflight responses were compared to those obtained during and immediately after the eight-day flight. No pre- or postflight subjective changes in taste or aroma or objective changes in olfactory function, taste recognition, or taste threshold were observed. Differences in flight length, susceptibility to motion sickness, or specific methodology may account for the conflict between the Skylab-4 and STS-41G results (Watt *et al.* 1985).

Diminished sensitivity to taste and odor could result from the passive nasal congestion reported in conjunction with the headward shift of fluid. Taste, particularly the non-volatile component mediated by the taste buds, may be susceptible to threshold shifts in microgravity because of mechanical factors related to reduced stimulation of the taste buds as a result of changes in convection in weightlessness (Rambaut & Johnson 1989). However, microgravity simulation studies using head-down best rest revealed no change in subjects' sensitivity to taste and odor (see Olabi *et al.* 2002 for review). Since sustained head-down tilt causes a shift of fluids from the lower to the upper body, as in microgravity conditions, the headward fluid shifts cannot be held responsible for the changes in taste and odor in astronauts during space flight.

Many other variables may affect the chemical senses during space flights: the physical and psychological stresses of the workload, sleep schedule, noise, confinement, and the adaptive physiological changes to the novel environment. It is quite possible that space motion sickness may cause a change in taste or smell perception of astronauts similar to the change in taste or odor perception that occurs as a result of terrestrial motion sickness or seasickness. *Conditioned taste aversion* (CTA) is known to occur as a result of motion sickness (Hu *et al.* 1996). One can readily imagine the occurrence of a mild conditioned taste aversion in astronauts who have experienced motion sickness symptoms and the possible effect of this CTA on the perception of the taste of foods.

Psychological variables could also have important influences on smell and taste. It is well known that responses to odors can be accentuated by the presence of visual cues. For example, during the Spacelab-3 mission that carried many animals in orbit, crewmembers complained of disturbing odors, which they attributed to the primates and test rats that were in view. In later missions, the animal cages were placed in visually separated areas and the astronauts mentioned no odor problems.

The smell function of astronauts is also likely to be affected by the presence of volatile trace contaminants present in the vehicle atmosphere, which can render the astronauts' smell fatigued or adapted to some odors.

Although the research on taste and odor perception in space is scare and the results of studies are somewhat contradictory, there seems to be changes in chemosensory perception during space flight. Most crewmembers desire foods with a higher flavor intensity. However, on the other hand, astronauts tend to avoid choosing foods with strong odors that will linger in the spacecraft atmosphere, such as oranges and their peel odor for example. Clearly, more research is needed in this area. According to Olabi *et al.* (2002): "In studying taste and odor alterations, the research path here should follow in the steps of other biomedical space research. The first step is to characterize the nature and extent of taste and odor alterations in space and to discover the important causes and mechanisms. The second step is to investigate the effects of taste and odor alterations on the mood and performance of astronauts, and the last step is to design and test appropriate countermeasures."

4   PROPRIOCEPTION

Somatosensation literally means 'body-sense'; it is the set of senses which originates from the entire body, including skin, bones, muscles, and tendons, not limited to the specialized sensory organs of the head. Somatosensation can be broken down into proprioception (perception of the physical body), kinesthesis (knowledge of movement), and the cutaneous senses, which include temperature, tactile (touch) and pain. The reason why all these seemingly different senses are collectively known as somatosensation isn't just because they are all body senses, but has more to do with the structure of the brain. Different areas of the brain are associated with different functions. The area related to vision is separate in structure and function to that of hearing, and they are both separate from somatosensation. However, the area responsible for somatosensation, the somatosensory cortex, receives input from structures responsible for pain, touch, proprioception, etc. Consequently, these senses may appear to be separate entities at first glance but it is their common destination, and subsequent similarity in processing, that has combined them into a whole.

Proprioception, used in this context, allows discrimination of position of body parts and discrimination of movement and amplitude of movement of body parts both passively and actively produced. Howard & Templeton (1966) suggest that the behavioral or perceptual responses connected with proprioception arise from stimuli associated with changes in the length of muscles, including tension, compression and shear forces (due to the effects of gravity), the relative movement of body parts, and muscular contraction.

It has been proposed that microgravity could impair the state of the proprioceptive sensory receptors, i.e., the neuromuscular spindles, Golgi tendon organs, tactile receptors, and joint receptors, because of atrophy of the antigravity muscles, a fluid shift, or the sudden release of a constant muscle tone (Lackner 1988), although daily sleep on Earth does not bring impairment to those mechanoreceptors.

Clearly, the virtual absence of gravity modifies the stimuli associated with proprioception and impacts knowledge of limb position. "The first night in space when I was drifting off to sleep," recalled one Apollo astronaut, "I suddenly realized that I had lost track of ... my arms and legs. For all my mind could tell, my limbs were not there. However, with a conscious command for an arm or leg to move, it instantly reappeared -- only to disappear again when I relaxed." Another astronaut from the Gemini program reported waking in the dark during a mission and seeing a disembodied glow-in-the-dark watch floating in front of him. Where had it come from? He realized moments later that the watch was around his own wrist!

Changes in the perception of mass, tactile sensitivity modifications, and difficulty acquiring targets during voluntary limb movement are other effects of microgravity exposure on proprioception. For example, rapid movements are found to be unaffected by the gravity level or arm orientation when blindfolded subjects make unsupported forearm movements of particular amplitudes and frequencies in a horizontal and a vertical plane in both hypo- and hypergravity. However, slow movements show a smaller amplitude and more frequent dynamic overshoots of final position in zero-g relative to one-g, both for horizontal and vertical arm orientations (Lackner & DiZio 1996). These findings are consistent with a decreased spindle gain in microgravity, which would also imply a decrease in position sense accuracy, such as that commonly reported by astronauts (see section below). The Golgi tendon organs are force transducers and in space, because of the apparent absence of gravity, they do not sense weight but only force due to an inertial load or acceleration acting on the body. There is a need for a reinterpretation of the messages given by these sensors, similar to that occurring for the otolith organs, which only provide information on head linear acceleration in space, not head tilt.

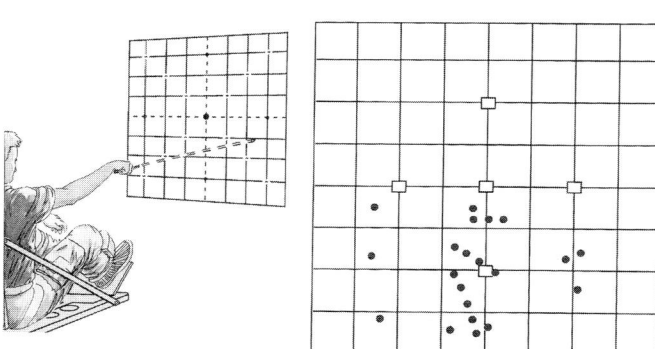

Figure 4-05. In this experiment, the astronauts looked at one of the five cardinal targets on a screen, then closed their eyes and pointed a laser beam at the remembered target position. The in-flight results for one subject are shown on the right. In microgravity, this subject exhibited a pronounced downward pointing bias. The divisions are in inches. Drawing courtesy of NASA.

4.1   Limb Position and Pointing Ability

During the Skylab flights, crewmembers reported that they were unaware of the position of their limbs in the dark. Other disturbances of perceived limb position during the Apollo flights have been anecdotally reported. Astronauts in microgravity reveal impaired ability to judge their elbow angles (Money & Cheung 1991, McCall *et al.* 2003). The astronauts studied by Young *et al.* (1984) revealed increased variability in the estimation of relaxed limb positions, suggesting reduced precision of the proprioceptive sense.

Objective measurement of limb position awareness was conducted during Spacelab-1 by asking crewmembers to point at remembered targets positions with the eyes closed both in-flight and immediately postflight (Young *et al.* 1984). Results of this experiment indicated that pointing accuracy was degraded during and immediately after flight. When crewmembers pointed at remembered target positions with they eyes closed, they made considerable errors and tended to point low (Figure 4-05). In an experiment on board STS-41G (Watt *et al.* 1985), changes in perception of the external world and arm position were studied in two blindfolded subjects. The purpose was to determine how accurately crewmembers could perform an active pointing task in the absence of vision and to assess changes in performance during the course of the flight and immediately afterward. Both crewmembers, who were very accurate prior to flight, showed considerable decrement in target acquisition early in-flight with one subject showing some improvement during the course of the eight-day flight. Immediately after flight, performance was less than that achieved preflight, but recovery was rapid and complete within a few hours after landing.

When crewmembers were asked to reproduce from memory the different positions of a handle, the accuracy of setting the handle to a given position was significantly lower with an error towards a decrease of handle deflection angle. Also, when trying to touch various body parts, they usually noted that their arms were not exactly where expected when vision was restored. The problem is that these examples are suggestive of either degradation in proprioceptive function, or an inaccurate external spatial map, or both (Young 1993, Watt 1997).

In microgravity, visual objects viewed in a dark or impoverished visual setting appear lower than they actually are, while initial attempts to reach for them tend to be too high (Clément *et al.* 1985, Young 1993, Watt 1997, Berger *et al.* 1997). The fact that these motor and visual effects work in opposite directions may account for some of the contradictions in the literature (Bock 1998). The opposite effects occur during hypergravity (Welch *et al.* 1996). In the microgravity phases of parabolic flight, non-visually guided arm movements are smaller in amplitude, with more frequent dynamic overshoots of final position than with one-g movements, whereas during the high-g phase these movements have larger amplitude and fewer dynamic overshoots than with one-g movements (Fisk *et al.* 1993). All of these eye-hand errors are prevented if observers are able to see their limbs or make their visual-motor responses slowly and deliberately. This suggests that rapid adaptation to altered gravity (both hypo- and hypergravity) can occur on the basis of proprioceptive feedback alone (Young *et al.* 1984, Welch *et al.* 1996).

4.2 Proprioceptive Illusions

An elegant way to evaluate changes in the proprioceptive function is to measure the subjective sensation generated by the stimulation of proprioceptive receptors. A classic technique consists in vibrating a muscle tendon to elicit illusory limb movement. As a result of spindle activation by vibration applied on lower leg muscle tendons, one can observe a contraction of the muscle being vibrated, i.e., a tilt of the body, or, if the muscle is immobilized, an illusion of body tilt. Adaptive properties of the proprioceptive system were studied during the French-Russian Aragatz mission using this technique (Roll *et al.* 1993). The perceptual effect of vibrating the soleus and anterior tibialis tendons was investigated while the subject was restrained with a back support or supported in-flight with elastic cords attached from the waist to the floor. Preflight, a 5-sec vibration episode at 70 Hz of the anterior tibialis induced a mild sensation of backward body tilt, while vibration of the soleus induced a forward tilt. Early in-flight sensations were similar, but of larger magnitude. After 19 to 20 days in flight, the tilting sensation associated with vibration of the anterior tibialis gave way to the sensation that the whole body was raised from the deck of the spacecraft while vibration of the soleus had no effect.

Another illustration of an alteration of proprioceptive inputs during the early exposure to microgravity is the impossibility for an astronaut to maintain a "vertical" posture perpendicular to the foot support in absence of visual information. Instead, the subjects tend to lean forward (Figure 4-06). The large body tilt observed in these conditions reveals an inaccuracy in the proprioceptive signals from the ankle joint or in their central interpretation. After flight day 3, however, the astronauts are able to maintain an upright posture, suggesting that adaptive processes take place quite rapidly (Clément *et al.* 1988).

Proprioceptive illusions have also been noted when making voluntary movements during and after several Space Shuttle flights. Watt *et al.* (1985) asked subjects on STS-41G to perform rhythmical deep knee bends and arm bends. In-flight the knee bends were performed with the feet held in foot restraints, and arm bends were accomplished with the subject grasping a handhold and alternately flexing and extending the arms. One crewperson experienced proprioceptive illusions: during arm flexions, the subject felt as though the wall moved toward him, and during leg bends, the illusion was of the floor moving up and down like a trampoline.

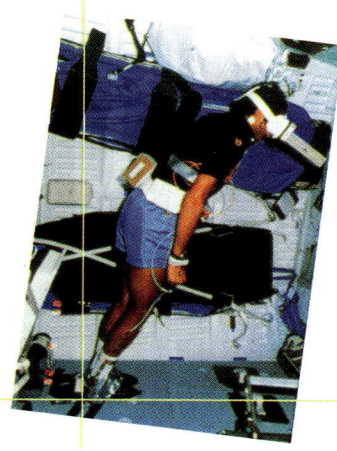

Figure 4-06. An astronaut with the feet attached to the floor of the Space Shuttle and placed in darkness using an occluding goggle is instructed to maintain an "upright" (i.e., perpendicular to the foot support) posture on flight day two. In absence of gravitational and visual inputs, his body is tilted forward, suggesting that the proprioceptive inputs from the ankle joint were misleading. Photo courtesy of NASA.

As a follow-up study, returning crewmembers were asked to make deep knee bends with the eyes open on Earth after a week in microgravity. When the knees were bent, there was a sensation that the knees were being bent more rapidly than intended, and that the floor came up to meet them as they went down (Figure 4-07). One aspect of this illusion is a change in position sense in that one does not correctly appreciate the rate of change or position of the ankle, knee and hip joints. Note that the antigravity musculature of the body is rich in muscle spindle receptors. On return to one-g, the weight of the body is felt to be greater than normal. This abnormal level of spindle activity will be interpreted as the antigravity muscle being longer than they actually are and this will be referred to the joints about which they act. Consequently, when lowering the body in a deep knee bend, an individual should experience his ankles, knees and hip as flexing more rapidly than intended. This will be referred to the deck having moved up under the feet thereby producing abnormally rapid flexion (Lackner & Graybiel 1981a).

Proprioceptive illusions similar to those recorded by Watt *et al.* (1986) were also reported by Spacelab-1 astronauts during passive drop testing primarily aimed at evaluating changes in the vestibulo-spinal motor neurons (Reschke *et al.* 1986). Three to four hours after landing, subjects reported that they did not have the sensation of falling after being unexpectedly dropped. Rather, they reported that the floor came up to meet their feet. In addition, the subjects frequently did not know where their feet were in relation to the floor and without restraints would have fallen backwards upon landing after the drop. These sensations continued for three days after flight and the instability remained until approximately 24 to 36 hours following the first exposure to the drop stimulus. Interestingly, the sensation of not knowing the position of the legs with respect to the trunk was present by the seventh day of flight, and the unexpected drop was followed by a backward rotation of the body.

Among the somatosensory systems projecting to the neurovestibular system, the position receptors of the cervical column, i.e., the neck receptors, also play an important role. During the Spacelab-D1 mission, the trunk of a crewmember was passively bent sideways or forward, while keeping his head fixed to the floor of Spacelab, thus stimulating the neck receptors without vestibular inputs. The crewmember reported an illusory rotation of a head-fixed target cross seen in the monitor of his helmet, which was entirely due to the stimulation of the cervical position receptors, since the otoliths were not stimulated.

The source of proprioceptive illusions is not known, but there are several possibilities. Watt's data suggests that proprioceptive function is altered in microgravity and is expressed as a sensory deficit. In addition, sensory motor programs could be altered due to the changed mode of locomotion in weightlessness. It is also possible that the system that compares motor commands with resulting sensory inputs may not, because of the stimulus rearrangement of microgravity and resulting reinterpretation of sensory input to the central nervous system, be capable of correctly distinguishing between self-motion and movement of the environment. The latter interpretation fits well with the theory of *otolith tilt-translation reinterpretation* (OTTR) (see Section 5.2), which has been proposed to account for illusions of self- or surround-motion experienced by returning astronauts (Young *et al.* 1984, Parker *et al.* 1985).

Preflight   Postflight

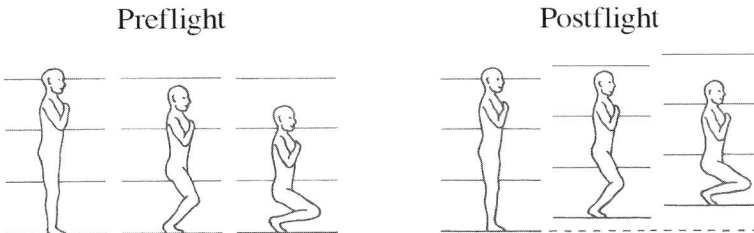

Figure 4-07. These schematics illustrate the perceived motion of a subject and his visual environment (represented by the horizontal lines) as he performs deep knee bends in one-g before and immediately after a space flight. After the flight, the subject experiences the floor coming up to meet him as he lowers his body. This illusion is partly due to an adaptation of the proprioceptive inputs (spindle receptors) to the microgravity conditions. Adapted from Lackner (1989).

Another interesting feature of microgravity is that it is possible to separate between two distinct physical concepts, mass and weight, which, on Earth, both produce similar sensations of heaviness. Indeed, on Earth, weight can be judged passively through the pressure receptors in the skin if the object is placed upon a supported limb. Weight can also be judged actively, if the object is held against the force of gravity by the muscular effort, or is repeatedly lifted. Mass can only be judged actively, derived from the force required to produce a given acceleration, or from the acceleration produced by imparting a given force. Thus, active weight perception usually includes mass perception. It is therefore difficult to investigate weight without mass during active movement, except in weightlessness. Using balls of various masses that the astronauts shook up and down moving their arms, it was found that discrimination in the mass of objects was poorer in microgravity than in normal gravity. Weight discrimination was impaired for two or three days postflight, while crewmembers felt their bodies and other objects to be extra heavy. The impairment in-flight was partly due to the loss of weight information (i.e., a reduction in the pressure stimulation), and probably also to incomplete adaptation to microgravity. The increase in apparent heaviness of objects reported for static weight judgments postflight suggests that some central rescaling of the static pressure systems had occurred (Ross *et al.* 1986, 1987).

The limited number of experiments mentioned above is an indication that the nature of proprioceptive changes in microgravity has been poorly studied. In particular, there is almost no space study of neck and joint angle sensors, and on the role of localized tactile cues in the perception of body verticality.

5   THE SENSE OF MOTION

Humans sense position and motion in a three-dimensional environment through the interaction of a variety of body proprioceptors, including muscles, tendons, joints, vision, touch, pressure, somesthesia, and the vestibular system. Afferent signals from these systems are interpreted by the brain as position and motion. Processing by the CNS enables us to determine body orientation, sense the direction and speed at which we are moving, and helps us maintain balance. It is interesting to note that the visual and vestibular systems are both sensory and motor systems. In the case of vision, the peripheral retina signals the position and movement of the head with respect to surrounding objects, and provides information about the direction of the vertical. As a motor system, the visual receptors that sense slipping of the retinal image supplement compensatory eye movements through a tracking mechanism called the optokinetic reflex. In its role as a sensory system, the vestibular system provides information about movement of the head and the position of the head with respect to gravity and any other acting inertial forces. As a motor system, the vestibular system plays an important role in posture control, that is, orienting to the vertical, controlling center of mass, and stabilizing the head. To this end, output from the vestibular system goes to the spinal cord to serve the vestibulo-spinal reflex. This reflex generates compensatory body movements to maintain head and postural stability. Output from the vestibular system also goes to the ocular muscles serving, in this case, the vestibular-ocular reflex that generates eye movements that enable clear vision while the head is in motion.

5.1   The Vestibular System

The vestibular system's main purpose is to create a stable platform for the eyes so that we can orient to the vertical—up is up and down is down—and move smoothly. The inner ear contains two balance-sensing organs: one is sensitive to linear acceleration, the other to angular acceleration.

The *linear* acceleration sensitive organ, comprised of the saccule and utricle, senses linear acceleration (translation) and provides information concerning changes in head position relative to gravity (tilt). The saccule and utricle are referred to collectively as the *otolith organs*. Each organ is basically composed two parts, the macula and the otolithic membrane. The macula is fixed to the bony labyrinth and includes supporting and sensory cells. Above the macula and its hair-like strands rests the otolithic membrane, which amounts to the moving mass. The otolithic membrane is a gelatinous substance embedded with small crystals of carbonate calcium, the *otoliths* (Figure 4-

08). The otolith crystals have a higher density (2.95) than the endolymph (1.02). Linear acceleration of the head causes a sliding displacement of the otolithic membrane relative to the macula, which in turn bends the hair cells of the macula. The bending of sensory cilia mechanically opens or closes (depending on the direction of bending) ion channels, which alter the electrical current of the respective sensory cell in the inner ear (Hudspeth and Gillespie 1994). The transformation to computable action potentials takes place at the level of the sensory cells. A signal transduction at the level of the CNS gives rise to a sensation of tilt or acceleration, and eventually causes a motor output, i.e., a behavioral response.

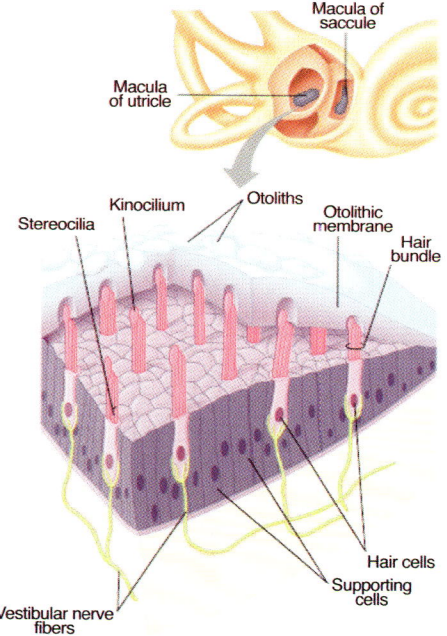

Figure 4-08. During horizontal linear acceleration, inertia causes the utricle's otolithic membrane and the embedded otoliths to deflect the hair-like extensions of the sensory hair cells. Deflection of the stereocilia toward or away from the kinocilium causes an increase or decrease in the firing rate of the sensory neuron at its basal surface. This information is provided to the brainstem via the vestibular nerve fibers.

On Earth, when the head is tilted to the left or right, forward or back, the otoliths tend to move along the gravity gradient (i.e., downwards). During translation, the crystals, being denser than the surrounding fluid, will tend to be left behind due to their inertia. It has been demonstrated that the resultant bending of the cilia causes cell excitation when the bending is toward the *kinocilium* (the longest hair cell), and inhibition when away from the kinocilium. During head motion, the weight and movement of the otoliths stimulate the nerve endings surrounding the hair cells and give the brain information on motion in a particular direction (up, down, forward, backward, right, left) or tilt in the sagittal (pitch) or the frontal (roll) plane (Figure 4-09).

The *angular* acceleration sensitive organ is comprised of three semicircular canals paired bilaterally and oriented in such a way that the plane of each canal is approximately orthogonal to the other two. The semicircular canals are embedded within a bony structure of the same shape. The central cavity of each canal is filled with a fluid called *endolymph*, which, by virtue of its inertia, flows through the canal whenever an angular acceleration in the plane of the canal is experienced by the head. Each endolymph-filled canal has an enlarged area near its base called an *ampulla*. Parts of the vestibular nerve penetrate the base of each ampulla and terminate in a tuft of specialized sensory hair cells. When the endolymph moves (or when the cupula moves and the fluid remains stationary), the gelatinous tip of the cupula and the hair cell extensions embedded within it are displaced to one side or the other. When the embedded hair cells bend, they send a signal via the vestibular nerve to the brain where the information is evaluated and appropriate action is initiated.

Each semicircular canal detects angular acceleration through the inertial movement of the endolymph within its plane. Collectively, they provide the brain with information about rotation about the three axes: yaw, pitch and roll. The semicircular canals do not react to the body's position with respect to gravity; they react to a change in the body's position. In other words, the semicircular canals do not measure motion itself, but change in motion. Not surprisingly, the semicircular canals are not affected by space flight, as shown by the absence of changes in the perception of rotation or in the compensatory eye movements in response to rotation both in-flight and after flight.

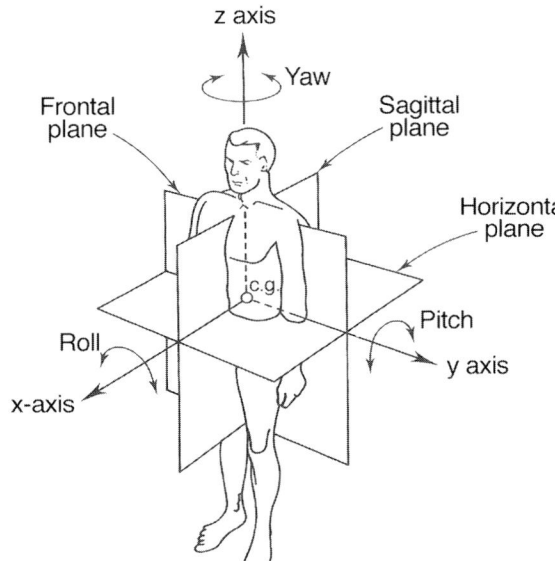

Figure 4-09. This diagram shows the three principal planes passing through the center of gravity (cg) of the body. The z-axis is formed by the intersection of the sagittal plane and the frontal plane; the y-axis, by the intersection of the frontal and horizontal planes; and the x-axis, by the intersection of the sagittal and horizontal planes. Rotations about the z-, y- and x-axes are called yaw, pitch and roll, respectively.

The findings of several researchers on the functional thresholds of the vestibular system have been summarized by Benson (1990). He reports that the mean threshold for angular accelerations of the head has been demonstrated as 0.32 deg/sec^2 with a range of 0.05 to 2.2 deg/sec^2. The perception of angular motion varies with frequency of motion, falling at around 0.2 log unit/decade between 0.1 and 1.0 Hz, and falling at minus 1 log unit/decade below 0.1 Hz. If the stimuli is of duration less than 15 seconds, the perception of angular motion is related to the time taken to detect angular acceleration, which has a mean constant value of 3.7 deg/sec. For sustained rotational stimulation with prolonged acceleration, the sensory threshold for angular rotation is determined by the magnitude of angular acceleration rather than velocity change. This type of stimulation can occur in an aircraft. The common peak angular velocity for passive nodding of the head, such as occurs during walking or running, is 10 deg/sec.

Volitional head movements usually exhibit a peak angular velocity of at least 100 deg/sec but may be as high as 500 deg/sec.

Thresholds for perception of linear acceleration have been obtained for motion along horizontal and along vertical tracks with the subject's x, y, or z body axis parallel to the direction of motion. Despite a large variability, the mean threshold was found to be around 0.10 m/sec^2, which is equivalent to a shear force of 0.01 g. The mean threshold for detection of static tilt is even lower, i.e., about 0.005 g or 0.25 deg of tilt from the vertical, with the head in the normal upright position. However, these values are far worse when the head is tilted or the body inverted. This decrease in precision is related to the decrease in the sensitivity of the utricles with increasing tilt of the body. It is also interesting to note that subjects are able to detect when a visual line is tilted as little as 0.5 deg from the vertical or the horizontal (Howard 1982, p. 368 & 415).

The vestibular system manifests a number of functional limitations. Transient movements lasting less than 10 sec with a change in angular velocity below roughly 2 deg/sec, or peak acceleration below roughly 0.05 m/sec^2, may not be detected. Misperceptions can be caused by prolonged rotation of the head (over about 15 sec) with cross-coupled stimulation of the semicircular canals. Prolonged linear acceleration or deceleration, lasting 40 to 60 sec, can cause misperceptions of attitude, or spatial disorientation episodes, particularly when the resultant effect of the imposed acceleration and head orientation is unaligned with the gravitational vertical. Head movements during linear accelerations over 1 g also cause misperceptions of the direction of the movements. The perception of tumbling occurs with head movements when the acceleration increases to more than 50 m/sec^2.

Illusions of passive self-motion have been studied for many years (Dichgans & Brandt 1978). Such perceptions can be generated by vestibular stimulation, visual stimulation, or a combination of both. In the presence of conflicting visual and vestibular cues, it seems that the vestibular cues dominate the short-term subjective determination of acceleration, whereas the visual cues dominate in the long-term sensation of velocity.

5.2 Linear Acceleration and Gravity

On the Earth's surface, two major sources of linear acceleration exist. One is related to the Earth's gravity. Gravity significantly affects most of our motor behavior (it has been estimated that about 60% of our musculature is devoted to opposing gravity), and it provides a constant reference for up and down. It is present under all conditions on Earth, and it forms one of the major pillars of spatial orientation (Howard 1982). Other sources of linear acceleration arise in the side-to-side, up-and-down, or front-to-back translations that commonly occur during walking or running, and from the centrifugal force that we feel when turning corners.

As Einstein noted, all linear acceleration is equivalent, whether it is produced by gravity or motion; and when we are in motion, the linear accelerations sum. The body responds to the resultant, and we tend to align our longitudinal body axis with the resultant linear acceleration vector, called the *gravito-inertial acceleration* (GIA) vector (see Figure 4-01). Unconsciously, the head, body, and eyes are oriented so that they tend to align with the GIA. The angle of tilt of these body parts depends on the speed of turning (Imai 2001). Put in simple terms, people align with gravity when standing upright and tilt into the direction of the turn when in motion. If they don't, they lose balance and fall.

When our head is horizontal the hair cells in the utricles are not bent and this stimulation is interpreted as signifying "normal posture." If our head is tilted forward, the otoliths shift downward under the action of gravity, bending the hair cells. Technically, it is the shearing force induced by gravity that stimulates the hair cells. If we translate backward, again there is a shift of the otoliths forward due to the inertial forces. Thus, an equivalent displacement of the otoliths (and consequently the same information is conveyed to the central nervous system) can be generated when the head is tilted 30 deg forward, or when the body is translating at 0.5 g backward (Figure 4-10). This example simply illustrates Einstein's principle stating that, on Earth, all linear accelerometers cannot distinguish between an actual linear acceleration and a head tilt relative to gravity.[8]

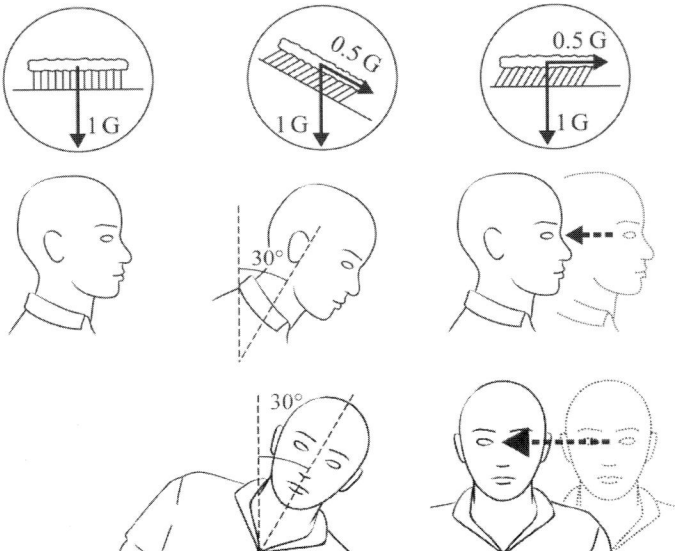

Figure 4-10. The otolithic membrane bends the hair cells of the utricles the same way when the head is maintained at a constant tilt angle of 30 deg relative to gravity and when the whole body in translated with an acceleration of 0.5 g in the opposite direction. Drawing courtesy of Philippe Tauzin (SCOM, Toulouse).

When there is no visual input as is common in many flight situations, we rely more heavily on our vestibular sense for the information of tilt or translation. However, in flight and in space, our vestibular system, which is designed to work on the ground in a one-g environment, often provides erroneous or disorienting information. For example, astronauts experience sensations of dizziness and disorientation during their first few days in the microgravity environment of space. Upon returning to Earth after prolonged exposure to microgravity, astronauts frequently have difficulty standing and walking upright, stabilizing their gaze, and walking or turning corners in a coordinated manner.

---

[8] In a normal situation, the CNS would easily distinguish between a tilt of the head relative to gravity and a head translation by comparing the sensory information from the otolith organs with that from the visual system or muscles proprioceptors. But in complete darkness, there could be a conflict between the proprioceptive input (e.g., signaling that the head is tilted) and the otolith input (e.g., signaling that the head translates).

5.3   Effects of Microgravity on the Vestibular System

Because the net, external contact force acting on the body is zero in weightless conditions, the otolith organs of the inner ear are unloaded. This means that the otoliths do not provide orientationally relevant signals in microgravity, except for voluntary or passively imposed head movements that will generate transient shear of the otolith membranes. The resting discharge pattern of the otoliths will not be influenced by static head orientation. Therefore, absent are the static otolith-spinal and otolith-oculomotor influences that are normally present and that depend on head orientation relative to gravity. The peripheral hydrodynamics of the semicircular canals are basically independent of linear acceleration and, therefore, are unaffected by weightless conditions (Wilson & Melvill Jones 1979).

Although it is difficult to measure changes in the vestibular end organs directly, several attempts have been made to determine whether exposure to microgravity produces anatomical or physiological changes in the vestibular-end organs and their primary afferents.

5.3.1   Anatomical Studies

The sensory epithelium of the vestibular organ of frogs returned from an eight-day stay on board the Russian Mir station remained basically intact (Suzuki *et al.* 1993). The morphology of the vestibular end organs of *Xenopus laevis* did not show major alterations following larval development in microgravity (Briegleb *et al.* 1988, Hertwig & Hentschel 1989).

Experiments on the Spacelab Life Sciences 1 (SLS-1) nine-day mission (STS-40) flown in June 1991 looked at anatomical changes in the otolith organs of rodents. The hypothesis was that otolith crystals might degenerate in microgravity because of changes in body calcium, carbohydrate and protein metabolism, body fluid redistribution, and hormone secretions. No deleterious effects were seen in the otoconia from rodents who flew as compared with those of ground controls (Ross 1993). This is in contrast to findings from a previous Kosmos-782 experiment, which suggested impairment of calcium exchange after a twenty-day space flight (Vinnikov *et al.* 1984). However, changes in the Kosmos experiment may have been influenced by accelerations during launch and landing. Changes in type-I (with single large afferent fiber) and type-II (with multiple afferent terminals) receptor cells were also noted in the Kosmos-782 experiment, but these changes were later attributed to the histological procedures used to remove and fix the animals' vestibular apparatus after landing.

An unexpected change found during the experiment on board SLS-1 was an increase by a factor of 12 in the number of synapses in type-II hair cells from the in-flight maculae as compared with the control data. Such an increase was confirmed during the SLS-2 mission (Ross & Tomko 1998). These findings suggest that mature utricular hair cells retain synaptic plasticity, permitting adaptation to an altered environment (Ross 1993). Consistent with these results are studies in animals centrifuged at 2 g for 14 days, which demonstrated that synapses were decreased by about 40% in type-II cells while type-I cells were unaffected. These data suggest that the maculae adapt to g-forces changes in either directions by up- or down-regulation of synaptic contacts in an attempt to modulate neural outflow to the CNS (Ross 1992). The functional meaning of the synaptic changes noted in the local microcircuitry synapses is, however, completely unknown (Ross & Tomko 1998).

More recent studies during the Neurolab mission also showed that snails reared in weightlessness developed more and 50% larger gravity-sensing crystals (similar to the otoliths in mammals) than their Earth-based counterparts (Wiederhold *et al.* 2003). In rats, the gravity sensors themselves developed normally, but the connections they made in the brain, specifically within the vestibular nuclei and cerebellum, were different (Raymond *et al.* 2003).

In conclusion, the functional significance of anatomical changes observed in vestibular sensory epithelia in animals during flights of several weeks and longer is unclear. There is no evidence that prolonged exposure to weightlessness lasting from months to years produces irreversible changes in the vestibular system. In fact, the observation that sensorimotor disturbances in astronauts returning from long-duration space flight disappear after days or weeks indicates that these changes are reversible. Nevertheless, because only half a dozen individuals have yet flown in space beyond eight months, vigilance is certainly appropriate. It is also known that radiation exposure affects some areas of the central vestibular system, including the hippocampus (Todorovic *et al.* 2005).

5.3.2 Electrophysiological Studies

The only direct experiment to test vestibular efferent discharge pattern and rate was conducted as early as 1970 in a fairly sophisticated experiment that monitored *in vivo* primary afferents of otolith receptors in the labyrinth of two orbiting bullfrogs (Bracchi *et al.* 1975). Primary afferent fibers of the vestibular nerve are the first components of the vestibular path, relaying the information originating at the cristae and maculae to the brainstem (within each nerve is also a system of efferent fibers from the CNS which provides neural feedback to modulate the activity of the peripheral organs). In the flight experiment, conducted over a period of seven days on just four units, a floating microelectrode continuously monitored changes in the resting activity of single otolith afferents and their response to centrifugal forces generated by a onboard centrifuge. The resting and (arguably) evoked unitary activity showed significant changes in microgravity compared to the ground. The fluctuation in unit activity was up to 20 times larger in-flight than any experienced on the ground (Gualtierotti 1987). However, this finding was heavily criticized for the lack of temperature control of the preparation during the flight and the fact that there was no way to ensure that the electrodes remained in the same cells throughout the flight.

More recently, a U.S.-Russian collaboration on the Kosmos-2044 biosatellite studied two Rhesus monkeys after a two-week space flight. They recorded extra cellular responses from single horizontal semicircular canals afferents pre- and postflight in two flown monkeys and three ground-based control monkeys (Correia *et al.* 1992). The spontaneous firing rate of an afferent, characterized by its mean frequency of firing and coefficient of variation, did not change pre- to postflight. However, the mean gain for nine afferents tested postflight during passive yaw rotation was twice that for 20 afferents tested during control studies. Studies of the eye movements in the same monkeys indicated that the gain and time constant of the horizontal vestibulo-ocular reflex changed in microgravity (Cohen *et al.* 1992), and that the modulation of vergence eye movements was reduced for 11 days after the flight (Dai *et al.* 1996).

Changes in afferents tested may be reflective of changes in the vestibular end organs themselves, changes in neural components such as hair cells, synapses or afferent terminals, or due to afferent feedback changes. Vestibular end organs changes may also be mediated through secondary mechanisms like calcium loss or through CNS pathways. Unfortunately, the adaptivity or maladaptivity of these changes cannot be determined without combining observations of anatomical changes with perceptual and behavioral responses.

5.3.3 Developmental Studies

Gravity-related behavior and their underlying neuronal networks are the most suitable models to study basic effects of altered gravitational input on the development of neuronal systems. An interesting feature of sensory and motor systems is their susceptibility to change depending on the physical or chemical stimuli during development. This discovery led to the formulation of *critical periods*, which defines the period of susceptibility to these changes during post-embryonal development.

Critical periods can be studied by altering the stimulus input for the gravity sensory system. Techniques include:

a.   Destruction of the vestibular end organ so that the gravitational acceleration no longer can be detected.
b.   Loading or unloading of body by weights or counterweights, respectively, which compensates for the gravitational pull in the muscles.
c.   Decrease or increase of the gravitational input by exposing animals to microgravity during space flight or hypergravity during centrifugation.

Using these techniques, significant changes in behavior and physiology were observed during the development of crickets, amphibians, fish, and rodents (see Clément & Klenzka 2006 for review). For example, behavioral responses including the compensatory eye or head movements induced by roll or pitch body tilt in tadpoles increased in microgravity and decreased in hypergravity. In crickets, the sensitivity of the position-sensitive neuron was reduced in microgravity. Several periods of development, such as hatching or when gravity-related reflexes first appear, were found to be particularly sensitive to altered gravity (Horn 2003). The mechanisms that might contribute to these alterations in microgravity include stimulus transduction within vestibular hair cells, activation of immediate early genes within central afferent and efferent vestibular nuclei, and modifications of cellular transcription factors' activity during early development (Horn 2006).

These structural changes may explain why animals that develop in space show definite differences in how their sensory and motor systems adapt to another gravito-inertial force level. An experiment during the Neurolab flight clearly showed differences in how the animals righted themselves when compared to Earth-raised rats. When an infant rat is held on its back and then dropped, it will right itself in whatever way it can. As it matures, it will learn to right itself in an efficient and smooth way: first the head, then the forelimbs, and then the rest of the body in one fluid motion. This righting reflex is related to gravity and is present in humans as well (see Figure 1-10).

During the Neurolab mission, a group of young rats were in space during the time when they would ordinarily have been learning to walk and right themselves. Upon release from being held on their backs, they floated up without ever trying to right themselves. No input from the gravity sensors told them they were upside down so they felt no need to do so. It turned out that the rats could right themselves after returning to Earth, but they never acquired the classic, smooth pattern or movement typical of adult rats. Their motor neurons involved fewer dendrites in postural control and righting.

These rats may have needed gravity to develop a normal righting reflex (Kalb *et al.* 2003).

Neurolab investigators also found that there were more numerous synapses in the parts of the brain related to hind limb movements in the rats that had been on the flight. One possible interpretation is that rather than being confined to one two-dimensional surface on which to move, the rats reared in space had three dimensions to move in and six cage surfaces to "walk" on. This provides more stimulation to the relevant areas of the brain, which could increase the number of synapses. The data from Neurolab suggest an intriguing possibility that the flight rats may be enriched in other areas as a result of the three-dimensional environment in which they grew up, even if they were deficient in their righting (Buckey & Lasley 2005).

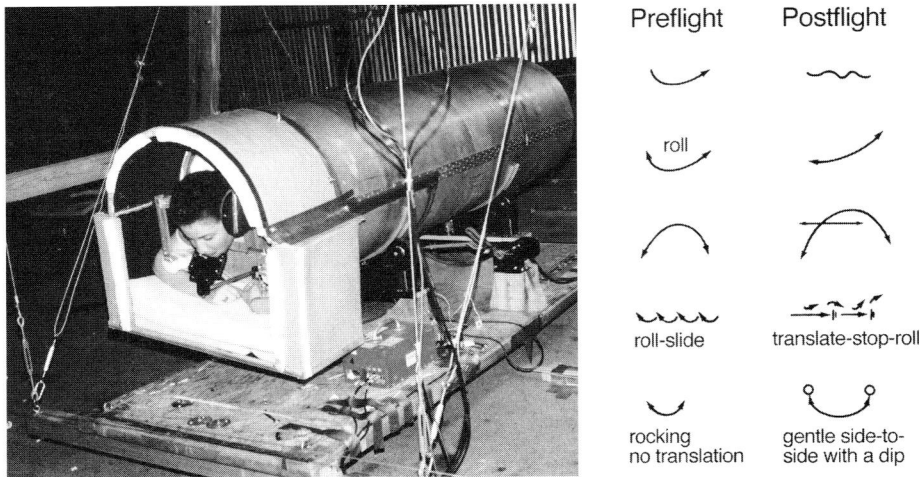

Figure 4-11. The Miami University Parallel Swing *either rolled or translated the subjects. During the runs, the subjects were in darkness. After the runs they drew on a sheet of paper their perceived motion path during the various stimuli. Preflight, all astronauts reported that cylinder roll produced primarily roll motion perception. Immediately postflight, they reported increased horizontal displacement during the roll stimulation. Photo courtesy of NASA.*

5.4   Perception Studies

The vestibular experiments on board Skylab clearly demonstrated that there was no alteration in the perception to angular rotation in yaw (Graybiel *et al.* 1977). This experiment used for the first time a rotating chair in orbit (see Figure 7-04). In one experimental protocol, the crewmembers reported their amplitude of turning sensation during small changes in passive angular acceleration in yaw. There was no change in the crewmembers ability to detect yaw accelerations either during the flight or postflight that differed from preflight values.

Changes in motion perception are, however, expected during head or body rotation in pitch or roll in microgravity, since head tilt about these axes are normally accompanied by changes in otolith inputs. In low Earth orbit, there is no apparent gravitational vertical. But the linear accelerations due to side-to-side, up-and-down, and front-to-back motions (translations) persist. Since tilt is meaningless in space (there is no vertical reference from gravity), it has been hypothesized that, during adaptation to weightlessness, the brain would reinterpret all otolith signals to indicate primarily translation, not tilt (Young *et al.* 1984, Parker *et al.* 1985). This *otolith tilt-translation reinterpretation* (OTTR) hypothesis has received some support from perceptual studies done after space flight.

5.4.1 Motion Perception

One of the earliest, and perhaps the most direct evidence of response incongruence following exposure to the stimulus rearrangement of microgravity showed that upon return to Earth, astronauts, perceived horizontal z-axis sinusoidal rotation as side-to-side translation (Parker *et al.* 1985, 1986, 1987). The motion apparatus employed was the Miami University parallel swing (Figure 4-11). The swing was a four-pole pendulum that produced translation oscillations when manually put in motion by the experimenter. The swing restraint system included an aluminum cylinder that was connected to a motor drive and rolled at an amplitude of ± 15 deg and a frequency of 0.26 Hz. The subjects were encased in a Styrofoam body mold inside of the aluminum cylinder. Their body was supine and their head dorsi-flexed about 50 deg, so that roll motion was about the head x-axis. Eye movements were recorded during motion.

Following the translation or tilt stimulation, the astronauts were instructed to describe their perceived motion and draw their motion path perception on a piece of paper. Preflight, the three astronauts reported that cylinder roll produced a sensation of nearly pure roll, which they illustrated by drawing a 'U' shape with arrows at the ends, and that swinging motion without roll of the restrained cylinder produced nearly pure apparent horizontal linear motion (Figure 4-11). Immediately postflight, roll stimulation was perceived as translation motion only with a small angular motion component. The verbal reports and drawings were congruent. Also, during roll stimulation, torsional eye movements preflight were primarily horizontal eye movements immediately postflight. In effect the eye movements were compensatory for the perception of translation self-motion (Parker *et al.* 1987).

If the simple OTTR hypothesis were true, the subjects would sense linear acceleration as translation postflight. Accordingly, the thresholds or changes in sensitivity to linear motion may or may not be affected. Several experiments performed during and immediately following Spacelab missions have tried to address this hypothesis. During the Spacelab-1 mission a floating-seat body-restraint system was used to provide rectilinear accelerations in response to metronome beats (Figure 4-12, left). The Spacelab D-1 mission used the European Space Agency's vestibular sled to extend this testing (Figure 4-12, right). Experiments using a linear sled were also performed before and after the Space Life Sciences (SLS) missions 1 and 2.

During Spacelab-1, the detection thresholds for linear-oscillatory motion at 0.3 Hz in all three orthogonal axes of the body were determined using the method of limits with a single staircase procedure. Thresholds for three crewmembers were higher (i.e., they were less sensitive) during flight than before; on Earth, thresholds were higher for vertical (z-axis) accelerations (0.077 m/sec^2) than for antero-posterior (x-axis) or lateral (y-axis) accelerations (0.029 m/sec^2) (Benson *et al.* 1986). Neurophysiological evidence supports such a lower gain for z-axis stimulation, as the sensitivity (spikes per second per g) in neurons signaling motion in the z-axis is 30% less than in neurons signaling motion in the x- and y-axes (Fernández & Goldberg 1976c). Threshold variability increased during and after flight, and thus the changes were not consistent. Threshold determinations typically require more data than can be collected in space experiments, which may account for this variability. Thus, the authors reasoned that a second measure of sensitivity, time to detect acceleration, might yield "cleaner" results.

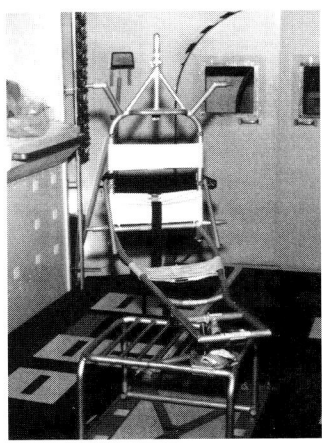 

Figure 4-12. Vestibular experiments on board Spacelab-1 and D-1 missions. Left: The ESA Body Restraint System (BRS) was a rotating chair with a harness to hold the test subject in place. The crewmember wore an accelerometer and electrodes to record his head motion and horizontal and vertical eye movement as the body rotated. Right: The ESA Linear Sled could passively translate the subjects along a 4-m linear track. Perception studies with these devices involved sinusoidal profiles with the subjects being manually spun or accelerated back and forth along each of the body axis. Photo and drawing courtesy of ESA.

Before and after the Spacelab-1 mission, four crewmembers used a joystick to indicate the onset and direction of acceleration during step acceleration followed by step deceleration in successive, discrete profiles of randomly determined amplitude (ranging from 0.001 to 0.08 g) and direction. No consistent pattern of change was noted in threshold or velocity constant (product of acceleration and time to detect) except for increased variability in response (Arrott & Young 1986). Four members of the Spacelab D-1 crew, however, had a consistently shorter time to detect acceleration in weightlessness, with no significant differences between the axes tested (Arrott *et al.* 1990). Before and after the SLS-1 mission, four crewmembers were also accelerated sinusoidally at 1 Hz and 0.25 Hz with a peak acceleration of 0.5 g and with a series of low acceleration steps. The results indicated a "pattern of confused and erratic responses to early postflight accelerations" (Young *et al.* 1993). There was an increased sensitivity to lateral (y-axis) linear acceleration, but a reduced sensitivity for longitudinal (z-axis) linear acceleration.

A closed-loop test used before and after Spacelab missions tested the OTTR hypothesis by having sixteen subjects try to null their movement on a sled. Movement of the sled included sinusoidal oscillations along the y- and z- axes at high (1 Hz) and low (0.25 Hz) frequencies, or pseudorandom linear motion (0.036-0.451 Hz) with peak acceleration ranging from 0.2 to 0.5 g. In about half of the subjects tested the ability to complete this task was generally improved postflight by comparison with the preflight performance, and this improvement decayed gradually over the ensuing week (Arrott *et al.* 1986, 1990, Merfeld *et al.* 1996b). This result supports the OTTR hypothesis. In this closed-loop test, the linear accelerations were large enough (0.2-0.5 g) to potentially generate a perception of tilt ranging from 15-27 deg preflight. This hilltop illusion could possibly interfere with the performance of nulling-out pure linear motion postflight. After adaptation to microgravity, however, tilt would no longer be perceived but reinterpreted as translation, and the performance of nulling-out pure linear acceleration would then improve (Merfeld *et al.* 1996b). This explanation remains a hypothesis, though, since simultaneous perception of tilt and translation was not recorded in these experiments.

Measurements of perception of motion were recently recorded when astronauts were exposed to *off-vertical axis rotation* (OVAR) in darkness before and after Shuttle missions. During OVAR, subjects are rotated in yaw about a rotation axis that is tilted relative to the direction of gravity (Figure 4-13, left). Initial rotation, driven by input from the semicircular canals, is sensed by the subject. This sensation is subject to exponential decay as the rotation approaches constant velocity. The otolith organs, however, are stimulated continuously by a rotating gravity component, which induces a sinusoidally varying linear stimulus along the utricular macula. The frequency of these sinusoidal variations in shearing force is proportional to the rotation velocity.

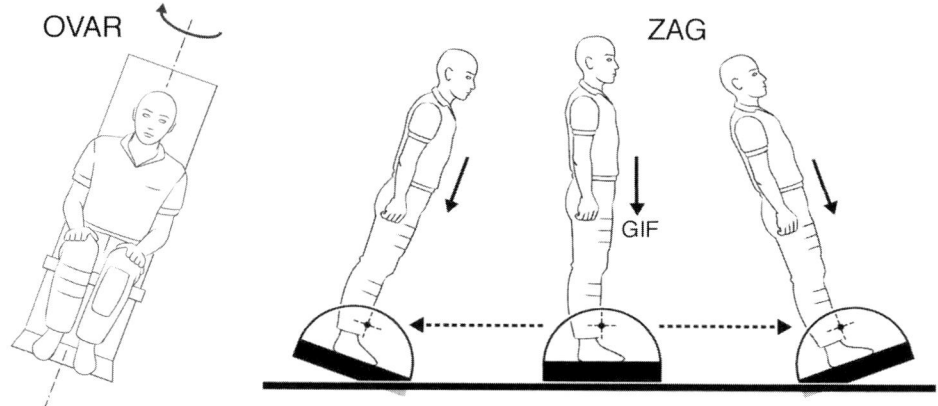

Figure 4-13. Left. A subject rotating at constant velocity about an axis tilted relative to the vertical (off-vertical axis rotation) has the illusion of describing either a conical or a cylindrical motion when rotation is at low or high velocity, respectively. Right. In the Z-axis Aligned Gravito-inertial force (ZAG) paradigm, the subject is sinusoidally translated while simultaneous tilted such as the gravito-inertial force (GIF) remains aligned with the longitudinal body axis. Both OVAR and ZAG allow to investigate the ambiguity between tilt and translation motion perception during stimulation of the otoliths on Earth. Drawings courtesy of Philippe Tauzin (SCOM, Toulouse).

On Earth, at low frequency constant velocity OVAR, subjects report the perception of progressing along the edge of a cone, whereas at high frequency they report the perception of translating along the edge of an upright cylinder (Miller & Graybiel 1973, Guedry 1974, Denise *et al.* 1988, Wood *et al.* 2007). Astronauts returning from short-duration missions generally experience a larger sense of tilt during OVAR at low frequency and a larger sense of translation during OVAR at high frequency than preflight (Clément *et al.* 1995, 2007, Wood *et al.* 2002b). This overestimation of tilt and translation during OVAR in returning astronauts is in general agreement with the results obtained using the human-rated centrifuge on board Neurolab (see Section 5.4.2).

An experiment is currently being conducted to evaluate perceived tilt and translation in astronauts returning from space flight when they are exposed to ambiguous inertial motion cues. During this *z-axis aligned gravito-inertial force* (ZAG) paradigm, the chair tilts (± 20 deg) within an enclosure that simultaneously translates so that the resultant GIA vector remains aligned with the longitudinal body axis (Figure 4-13, right). This condition provides a space flight analog in that tilt signals from the canals conflict with otolith signals that do not indicate tilt (Golding *et al.* 2003). It is expected that, in agreement with the OTTR hypothesis, the crewmembers should perceive larger translational motion immediately postflight compared to preflight.

Postflight measurements in this experiment (first astronauts will be tested in April 2008) will also include a closed-loop nulling task, where subjects will use a joystick to null-out tilt motion disturbances with or without concomitant translation motion. Finally, we will evaluate how a tactile display countermeasure can aid piloting performance when sensorimotor function is compromised after tilt-translation adaptation following exposure to microgravity. The tactile display will consist of a matrix of electromechanical tactile stimulators applied on the subjects' torso (see Figure 3-09). These stimulators will convey orientation cues to the skin, such as the individual's amplitude of body tilt relative to gravity. By overcoming the limitations of multi-sensory integration when sensorimotor function is compromised during unusual acceleration environments, such aid is a promising tool for reducing spatial disorientation mishaps (Rupert 2000).

5.4.2 Tilt Perception

Humans are fully aware of what is "up" and what is "down", even with the eyes closed. It is not known why this is so and why the outer world is not perceived tilted or event inverted (Howard 1982). We do know that several senses are involved. Vision is important, because trees and houses are naturally oriented vertically and their substrates are mostly horizontal, for example (see Figure 1-08). Our vestibular system senses how we move and if we are tilted with respect to gravity. The signals from the otolith organs, as well as from somatosensory receptors (e.g., ankle joints, abdominal graviceptors, neck muscle afferents, and "seat of the pants") all typically point at the direction of the GIA. Depending on how it manifests itself, the perceived vertical (percept of *g*) has been given various adjectives such as visual, kinaesthetic, postural, subjective, apparent, or gravitational (Gibson 1966, Mittelstaedt 1983, 1992, Bos & Bles 2002). In this section, we will review the effects of adaptation to microgravity on the perception of tilt in absence of visual cues. The influence of visual cues on spatial orientation will be more detailed in Chapter 7, Section 3.2.

It is known that, on Earth, the subjective vertical of tilted subjects in darkness typically deviates in the direction of head tilt for roll tilts beyond 60 deg, as if body tilt is underestimated, an error known as the *Aubert* or *A-effect* (Schöne 1964, Udo de Haes 1970). It is argued that at such large angles of body tilt, the utricles become less effective and changes in somesthetic and kinesthetic stimuli largely affect the precision and accuracy of judgments of the visual vertical (Howard 1982). It has also been proposed that the additive effect of an idiotropic head-fixed vector biases the subjective vertical toward the head z-axis, and reduces subjective vertical errors in the commonly used working range of small tilts (Mittelstaedt 1983).

The experiments performed during and immediately after space flight in astronauts show that the perception of self-orientation with respect to the environment is altered (Clément *et al.* 1987, Glasauer & Mittelstead 1998). As expected, in the absence of visual and graviceptive cues, free-floating astronauts are unable to accurately report the orientation of the local (spacecraft) vertical, i.e., the error in their estimate of tilt ranges between 0 and 180 deg (van Erp & van Veen 2006). Within a few days, however, they get better at recognizing the amplitude of passive rotation about roll, and pitch axes (Clément *et al.* 1987) (Figure 4-14, left). Upon return to Earth though, the perception of tilt is again altered. In a series of experiments carried out after several Shuttle missions, we were able to compare static tilt perception in seven astronauts who were tilted with respect to the gravitational vertical on a tilt table (Clément *et al.* 2001, 2007). Results showed that the sense of tilt was significantly overestimated a few days after landing (Figure 4-14, right).

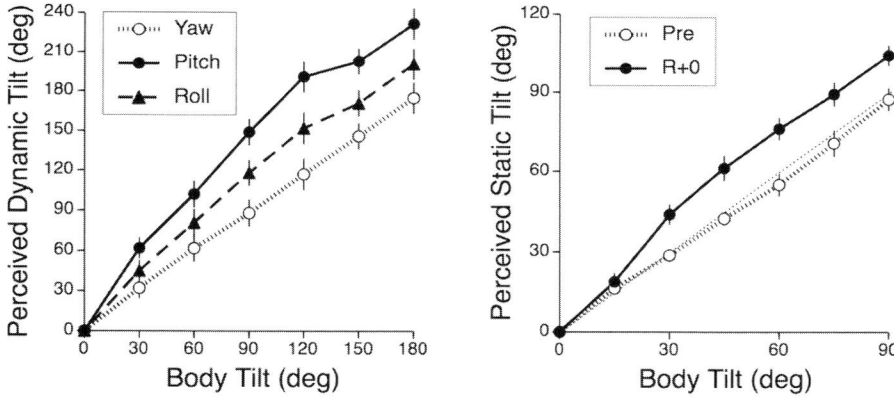

Figure 4-14. Perceived tilt during and after space flight. Left: When free-floating astronauts are being rotated around their center of gravity in complete darkness, they perceive larger amplitude of tilt during rotations in pitch and roll than during rotation in yaw. Right: This overestimation carries over to the postflight period. When they are statically tilted in roll relative to the vertical, they overestimate the amplitude of this roll tilt. Adapted from Clément et al. (1987) and Clément et al. (2007)

Three space experiments flown on Space Shuttle Columbia have used controlled rotation devices generating passive motion, which also induced a sense of tilt. In the first experiment flown on STS-42 in 1982, the *Microgravity Vestibular Investigations* (MVI) project used a computer-controlled rotating chair that could rotate subjects in yaw, pitch, or roll. During the pitch and roll orientations, the subject's head was located at about 45 cm from the center of rotation (Figure 4-15, left). Therefore, during rotation at constant velocity a low magnitude centripetal acceleration was generated that stimulated the vestibular system. This offered exceptional opportunity for study of perception of tilt under conditions that only occur in orbit. Some motion profiles included a ramp to a constant velocity of 20 rpm that was sustained for 60 sec. During spin-up, the centripetal acceleration increased up to 0.22 g and remained constant for the entire 60 sec of constant velocity. On Earth, this 0.22 g centripetal acceleration at the head level was combined with gravitational acceleration and generated a resultant vector that was tilted by 11.3 deg relative to gravity. Therefore, subjects generally felt inclined slightly head down during constant velocity rotation in pitch or roll.

In orbit, however, the gravity vector is not effective so the resultant vector was aligned with the subject's longitudinal body axis (z-axis) in a direction that could induce a sensation of being inverted. Nevertheless, none of the four subjects tested throughout the seven-day mission felt inclined during constant velocity rotation (Benson *et al.* 1997). One possibility is because the lower body was on the opposite side of the rotation axis and a centripetal acceleration of 0.36 g was simultaneously generated at the feet level. Another interpretation is that otolith stimulation of 0.22 g is not of sufficient magnitude to provide an artificial gravity perceptual reference. This is a very important issue because it relates to questions concerning spatial orientation on the Moon's surface (where the gravity level, 0.16 g, is less than the 0.22 g utilized in this experiment) and the use of artificial gravity for long-duration missions (Clément & Bukley 2007).

Figure 4-15. Left. In the rotator used for the Microgravity Vestibular Investigations *on board IML-1, the subject could be rotated in yaw, pitch or roll. In the pitch orientation seen here, the subject's head was 65 cm from the axis of rotation. Constant velocity rotation at 120 deg/sec generated a centripetal acceleration of 0.22 g along the head y-axis. Right. The off-axis rotator flown during the Neurolab mission. The subject's head was 50 cm from the axis of rotation. Rotation at constant velocity of 253 deg/sec and 158 deg/sec generated linear acceleration of 1 g and 0.5 g, respectively, along the head y-axis. Centripetal acceleration at 0.5 g and 1 g provided the subjects a sense of tilt both on the ground and in-flight. Photos courtesy of NASA.*

A second experiment flown on Spacelab-D1 in 1985 used a sled moving along 4-m long rails fixed to the floor of the Spacelab (see Figure 4-12, right). Seated test subject could be moved backward and forward with precisely controlled accelerations. On Earth, during lateral sinusoidal oscillations at low frequency (below 0.3 Hz), about half of the subjects have the perception of being transported not on an Earth-horizontal path, but over a small hill. This *hill-top illusion* is explained by the partial reinterpretation of the linear acceleration signal as a tilt from the gravitational vertical (Arrott *et al.* 1990). Like for the MVI off-axis rotator experiment described, the subjects did not experience tilt in orbit, but a pure horizontal self-motion. The peak linear acceleration of the sled was only 0.2 g, so it is possible that this level was too low to produce a perceived sensation of tilt relative to an artificial gravity reference. It is interesting to note that those subjects who experienced the "hilltop" illusion preflight never reported a tilt illusion immediately after the flight (Young *et al.* 1993).

The third experiment involved a short-radius centrifuge flown on STS-90 during the Neurolab mission in 1998 (Figure 4-15, right). When a subject fixed at the end of a centrifuge arm (a distance *r* from the center of rotation) is rapidly brought to constant angular velocity (ω) about an Earth vertical axis, he will experience a centrifugal force ($\omega^2 r$) that is perpendicular to the gravitational force. The resultant gravito-inertial force hence tilts with respect to the subject (Figure 4-16, left). For example, when the centrifugal force is equivalent to 1 g, on Earth, subjects who are upright and facing the direction of motion ("nose in the wind") feel tilt 45 deg outwards. This apparent tilt during centrifugation was termed the *somatogravic illusion* (Gillingham & Wolfe 1985). Labyrinthine-defective subjects are less susceptible to this illusion (Clark & Graybiel 1966), indicating the significance of the vestibular apparatus regarding this effect. The somatogravic illusion illustrates that we do *not* employ a veridical "sense" of verticality, but adapt to the direction of the resultant vector instead.

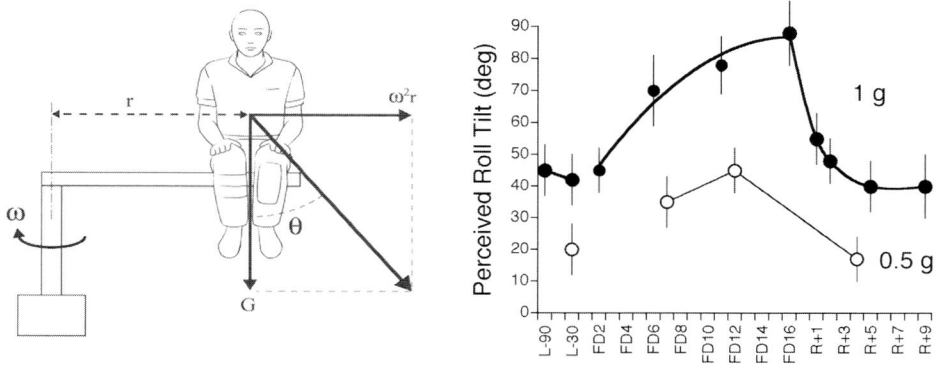

Figure 4-16. Principle of centrifugation along the subject's interaural axis (left). Subjects were being accelerated on the Neurolab centrifuge in complete darkness up to a constant velocity of 44 rpm. After the perception of rotation ceased (after about 40 sec the onset of rotation) the subjects were then prompted by an operator to verbally report whether they had a perceived sensation of tilt or linear motion during steady-state centrifugation. All subjects reported tilt before, during and after the mission. The graph on the right shows the amplitude of body tilt in roll that was perceived by the subjects for centripetal linear accelerations of 1 g and 0.5 g along the interaural axis throughout the mission.

In orbit on the second day of flight, the four subjects tested on this short-radius centrifuge continued to perceive a 45-deg tilt relative to their position when the chair was not in motion when the centripetal acceleration was 1 g (Clément *et al.* 2001). The apparent tilt increased throughout the flight and on flight day 16 all four subjects perceived a 90-deg tilt (Figure 4-16, right). In-flight tilt perception during 0.5-g centrifugation on flight days 7 and 12 was approximately half that reported during 1-g centrifugation, and higher than the 27-deg tilt reported preflight. Like for the static condition (see Figure 4-14, right), tilt was overestimated during centrifugation on return to Earth. At no point during or after the mission did the subjects perceive translation during constant-velocity centrifugation.

The results of this simple experiment showed that the astronauts perceived a body tilt relative to a perceived spatial vertical when exposed to 0.5-g and 1-g artificial gravity, and that the magnitude of this perception adapted throughout the mission. After two weeks in space, the subjects perceived an almost 90-deg tilt when they received a 1-g sideways linear acceleration in space, and about half of this when they received a 0.5-g acceleration. Although they had never encountered this stimulus before, their perception was essentially veridical in that it represented the actual levels of linear acceleration experienced by the graviceptors. It suggests that the otoliths are operating normally in space when exposed to 0.5-g and 1-g steady-state linear acceleration, after the initial period of adaptation.

The finding that none of the astronauts felt translation instead of tilt in response to the 0.5-g or 1-g constant linear accelerations in space does not support the OTTR hypothesis. Tilt is perceived as tilt, regardless of whether the subjects are in microgravity or the one-g environment of Earth, and is not sensed as translation. The underestimation of tilt at the beginning of the flight suggests that the subjects continued to use an internal representation of the magnitude and direction of gravity while in space, despite of the apparent absence of gravitational force. As the flight progressed, the "weight" of this internal model of gravity gradually decreased, and the subjects finally adopted the centripetal acceleration as the new spatial vertical. On return to Earth, perceived body tilt was larger than preflight. This overestimation of body tilt can be interpreted as the result of the continued small weighting of the internal model of gravity after adaptation to the weightless environment. Thus, the underestimation of tilt during the period immediately following orbital insertion, and the exaggerated sense of tilt on return, could both be due to the lag in readjusting the weight of the sense of the internal model of gravity in determining the perceived spatial vertical reference (Clément *et al.* 2001, 2003).

One consequence of this finding is that if low-frequency linear acceleration (of at least 0.5 g) is always perceived as tilt when subjects are in weightlessness, long-duration missions can proceed with the expectation that the astronauts will respond normally to artificial gravity or to the reduced gravitational fields of other planets. The results of these experiments must, however, be interpreted with caution. When using a rotator for generating centripetal linear acceleration there are subtle canal-otolith interactions during the onset acceleration and during the deceleration when the subject's head is fixed and off center. Ground-based studies have demonstrated that these interactions affect the perceptual responses, particularly their time of onset and magnitude (Lackner & DiZio 1993) In addition, the canal-otolith interaction during the onset acceleration differs from the interactions during deceleration. On a large-radius centrifuge on Earth, the onset acceleration generates the illusion of eccentric rotation, whereas the deceleration engenders a perceived turn about the body z-axis and a more pronounced sense of turning (Guedry 1974). Perception is different on a small-radius centrifuge, with a pronounced sense of pitch during back-to-motion profiles (Clément *et al.* 2002). These differences may be exacerbated in orbit and confound the subjective perceptual responses.

Figure 4-17. The Body Restraint System *(BRS) flown on board the first Spacelab mission was used to evaluate changes in the threshold of perception of linear acceleration during adaptation to microgravity. An operator manually moved the BRS, while the subject reported the beginning and direction of perceived motion using a joystick. Photo courtesy of NASA.*

6   SUMMARY

There are many anecdotal reports suggesting that visual acuity, binaural hearing, taste and olfaction, and limb proprioception are affected by exposure to weightlessness. Clearly, more research is needed in these areas, especially to investigate potential changes during long-duration space flights.

In fact, the human body has seven sensory systems – not five. The sixth system indicates whether and how fast a person is spinning and the seventh indicates if the individual is tilted relative to gravity as well as when he or she starts and stops along a straight line. The seventh system no longer provides tilt information in the absence of perceived gravity; however, it does continue to signal translation, so the afferent signals to the CNS are confusing. The experience of living and working in space alters the manner in which the CNS deals with otolith organ signals during linear acceleration. Although the perception is fairly accurate when subjects are exposed to angular acceleration in yaw in-flight, there are disturbances during angular rotation in pitch and roll, and during linear acceleration along the body y- and z-axes.

Perception of body motion is also altered during the same stimulations immediately after landing. There is an adaptation to microgravity that carries over to postflight reactions to linear acceleration.

An adaptive change in the way that the CNS interprets tilt cues in microgravity is evident because of the observed error in tilt perception. The amplitude of the internal estimate of gravitational vertical in microgravity decreases to zero in a few days. This adaptation to microgravity carries over to the early postflight period. The exaggerated sense of tilt upon return is then because of the lag in readjusting the amplitude of the internal estimate of gravitational vertical in determining the perceived spatial vertical reference. In the next three chapters we will review how these perceptual changes are reflected in posture, eye movements, and spatial orientation.

Chapter 5

POSTURE, MOVEMENT AND LOCOMOTION

Postural activity is the complex result of integrated orientation and motion information from visual, vestibular and somatosensory inputs. These inputs collectively contribute to a sense of body orientation relative to the Earth or other support surface and, additionally, coordinate body muscle activities that are largely automatic and independent of conscious perception and, in some cases, voluntary control (Figure 5-01).

Brain structures concerning posture, movement and locomotion are hierarchically organized. Whereas local reflexes take place in local interneural circuits in the spinal cord, standing posture and equilibrium are achieved by excitation of the brain vestibular system in the midbrain and alpha and gamma motor neurons units of extensor muscles. Complex movement sequences and gait result from the activation of forebrain structures, such as thalamus and premotor or motor areas of cerebral cortex. Both voluntary and reflex pathways participate in the control of excitation of synergist and inhibition of antagonist motor neurons. Also, while the movement is underway, feedback from proprioceptors influences subsequent neuronal activity in motor centers to effect the desired movement. The cerebellum influences neuronal activity in the initiating motor centers and continuously modulates neuronal activity, based on information about motor commands and proprioceptive feedback about position and acceleration.

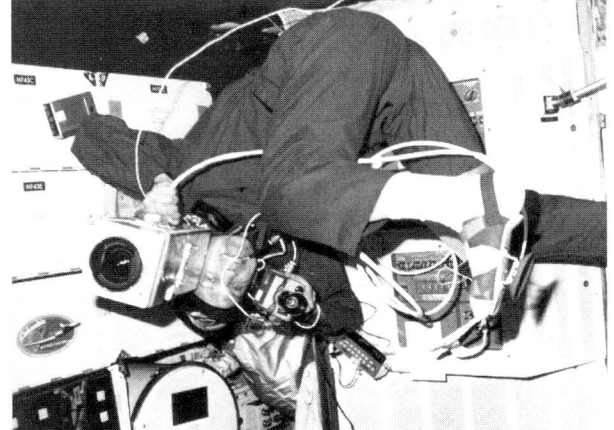

Figure 5-01. An astronaut in the mid-deck of the Space Shuttle is wearing sandals with suction cups to help stabilizing his body relative to the spacecraft walls when operating high-resolution photographic equipment. Photo courtesy of NASA.

Exposure to the microgravity conditions of space flight induces adaptive modification in the central processing of sensory input to produce motor responses appropriate for the prevailing gravito-inertial environment. As a result, terrestrial motor strategies are progressively abandoned, as astronauts adapt to the new demands of the zero-g environment. This is particularly true for the major postural muscles found in the lower legs. The plastic modifications in posture, movement and gait functions acquired during space flight are then inappropriate for a one-g environment upon return to Earth.

G. Clément, M.F. Reschke, *Neuroscience in Space*.
DOI: 10.1007/978-0-387-78950-7_5, © Springer Science+Business Media, LLC 2008

Difficulties with standing, walking, turning corners, and climbing stairs are experienced as astronauts re-adapt to a one-g environment, until terrestrial motor strategies are fully reacquired. These difficulties can have adverse consequences for an astronauts' ability to stand up, bail out, or escape from the vehicle during emergencies and to function effectively immediately after leaving the spacecraft after flight. Thus it is important to understand the cause of these profound impairments of posture and locomotion stability, and develop countermeasures.

1   INTRODUCTION

Homo erectus' ability to maintain stable upright posture and to move in Earth's gravitational field has evolved over many millennia. As part of this evolutionary process, numerous neurosensory and neuromuscular systems have developed to sense the orientation of the individual with respect to the gravitational field and to support and move the individual's body mass through this gravito-inertial environment. In weightlessness, the structural and anatomical systems that provide for upright orientation and movement on Earth are, at best, not required and, at worst, not appropriate for orientation and movement. As a result, part of the adaptation process involves elimination, reinterpretation, or modification of the information and control provided by these systems. One consequence is that, upon return to Earth following a sufficiently long space flight, the orientation and movement control systems of the crewmember are no longer optimized for terrestrial gravity. Indeed, disturbances in postural equilibrium and gait upon return from flight have been among the most consistently observed and reported responses associated with space flight. Careful study of these changes may provide a key to understanding how sensorimotor systems adapt to the unique environment of microgravity.

Over the past forty five years, returning crewmembers from both Russian and American space missions have reported one or more of four basic unusual sensations associated with posture or locomotion during the first few hours after landing. The first of these is the sensation of turning or lateral deviation while attempting to walk a straight path. Overcompensating for this, many crewmembers actually walk in a curved path in the opposite direction. Second is a sudden loss of postural stability, much as though the crewmember has been pushed to one side by a "giant hand", usually experienced while attempting to walk around corners. Third is the perception that the pitching and rolling head motions that accompany normal walking are greatly exaggerated. Finally, in an environment with no clear visual vertical, some crewmembers experience a sudden loss of orientation and pitch forward, or fall to the side before position awareness is regained (Young 1993).

Maintenance of a stable postural equilibrium requires constant interaction between sensory input and motor output. Disturbance of either can result in inappropriate postural responses, which lead to postural instabilities. The sensory inputs required to maintain postural stability on Earth are provided by visual, vestibular, somatosensory and proprioceptive receptors. Adaptation to microgravity apparently results in elimination, reinterpretation, or modification of the weighting of sensory information from these receptors.

The effects of weightlessness on postural stability have been examined using a number of different methods including:

   a.   Crewmembers' reports of changes in sensations (illusionary movement).

b.   Performance on balance rails.

c.   Performance on moving platforms.

d.   Performance while standing following a voluntary movement (raising the arm from the side, tiptoeing, or bending at the waist) or involuntary movement (push to the chest) designed to perturb the body's center of gravity.

e.   Measurement of muscle potentials from the major antigravity and weight-bearing postural muscles.

f.   Application of a vibrator to selected muscles so as to elicit postural responses.

g.   Performance during complex postural tests designed to selectively eliminate visual, proprioceptive or vestibular information.

Although some in-flight data have been collected (Figure 5-02), most postural stability testing has been limited to the comparison between preflight and postflight performance.

Figure 5-02. One of the engineering and technology experiments on board Skylab was a special suit instrumented to measure body motions as the wearer went through typical tasks on board the space station. One Skylab astronaut inspects such a vest during a training session on Earth. Photo courtesy of NASA.

2   IN-FLIGHT POSTURE STUDIES IN ANIMALS

The first microgravity experiment on spatial orientation and postural control was conducted some three decades ago using fish as test subjects (von Baumgarten *et al.* 1972). During the microgravity phase of parabolic flight, the animals exhibited a continuing diving response, i.e., swimming inward looping. This behavior was called "looping response." Other individuals performed spinning movements around their longitudinal body axis (von Baumgarten *et al.* 1972, DeJong *et al.* 1996). The looping behavior would result from the absence of any otolith feedback to the animals indicating completion of the maneuver. Long-axis rotation appears to be the result of the repetitive execution of the righting response, such as that observed in a falling cat (see Figure 1-09), in a situation where it is non-effective. These early experiments clearly showed that fish face severe orientation problems in a microgravity environment. Fish have not been observed to vomit under microgravity, and they may therefore be presumed not to suffer true motion sickness. Therefore, regarding fish the term *kinetosis* is more appropriate than *motion sickness*, since they do demonstrate gastro-intestinal symptoms (increased fecal output) when exposed to unusual motions outside of the water (Money 1970).

Like humans, fish also use basic visual and vestibular cues for postural equilibrium maintenance and orientation (Allum *et al.* 1976). As early as 1935 the so-called *dorsal light response* (DLR) was described (von Holst 1935). When illuminated from the side at one-g, a fish tilts its back towards the light source, but under microgravity conditions the tilt is guided by light alone. The DLR thus expresses a balance between the tilting force induced by visual information and the vestibular righting response (Watanabe *et al.* 1991) induced by tonic vestibular information. It has been suggested that the intensity of the DLR is species specific based on the finding that particular genetic strains of medakas (Japanese ricefish *Oryzias latipes*) differ in their DLR performance (Ijiri 1995). However, recent investigations have clearly demonstrated that the DLR depends on the specific ability of an individual, thus suggesting that some individuals are more "vestibular" and others more "visual," as is the case in humans (Harm & Parker 1993, Isableu *et al.* 1997).

Like the fish, the midwater tadpoles of the African clawed-frog (Xenopus laevis) make forward somersaults when subjected to microgravity (Wassersug & Souza 1990, Wassersug 1992). However, aquatic amphibians either float randomly or make reverse (backward) somersaults when abruptly exposed to microgravity (Mori 1995). Adult non-arboreal frogs and salamander larvae rotate along their rostral-caudal axis in response to microgravity. This long axis rotation is similar to their righting reflex when inverted in normal gravity (Wassersug *et al.* 1991, Wassersug *et al.* 1993).

Arboreal frogs take up an extended-limb gliding or parachuting posture when suspended in air during microgravity (Izumi-Kurotani *et al.* 1992). A semi-arboreal snake was observed taking a stereotypic defensive posture in microgravity during parabolic flight. Pond turtles in microgravity extend their neck and limbs in an asymmetric fashion identical to their righting response when placed upside-down in normal gravity (Wassersug & Izumi-Kurotani 1993). Birds adopt a flying behavior (Oosterveld & Greven 1975) and mammals, such as hamsters or rats, frequently extend their extremities and back in-flight, similar to a flying squirrel, or spiral along their long body axis (Kalb *et al.* 2003). Some of these responses, such as in the snake, appear to be extensions of escape behavior in response to stress. However, the extension of the extremities would be the consequence from the release of the inhibitory influence exerted by the otolith organs on the antigravity muscles, the extensors (Clément *et al.* 1984, Wassersug *et al.* 1993).

The various types of neurobiological data (behavioral, morphometrical, histochemical, biochemical, and electron microscopic) using animal models for studying the signal-response chain of graviperception favor the following concept of interactions: Sudden exposure to altered gravity can induce transitionally aberrant behavior due to malfunction of the inner ear originating from asymmetric otolithic loading or, generally, from a mismatch between otolith afferents and the other sensory inputs that also provide orientation information. This aberrant behavior in different gravito-inertial environments vanishes due to a re-weighting of sensory inputs and vestibular offset and/or gain compensation, probably on a bioelectrical basis. During steady-state exposure to altered gravity, step-by-step neuroplastic reactions on a molecular basis (i.e., molecular facilitation) in the brain and inner ear possibly activate feedback mechanisms between the CNS and the vestibular organs for the regain of normal behavior (Anken & Rahmann 1999).

3   IN-FLIGHT POSTURE STUDIES IN HUMANS

In-flight studies of postural control changes associated with exposure to weightlessness in humans have been performed on both Russian and American missions. In these studies, postural equilibrium was disturbed either by a voluntary (subject initiated) movement or by an involuntary (externally initiated) movement and the postural responses to that disturbance were measured. Comparisons between in-flight and control (preflight and postflight) measurements were used to identify adaptive changes in posture control. Voluntary movement posture control paradigms included requiring crewmembers to respond when their resting position was disturbed with either a rapid arm movement, elevation of the whole body (voluntary tiptoeing), bending at the waist, or squatting (Clément *et al.* 1984, Massion *et al.* 1993). Involuntary movement posture control paradigms included displacement with a foot support platform capable of providing a forward step velocity, vibration of select muscle groups, and sudden "falls" where crewmembers were pulled to the floor of the spacecraft with elastic cords (Clément *et al.* 1984, Clément & Lestienne 1988, Roll *et al.* 1993, Reschke *et al.* 1986, Watt *et al.* 1986). The results of these investigations are described below.

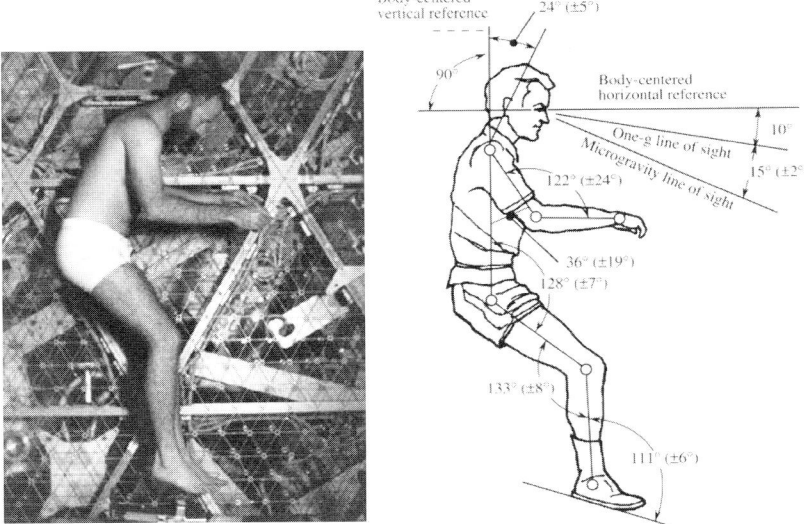

Figure 5-03. Photograph of an astronaut free-floating on board the Skylab space station (left). A series of photographs was used to construct a model of the neutral body position in weightlessness (right). Photo courtesy of NASA.

3.1   Rest Posture

Human factor studies, after investigating photographs taken during Skylab missions have led to the NASA *neutral body posture* model (Figures 1-10 and 5-03).

This model is characterized by a forward tilt of the head (with the line of sight 25 deg lower than the body-centered horizontal reference), shoulders up (like a shrug), and arms afloat, up and forward with hands chest high (Thornton 1978). However, recent investigations, taking into account body size, gender and mission duration suggest that the neutral body posture model is too generalized. Data collected on a larger number of astronauts showed that arm and shoulder positions were less bent, and there were straighter leg positions at the hip and knee than expected from the neutral body model (Mount & Foley 1999). Further studies should be made of posture in zero-g to better define not only the differences in postural response in microgravity, but to seek a more normalized picture of crew responses over different lengths of flight. Also, it is unclear how the direction of the line of sight has been evaluated from the Skylab photographs. The downward deviation of gaze in microgravity in the neutral body model is in contradiction with the results of several space experiments that actually measured the eye deviation during space flight (Clément 1998).

Frame-by-frame analysis of video recordings made during various on-board activities of crewmembers have allowed researchers to characterize prevalent orientations and stereotyped motor acts (Tafforin & Lambin 1993). Results revealed that head and body movements in yaw were more frequent in space than on the ground, and that the astronauts quickly learn to anchor their feet and use handgrips for stabilizing their posture. Head-down orientation increases in frequency as flight progresses, presumably in phase with the development of a new internal representation of the environment and the location of objects (see Chapter 7, Section 4.4).

During a standing posture in microgravity, dorsi-flexor muscles (e.g., the *anterior tibialis* leg muscle) assume a larger role in space than on Earth in regulating the orientation of the individual relative to his/her support. This is in contrast with the general use of muscle extensors on Earth, which are used to counteract gravity. This transfer of motor strategies from one muscle group to another explains the forward tilted posture of crewmembers placed in darkness when instructed to maintain a posture perpendicular to the foot support (see Figure 4-06) (Clément *et al.* 1984).

Why is there an activity in the flexor muscles in weightlessness? One explanation is that it is the result of a sudden disinhibition from the normal excitatory drive exerted by the otolith inputs (perhaps the saccule) on the extensor muscles under the influence of gravity. Another explanation has been proposed by Clément *et al.* (1988), namely that this activity is compensatory for passive resistance. In other words, the normal biomechanically neutral posture of the ankle is when the foot is slightly extended, which would bring the body backward. Therefore, in absence of apparent gravity, in order to have the feet at right angle with the leg, a small flexor tone has to be generated.

Massion *et al.* (1997) proposed that this flexor tone is aimed at maintaining a virtual vertical projection of the body's center of mass on the polygon of sustentation created by the feet. In other words, the CNS would try to recreate in weightlessness a condition similar to Earth. This interpretation is in agreement with the idea of an internal model of gravity that is oriented along the longitudinal body axis or an idiotropic vector (Mittelstaedt & Glasauer 1993, Clément *et al.* 2001). This model would allow a coherent mental representation of the body with an alignment of the longitudinal head and body axes. This internal model of gravity would also serve as a reference frame for movement, as demonstrated by the experiments detailed in the next sections.

3.2   In-Flight Postural Responses to Voluntary Movements

On the Earth's surface, gravity significantly affects most of our motor behavior. For example, when making limb movements during static balance, anticipatory responses from the leg muscles compensate for the impending reaction torques and the changes in location and projection of the center of mass associated with these movements. Similar patterns of anticipatory compensations are seen in-flight, although they are functionally unnecessary (you can not lose your balance or fall). Also, rapidly bending the trunk forward and backward at the waist is accompanied on Earth by backward and forward displacements of hips and knees to maintain balance. Since the effective torques observed in a normal gravitational environment are absent during space flight, the motor responses necessary to achieve these synergies in weightlessness are different from those needed on Earth. Consequently, movements executed in-flight must reflect reorganized patterns of muscle activation.

3.2.1   Arm Raising and Tiptoe Raising

In a joint French-Russian experiment on board Salyut-7, control of upright posture was examined during voluntary upward movement of the arm and voluntary raising on tiptoe (Clément *et al.* 1984, 1985). Early in-flight postural attitude was similar to that on Earth, but as the flight progressed, there was a forward inclination of the body, which increased when vision was stabilized, i.e., when the eyes were open but with no vision of the surrounding spacecraft. Muscle responses to sudden voluntary perturbations (raising the arm rapidly) indicated a redistribution of tonic activity between extensor and ankle flexor muscles, and a general reduction of extensor tone (Figure 5-04).

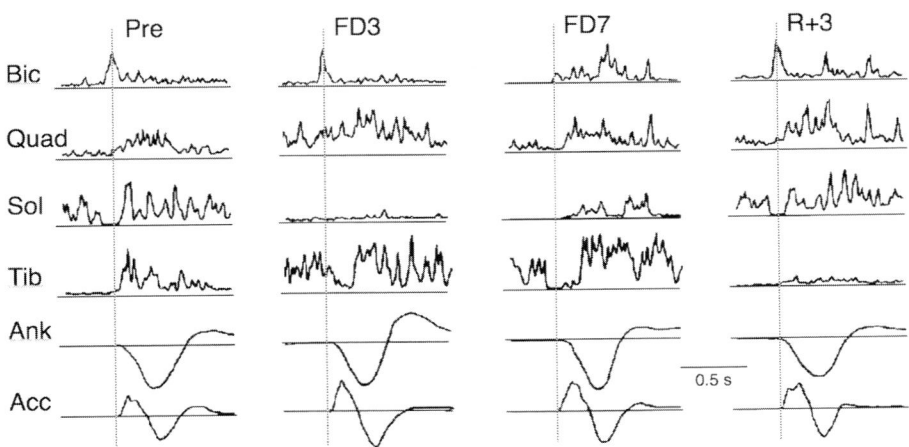

Figure 5-04. Electro-myographic (EMG) activity of leg muscles (Bic: biceps femoris; Quad: quadriceps; Sol: soleus; Tib: anterior tibialis), ankle displacement (Ank) and arm acceleration (Acc) during arm raising in one astronaut before flight (Pre), on flight days 3 (FD3) and 7 (FD7) and 3 days after (R+3) a seven-day space flight. The dashed line indicates the beginning of arm raising. On Earth, the soleus tonic activity decreases before the arm moves. In-flight, this anticipatory deactivation is seen on the tibialis muscle, which maintains the postural tone required for an upright posture in weightlessness. The EMG activity of the biceps femoris and quadriceps is not fundamentally changed in-flight. Adapted from Clément et al. (1984).

Another experiment looked at motor strategies during a rapid toe rise from a standing position. In trials conducted in normal gravity conditions, the temporal characteristics of the anticipatory activity of the postural muscles preceding the elevation to the toes showed an inhibition of the spontaneous activity of the soleus muscle followed by a burst of activity in the anterior tibialis muscle, which continued as long as the subject remained standing on his toes (Lipshits *et al.* 1981). In contrast to this rise and hold technique, if the subject immediately returned to the initial position then the anticipatory activity in the tibialis was absent. Lipshits and his colleagues (1981) proposed that this anticipatory activity functions to displace the body's center of gravity into a new stable position. When a similar toe rise experiment was conducted in-flight, results typical of those observed preflight were found on the third flight day. The finding that the sequence of motor patterns were preserved in-flight is significant and suggests that terrestrial postural programs continue to operate for a relatively long period of time in weightlessness, independent of how sensory inputs are modified (Clément *et al.* 1985).

3.2.2   Bending at the Waist

Rapid voluntary pitch movements at the waist (forward and backward) were made while the crewmember's feet were fastened to the wall of the spacecraft with Velcro bands (Massion *et al.* 1993). Kinematic analysis, in addition to confirming the forward tilt posture reported by Clément *et al.* (1985), showed that upper trunk movements were accompanied by hip and knee movements in the opposite direction, and that there was little difference between in-flight measurements and those obtained both pre- and postflight. The results of EMG analysis, like that observed during the Salyut-7 flight (Clément *et al.* 1985), showed that the early activation of the soleus muscle group observed under terrestrial conditions was replaced in-flight by an early activation of the anterior tibialis. This in-flight motor strategy was still in evidence five days following the flight.

3.2.3   Squatting

Under terrestrial conditions, upright posture is maintained primarily through tonic activity in the extensor muscles. In microgravity, simultaneous recordings of EMG activity in the tibialis and soleus muscles while the crewmember's feet were fixed to the floor of the spacecraft demonstrated that upright posture was maintained through tonic activity in the flexor muscle (Clément *et al.* 1984). This reported change prompted an investigation on STS-51G into the relationship between conscious appreciation of limb position and body position in space and muscle afferent activity. Two crewmembers were asked to lower their bodies into a squatting position, pause and then rise to a fully erect position. By the third day in-flight, the subjects reported illusions of floor motion during execution of the deep knee bends (see Figure 4-07). Similar illusions occurred following the Spacelab-1 flight (Watt *et al.* 1986, Reschke *et al.* 1986), during STS-41G (Watt *et al.* 1985), as well as during parabolic flight (Lackner & Graybiel 1981).

3.3   In-Flight Postural Responses to Involuntary Movements

3.3.1   Support Surface Translation

Using a foot support platform designed to provide sudden forward translation, postural responses to involuntary body displacements were also investigated during the

Salyut-7 flight (Clément *et al.* 1985). Preflight, when the platform was unexpectedly moved forward, the ankle joint extended (plantar flexion) and then returned to its initial position. The motor pattern in response to the sudden plantar flexion showed an initial tibialis muscle burst with latency between 80 and 120 msec. When the test was repeated (up to six times), this early burst of activity from the anterior tibialis was reduced by approximately 40%. On the second day of the flight, the initial burst of tibialis activity was similar to that observed preflight, but the level of tonic activity in the tibialis was greater than that observed on the ground. The tibialis burst of activity decreased quickly in amplitude with the repetition of the trials (Figure 5-05). On the third day after landing, the tibialis motor response returned to baseline, but the ankle rotation trajectory suggested postural destabilization. In discussing these results, the authors suggest that the early tibialis burst resembles the EMG activity of a "functional stretch reflex" mediated by supraspinal centers (Melvill-Jones & Watt 1971), and that the changes in overall EMG amplitude during flight reflect reduced output from the otoliths. These results are consistent with the findings from the Hoffmann reflex experiment (Reschke *et al.* 1986) and the otolith-spinal reflex measurements (Watt *et al.* 1986) performed on Spacelab-1 and described below.

Figure 5-05. EMG reflex activity of anterior tibialis in one astronaut during six consecutive support surface forward translations before (Pre) and during space flight (flight day 2, FD2). The vertical line indicates the beginning of ankle extension. The reduction in the amplitude of the tibilialis activity burst reflex in response to this ankle extension was faster during the flight. Adapted from Clément et al. (1985).

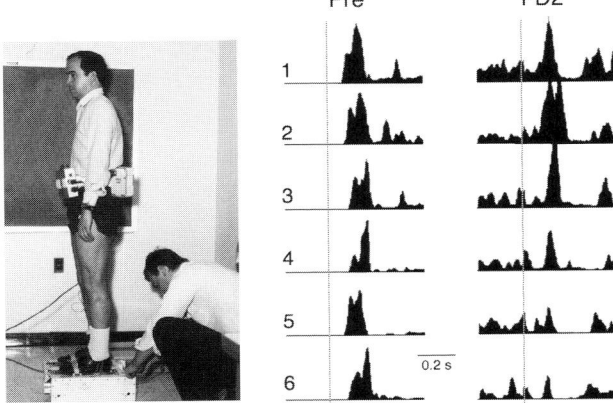

3.3.2 Sudden Drop

In an experiment performed on board Spacelab-1, Reschke *et al.* (1986) examined the effects of weightlessness on the *Hoffmann reflex* (H-reflex). This reflex takes advantage of the anatomical pathways that link the otoliths and spinal motoneurons. Therefore, it can be used as a method of monosynaptic spinal reflex testing to assess otolith-induced changes in postural muscles.

By contrast to doctor tapping a patient's knee to produce the proverbial "knee jerk" reflex (i.e., a mechanically induced spinal stretch reflex), during H-reflex the stimulus is an electrical shock to sensory fibers coming from stretch receptors in the calf (soleus) muscle, and the response is the electrical activity mediated by the muscle motor neurons through the spinal cord and recorded from the muscle. Each time a subject is tested, the number of motoneurons that have been excited by a standard volley of sensory impulses is counted. That number is an indicator of spinal cord excitability as established by the descending vestibular output. The H-reflex data can also be related to EMG from the calf muscle (the M-wave) and self-motion reports.

Activity in this otolith pathway was elicited by exposing the subjects to unexpected drops (falls) (Figure 5-06, left). It was hypothesized that exposure to free fall would reduce the necessity for postural reflexes in the major leg muscles, and that postural modifications would reflect a change, not in the peripheral vestibular organs, but more centrally. This postural adjustment would reflect a sensorimotor rearrangement in which otolith receptor input was reinterpreted to provide an environmentally appropriate response. Early in-flight H-reflex amplitude was similar to that recorded preflight, but measurements obtained on the seventh day of flight did not show a change in potentiation as a function of the drop-to-shock intervals (Reschke *et al.* 1986). Immediate postflight H-reflex response in three of four astronauts tested showed a rebound effect. This effect returned rapidly to baseline.

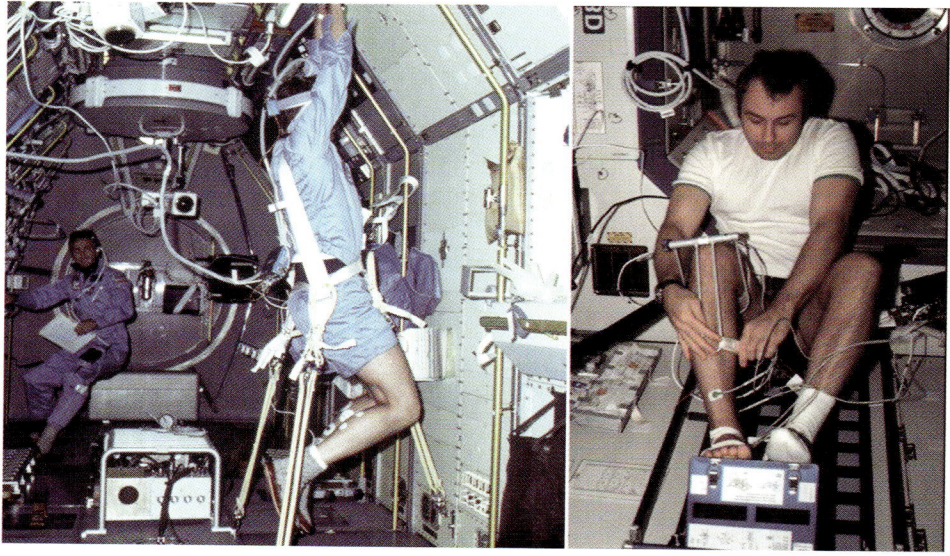

Figure 5-06. Left. H-reflex experiments on board Spacelab (left) and the ISS (right). On Spacelab, subjects were suddenly released and dropped to the floor my means of bungee cords. On Earth, during such drop the otoliths signal the muscles to prepare for jolts associated with falling. This anticipation was partially inhibited early in flight, and declined further as the flight progressed, suggesting that the brain ignored or reinterpreted otolith signals during space flight. The response returned to normal immediately after landing. During the flight, crewmembers also reported a lack of awareness of position and location of feet, difficulty in maintaining balance after hitting the floor, and a perception that falls were more sudden, faster, and harder than similar drops experienced preflight. Photo courtesy of NASA.

In-flight self-motion perception reports suggested that the early in-flight drops were perceived like those preflight. Drops later in-flight were described as sudden, fast, hard, and translational in nature. Immediately postflight, the drops were perceived like those late in-flight, with the astronauts reporting that they did not feel as though they were falling, but rather that the floor came up to meet them.

In a related Spacelab-1 experiment (Watt *et al.* 1986), otolith-spinal reflexes were elicited by sudden, unexpected Earth vertical falls. Like the H-reflex experiment, falls were executed in-flight by pulling subjects to the deck of the Spacelab using elastic cords. EMG activity recorded early in-flight from the gastrocnemius-soleus complex during the fall was of lower amplitude than that observed preflight and continued to decline as the flight progressed. These results agree with the results of the H-reflex experiment showing little or no potentiation of the monosynaptic reflex as a function of a vertical fall late in-flight (Reschke *et al.* 1986).

In astronauts tested on board the ISS (Figure 5-06, right), the spinal cord excitability decreased by about 35% in microgravity and stayed at this new level for the duration of 3-6 month missions. Although there was notable improvement in the H-reflex response the day after landing, it took about ten days back on Earth for astronauts to fully recover their muscle strength and spinal cord excitability (Watt & Lefebvre 2001, Watt 2003). This difference in excitability means that only a portion of muscle fiber units are contracting in response to signals from the nervous system and explains functionally why muscle mass declines in weightlessness, even with exercise. Reduced excitability means that there might be limits on the degree to which heart muscle strength, leg muscle tone, and bone density (for which muscle contraction is an important regulating factor) can be maintained through exercise on long-duration missions. Because this decrease in excitability is only observed on orbit and not during bed rest, an analogue for weightless space travel, the results highlight the possibility that reduced excitability with corresponding loss of muscle and bone might be partly a CNS response and not simply due to disuse of the legs (Watt 2007).

3.3.3 Muscle Vibration

The role of muscle proprioceptive receptors in control of upright posture was investigated by vibratory stimulation of the soleus and anterior tibialis muscle tendons during the Mir Aragatz mission (Roll *et al.* 1993). Two subjects participated in the experiment; one remained on-orbit for four months and was joined by a second who remained in the Mir station approximately five months.

Before flight, vibratory stimulation of the soleus resulted in backward sway about the ankle joint, whereas stimulation of the anterior tibialis resulted in forward sway. During flight, the postural responses developed differently depending on which muscle group was stimulated. Sway during stimulation of the anterior tibialis either decreased or disappeared (depending upon the subject), whereas the response to the soleus remained normal (somewhat decreased in one subject) throughout the twenty-day in-flight test period. In addition, the compensatory EMG recorded preflight disappeared in-flight even though muscle activity concomitant with the vibration was observed in both soleus and anterior tibialis (similar to the classic tonic vibratory response). No testing was possible until two days after landing. At that time, the responses of the subject who spent the least time on orbit were comparable to those obtained before flight. The same appeared to be true of the second subject, but no objective measurements were made (Roll *et al.* 1993).

The authors concluded that muscle proprioception remained intact after prolonged flight, since it was still possible to activate the muscle spindle with vibration. However, the characteristics of the response to muscle vibration changed in microgravity, indicating that adaptive sensorimotor responses occur and that these new responses were appropriate to the environment.

4   PRE- AND POSTFLIGHT POSTURE STUDIES

Owing to both the physical difficulties and constraints of performing posture studies in-flight, many investigators have chosen to test crewmembers before and immediately after flight (presumably before significant re-adaptation to one-g has occurred) in order to better understand in-flight adaptation. The first studies designed to quantify postflight postural ataxia in this fashion required astronauts, upon landing, to tandemly stand on narrow rails of various widths with their eyes either open or closed and arms folded across their chests (Berry & Homick 1973, Homick & Miller 1975, Homick *et al.* 1977, Kenyon & Young 1986). Other studies have used static force plates for stabilometry and simpler tests, such as the clinical Romberg test, a sharpened (toe-to-heel) Romberg test, and vertical posture with varying head positions, to assess postural ataxia immediately after flight (Yegorov 1979, Bryanov *et al.* 1976). Later postural performance studies have relied on dynamic posture platforms that translate the subject (Reschke *et al.* 1984, Clément *et al.* 1985, Anderson *et al.* 1986), tilt the subject (Kenyon & Young 1986, Reschke *et al.* 1991), or provide more sophisticated posture control tasks such as stabilization of ankle rotation and/or vision (Paloski *et al.* 1993). Pre- and postflight studies of vestibulo-spinal reflexes (Baker *et al.* 1977, Reschke *et al.* 1984, Kozlovskaya *et al.* 1984, Watt *et al.* 1986) and postural responses to voluntary body movements (Reschke & Parker 1987) have also been performed. A summary of the results of these studies follows.

4.1   Rail Tests

Early measurements of postural ataxia were based on the hypothesis that prolonged exposure to a weightless environment would result in changes in the sensory systems (with the possible exception of vision) necessary for the maintenance of postural stability. It was postulated that these changes would most likely originate at the periphery and involve modification of input from the receptors serving kinesthesia, touch, pressure, and otolith functions. Furthermore, as exposure time increased, adaptive responses appropriate to the new inertial environment were expected to occur at a central level. Upon return to Earth, postural instability would be manifested as a result of the in-flight neural reorganization.

The first tests of this hypothesis were performed following the Apollo-16 mission (Homick & Miller 1975) and the Skylab-2, -3 and -4 flights (Homick & Reschke 1977). Ataxia was evaluated using a modified version of a standard laboratory test developed by Graybiel & Fregly (1965). Metal rails of varying widths were provided for the crewmembers to stand on in a sharpened Romberg position (feet, heel-to-toe; arms crossed and folded across the chest) with eyes opened or eyes closed (Figure 5-07). Time before stepping (or timeout) was the performance measure of postural stability. Postflight decrements in postural stability during the eyes open tests ranged from none to moderate. However, during the eyes closed tests, postural stability was considerably decreased in all crewmembers tested. The magnitude of the change was greatest during the first postflight test. Since the Apollo and Skylab tests were not performed until the fourth and second day after landing, respectively, the magnitude of ataxia immediately postflight is believed to have been even greater than that observed at the first postflight test. As it was, one Skylab crewmember had difficulty maintaining balance with his eyes closed while standing on the floor. Improvement was slow and appeared to be related to the length of the mission.

Rail tests were repeated by another group of investigators as part of the complement of vestibular tests performed with the crew of the Spacelab-1 mission (Kenyon & Young 1986). With open eyes, performance on the narrow rail width (1.90 cm) was found to be considerably reduced postflight and did not return to preflight levels before the last test session, seven days after landing. With the eyes closed, all four crewmembers tested exhibited a significant decrement in performance immediately postflight, even while standing on the 5.72 cm wide rail. In at least one case, return to baseline had not occurred by the seventh day postflight. In addition to the static rail-standing task, crewmembers were asked to walk on the 1.90 cm rail. All subjects adopted a strategy of speed, trying to complete the test trials as quickly as possible and minimize instability. Postflight performance was in all cases below that of preflight data, but was only consistently reduced for one subject.

Figure 5-07. Astronauts John Glenn and Scott Carpenter during the posture rail tests performed before and immediately after their Mercury missions. Similar tests were done on Apollo, Skylab, and early Space Shuttle crewmembers. Photos courtesy of NASA.

4.2   Stabilometry

Russian investigators obtained their earliest quantification of postflight postural ataxia in a unique investigation associated with the Soyuz program. Operating under a hypothesis similar to that of their American counterparts, the Russians stressed that postural activity observed in human is based on biomechanical (support), physiological, neurological (vestibular, muscle tonus, tonic activity, coordination of movement, etc.), and psychological (perception, need, etc.) components. They postulated that space flight produces a reorganization of these components and that the subsequent return to Earth requires conscious control of these components for their restoration (Bryanov *et al.* 1976).

For many of the shorter Soyuz missions (Soyuz-3 through Soyuz-8), postural stability was measured using stabilometry 30 to 40 days before flight and at various times (9, 18, 27 hours) after flight. For longer duration flights (Soyuz-9 and Soyuz-17), additional repeated observations were collected postflight. Stabilograms were recorded for periods of one and two minutes during predetermined postural stances, including standing with the head erect (eyes open or closed), standing in the Romberg posture, and standing with the head tilted either forward or backward. Primary measures obtained from the stabilogram were the average frequency and amplitude of sway of the derived body center of gravity in both the sagittal and frontal planes. The postflight stabilographic data in all assumed postural stances were characterized by an increase of sway amplitude primarily in the frontal plane coupled with, in most crewmembers, a decrease in oscillation frequency. The magnitude of change was coupled with the length of flight, with significant changes occurring following the Soyuz-9 flight (Bryanov *et al.* 1976).

In a later study, the prime crew of the Mir Kvant expedition also participated in postural stability tests using the stabilogram technique. In this study, normal upright posture was perturbed by a calibrated force that was momentarily applied to the subject's chest. In fact, the operator pushed the subject with a stick coupled with a force transducer. Three cosmonauts participated in this study; two had been on-orbit for 151 days and the third for 241 days. Postflight testing was not initiated until six days after return of the crew. In all but one crewmember, less force was required to perturb vertical posture postflight, and in all crewmembers the time to recover from the applied perturbation increased postflight. Overall muscle activity required to maintain upright posture following the perturbation was also increased postflight. All changes observed on the sixth postflight day were still present on the eleventh day, but to a lesser degree, and were reportedly similar to those observed following other missions of comparable length (Grigoriev & Yegorov 1990).

4.3 Moving Platform Tests

4.3.1 Support Base Translation

Pre- and postflight postural stability measurements were made on four crewmembers from the Spacelab-1 mission using a dynamic posture platform that could be moved parallel to the floor in both predictable and unpredictable patterns, including sinusoids, pseudorandom and velocity steps (Anderson *et al.* 1986). In these studies, the subject attempted to maintain a normal upright stance with eyes either open or closed as the moving platform perturbed his base of support. EMG data obtained from the soleus and anterior tibialis muscles and the hip and shoulder displacements relative to the moving platform were recorded with edge detection cameras throughout the testing period. Postflight, when the subject's eyes were open and the platform was moved with a backward step function, the subject's response showed an overshoot with the shoulders and an undershoot with the hips relative to his preflight response. Also, the time required to assume a new stable position was greater after flight than before. The EMG data indicated that soleus muscle latency was greater postflight. It is interesting to note, in contrast to other posture tests, that vision appeared to degrade performance.

Another interpretation, however, would suggest that visual stabilization (i.e., gaze) was the important parameter, and that shoulder (in lieu of actual head movement measurements) tracking of the stimulus would reflect a decrease in head stability (and gaze by inference). This interesting result has never been verified with additional testing.

4.3.2 Support Base Rotation

In another dynamic posture platform study on Spacelab-1 (Kenyon & Young 1986), the crewmember's erect posture was perturbed by pitching the platform base unexpectedly about the ankle joint. EMG activity from the anterior tibialis and Gastrocnemius muscles was measured with the eyes open and closed. Postflight, the early EMG response (first 500 ms) did not change in latency or amplitude when the platform was pitched. However, the late EMG response (after the first 500 ms) was found to be higher in amplitude than that obtained preflight.

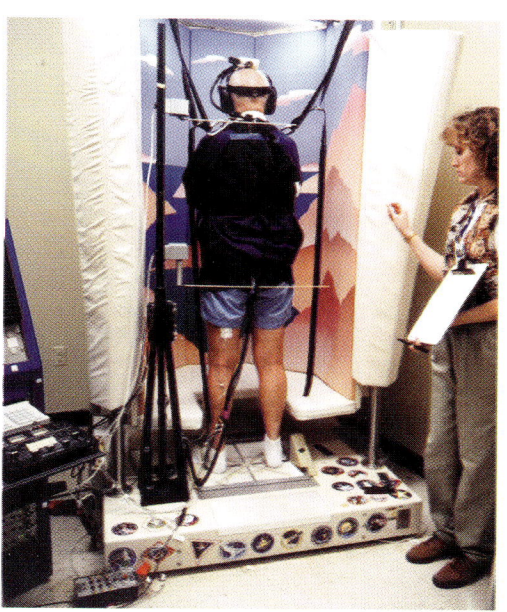

Figure 5-08. Subjects ability to stand as still as possible is investigated while standing on a platform inside a booth. The platform and the booth are designed to isolate the visual, vestibular and proprioceptive information used for balance control. For example, the booth is slaved to the body sway to prevent changes in visual information (sway-referenced vision). Similarly, information from proprioceptive receptors in the ankles is cancelled by moving the foot platform in phase with the displacements of the center of gravity (sway-referenced support). Measurements include displacements of the center of gravity, hip and shoulder; angular velocity of the head in pitch yaw and roll; and EMG activity of leg muscles. Photo courtesy of NASA.

4.4 Complex Visual, Vestibular and Proprioceptive Tests

The relative importance of visual, vestibular and somatosensory information to control of postural stability was studied before and after the seven-day Spacelab D-1 mission using a tilting room (von Baumgarten *et al.* 1986). Crewmembers stood with their feet on an Earth-fixed stabilometer anchored to the floor beneath the tilting room while body sway was measured under conditions of no visual input (eyes closed), conflicting visual-vestibular input (eyes open, room tilted with a sinusoidal motion), normal vision (eyes open, room upright), or reduced somatosensory input (foam rubber placed between the stabilometer and the astronaut's feet). Immediately postflight (a few hours after landing), two crewmembers showed an increased reliance on visual feedback for maintenance of upright postural equilibrium; stability was decreased when the room was oscillating or when eyes were closed. By the second day after landing, when three additional subjects were tested, stability under the oscillating room condition was analogous to that observed before flight. On the other hand, postural stability remained impaired for up to five days postflight when the crewmembers stood on the foam rubber or closed their eyes.

Later, a clinical posturography system (Equitest, Neurocom International, Clackamas, OR, USA) was used to assess the magnitude and recovery time course of postflight postural instabilities in returning astronauts from Space Shuttle missions (Paloski *et al.* 1993). This system consists of a platform and a visual surround scene, both of which are motorized to allow either a step input to the subject or servo-slave the platform and the scene to the subjects sway motion (Figure 5-08). Subjects complete multiple tests before and after the flight to establish stable individual performance levels and the time required recovering them. Two balance control performance tests are administered. The first test examines the subject's responses to sudden, balance-threatening movements of the platform. Computer-controlled platform motors produce sequences of rotations (toes-up and toes-down) and translations (backward and forward) to perturb the subject's balance. The second test examines the subject's ability to stay upright when visual or ankle muscle and joint information is modified mechanically. A battery of six sensory organization tests is used to assess a subject's ability to maintain postural equilibrium under normal and reduced sensory feedback conditions. The basic paradigm involves measuring hip, shoulder, head, and center of mass sway over 20-sec periods while the subject attempts to maintain a stable upright stance. Sway measurements are made three times under each of six randomly presented test conditions, including an eyes-open Romberg test, an eyes-closed Romberg test, and four other tests in which vision and/or ankle proprioceptive inputs are selectively eliminated by having the subject close his eyes or by servo-controlling the visual surround and/or support surface to the subject's center of mass sway.

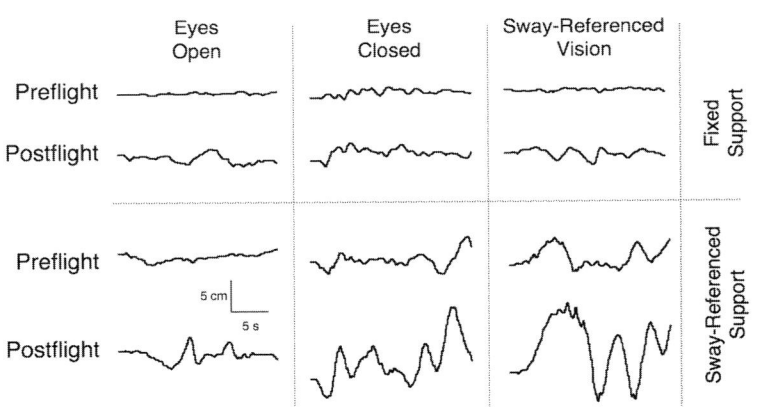

Figure 5-09. Pre- and post-flight anterior-posterior sway for a subject standing on a force platform for 20 sec. Each column and row re-presents a different visual and support surface con-dition, respec-tively.

The upper traces in each panel represent the preflight performances and the lower traces represent the postflight performances. After flight the subject's anterior-posterior sway amplitude increased under all test conditions compared to preflight. The increased amplitudes observed under sway-referenced support were balance threatening. When both visual and proprioceptive cues were sway-referenced, this subject's center-of-gravity oscillated between his/her forward and backward stability limits. Adapted from Paloski et al. (1993).

This test has the advantage of a huge clinical database against which the effects of space flight can be evaluated. Postflight measurements using this system revealed significant deviations from the results obtained before flight (Figure 5-09). The strategy used by the individuals for balance on the moving platform is modified and their behavior indicates a decrease in awareness of the direction and magnitude of the motion. On landing day, every subject exhibited a substantial decrease in postural stability. Some had clinically abnormal scores, being below the normative population 5th percentile. As dramatic as these results are, testing postural performance with the head upright may underestimate the level of disequilibrium following space flight. During a recent study, Paloski and his colleagues (2004) observed 90% fall incidence on landing day in trials during which crewmembers performed active pitch head tilts versus no falls during trials with the head held upright.

Significant differences in this complex posture test were identified between rookie (first-time space travelers) and veteran (experienced space travelers), suggesting that something learned in the adaptation/re-adaptation process is retained from one flight to the next. It was suggested that experienced space travelers are better able to use vestibular information immediately after flight than first-time fliers (Paloski *et al.* 1999). Since experienced astronauts have previously made the one-g to zero-g to one-g transitions, they may be partially dual-adapted and able to more readily transition from one set of internal models to the other.

It was also found that the recovery time course followed a double exponential path. In the 34 astronauts tested after 10-12 day Shuttle flights, an initial rapid improvement in stability during the first eight to ten hours was followed by a more gradual return to preflight stability levels over the next four to eight days (Figure 5-10). It was concluded that postflight postural instability appears to be mediated primarily by alterations in the vestibular (presumably otolithic) feedback loop and, secondarily, by alterations in ankle proprioceptive feedback, at least in some subjects (Paloski *et al.* 1999). It also appears that increased reliance on vision may partially compensate for the degraded performance of the other two feedback systems.

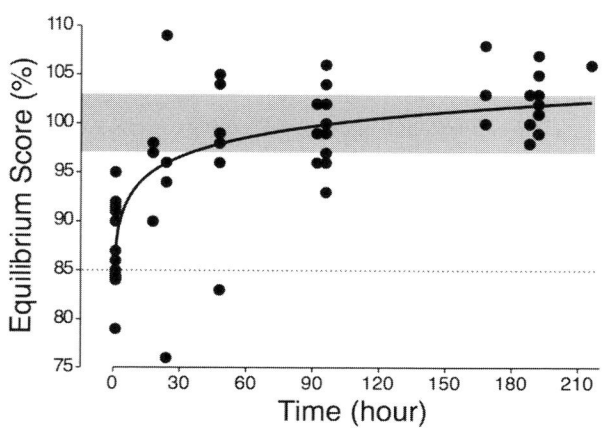

Figure 5-10. Sum of the equilibrium scores from the various sensory tests performed on astronauts after landing relative to preflight. The grey area indicates the mean (100%) and standard error measured preflight on the same subjects. A few hours after landing, the average returning crewmember was below the limit of clinical normality (dashed line). After flights ranging from 5-13 days, postflight re-adaptation took place in about four days and could be modeled as a double-exponential process, with an initial rapid phase lasting about 2.7 hours and a secondary slower phase lasting about 100 hours. Adapted from Paloski et al. *(1999).*

In a related study, Speers and his co-investigators (1998) examined changes associated with space flight in postural strategies employed by a 10-subject subset of the original 34 subjects. Using a multivariate approach, they found an increase in the relative utilization of hip sway strategy after flight and they conclude that these changes are consistent with re-weighting of vestibular inputs and changes in control strategy in the multivariable posture control system.

Postflight postural instability appears to be mediated primarily by alterations in the vestibular (presumably otolithic) feedback loop and, secondarily, by alterations in ankle proprioceptive feedback, at least in some subjects. The effects of demographic factors like age, gender and longer mission duration on these responses are currently being evaluated.

4.5   Tests of Vestibulo-Spinal Reflexes

In a number of pre- and postflight studies using the *Hoffmann reflex* (H-reflex) and the *tendon reflex* (T-reflex) techniques, it has been shown that both the alpha and gamma motor systems can be altered by space flight. In crewmembers studied during the Skylab program, the postflight T-reflex elicited by Achilles tendon percussions (mechanical stimulation) showed considerable potentiation over preflight baseline values (Baker *et al.* 1977). Before and after the Spacelab-1 mission, the H-reflex (electrical stimulation) showed a potentiation related to the selected drop-to-shock interval (see Section 3.3.2) for up to seven days postflight (Reschke *et al.* 1984).

Similar results have been obtained in the Russian space program. Kozlovskaya *et al.* (1984) demonstrated that two days after landing, the H-reflex in monkeys required a lower stimulus threshold and was potentiated over that observed before flight. In this study, it was observed that a single shock could elicit a response where the usual protocol required a double shock technique. When the double shock was employed, the second response following the conditioning shock (by 100 ms) was not inhibited, but rather enhanced. More recently, Grigoriev & Yegorov (1990) reported that the T-reflex in crewmembers who had been in orbit for up to 241 days on board Mir was characterized by a decrease in threshold and a three- to fourfold increase in amplitude over preflight values even on the sixth day postflight.

Coupled with the gradual decrease in the vestibulo-spinal reflex amplitude observed in-flight (Reschke *et al.* 1984, 1986, Watt *et al.* 1986, Watt 2001), the postflight potentiation of the H-reflex and T-reflex suggests response mediation via descending vestibular (otolith) pathways and a reinterpretation of otolith function via adaptation within the CNS in response to the stimulus rearrangement of orbital flight.

4.6   Postural Response to Voluntary Movements

In a simple test following two short-duration Shuttle missions, crewmembers provided with a visual reference (imaginary with eyes closed) were ask to bend at the waist in roll or pitch in an attempt to match a 20-deg angle (Reschke & Parker 1987). It was reasoned that if visual signals were eliminated and the otolith output was not interpreted as tilt immediately postflight, then the magnitude of the feedback signal during voluntary tilting would be reduced. Consequently, the astronauts would be expected to bend too far as they attempted to perform the roll or pitch movements. Kinematic analysis obtained from video recordings showed no significant difference in estimating the magnitude of tilt between pre- and postflight bending in either the roll or pitch plane planes. However, the basic premise may still be correct. Following the STS-41B Shuttle mission, one crewmember, attempting to test his limits of stability, demonstrated that by tilting in the roll plane there was a consistent angle at which he would lose his balance. When it was pointed out to him that most individuals would lose balance at that tilt angle, his comment was: "Yes, but I am unaware of the angle that I am leaning, even with my eyes open" (Reschke *et al.* 1991).

Another interpretation of the results is possible. While accuracy in achieving a specified tilt angle was unchanged immediately postflight, the strategy employed to maintain this accuracy appeared to involve a change in the use of the hips and shoulders. The hips were thrust backward more postflight and the angle of the head indicated that there was an attempt (or strategy) to stabilize the head in space, thus ensuring that gaze was maintained. This finding is related to the results on the translation platform, the more complex measures of postural stability and the locomotion studies described below.

4.7   Clinical Benefits

The results of space experiments on posture have put forward the remarkable plasticity in the organization of postural reactions. Prolonged exposure to a weightless environment results in changes in the sensory systems (with the possible exception of vision) necessary for the maintenance of postural stability. These changes, driven by the new complex of stimuli of microgravity, originate at the periphery and involve a subsequent reinterpretation of the sensory input from the receptors serving kinesthesia, touch, pressure, and otolith functions. Furthermore, as in-flight time increases, habituation of responses appropriate to the new inertial environment occurs at a central level, but the terrestrial motor programs are maintained. Upon return to Earth, postural instability, to a point that borders on clinical ataxia, is manifested as a result of this in-flight neural reorganization. Postflight recovery of posture is then probably related to the time it takes for the CNS to re-adapt to the appropriate interpretation of graviceptor signals. The faster re-adaptation observed in veteran astronauts on their subsequent flights opens interesting perspectives for the rehabilitation of patients after lesions of the vestibular system and countermeasures that may be developed for planetary exploration missions.

Information obtained from these investigations is promising for ground-based clinical research. A relatively large number of individuals on Earth suffer from prolonged, frequently life-long, clinical balance disorders. Disorders like Ménière's disease and traumatic injuries to the inner ear can severely influence quality of life. Falls are the leading cause of injury-related deaths in the elderly and these numbers continue to grow. Inner ear disorders are thought to account for 10-50% of falls among senior citizens. Currently, human space flight is the only means available for studying the response to sustained loss and recovery of inner ear information. Comparison between data from astronaut-subjects and similar data from patients and elderly subjects demonstrates similarities between these balance disorders. One sensible difference is that the posture problems recover in a few days for the astronauts, whereas it can take weeks (or never recover) in the patients. It is hoped that a better understanding of the strategies used during the recovery process in the astronauts and of the plasticity of this system in general, will help to improve rehabilitation treatments for patients with balance disorders on Earth.

5   LOCOMOTION STUDIES

Erect walking is a unique feature of human locomotion. Its evolutionary history indicates highly specific adaptations of the skeletal and muscular apparatus. Also, erect posture is mechanically efficient in humans because the center of body mass vaults over the supporting limb like an inverted pendulum, thereby limiting energy expenditure by means of an exchange of the forward kinetic energy with the gravitational potential energy (Cavagna *et al.* 1977).

Normal gait control depends on the acquisition of pre-programmed patterns of muscle activation and requires the continuous monitoring of external sensory input and internal reafferent signals. Locomotion pattern generators in the CNS are subject to overriding control from higher neural centers (Brooks 1986). Peripheral sensory and internal reafferent feedbacks modify patterns of activation emitted by pattern generators to improve ongoing motor performance.

Detailed postflight locomotor studies indicate that the relationship between sensory input and motor output is altered in the microgravity environment. During prolonged missions, neural adaptive processes come into play to permit new locomotion strategies to emerge in this novel sensory environment. This recalibration is associated with a time constant of acquisition and decay. The adaptive state achieved on-orbit is inappropriate for a one-g environment, leading to gait instabilities on return to Earth.

5.1   In-flight Observations

The cautious gait of astronauts descending the stairs of the "white room" docked with the Space Shuttle and walking on the runway is an obvious example of changes in sensorimotor coordination. Typically, locomotion in microgravity poses no problem and is quickly learned. However, adaptation continues for about a month. The astronauts who just visit the ISS note that the long-duration crewmembers move more gracefully, with no unnecessary motion. They can hover freely in front of a display when the new comers would be constantly touching something to hold their position (Clément 2005).

When moving about in space, the astronauts stop using the legs as they do on Earth. Instead they will increase the use the arms or fingers to push or pull themselves within the available space. For clean one-directional movements, push must be applied through the center of gravity, i.e., just above the hips for a stretched-out body. When translating though, the natural place for the arms is overhead to grab onto and push off from things as they come whizzing by. This is the worst possible place from the physics of pushing and pulling for clean movements, for by exerting forces with arms overhead, some unwanted rotations will invariably occur, which have to be compensated with ever more pushes and pulls, giving an awkward look to the whole movement. "To cleanly translate, I found it is best to keep the hands by your hips when exerting forces and boldly go headfirst. This way your pushing and pulling is directed through your body's center of gravity and gives nice controlled motions without unwanted rotations" (Pettit 2003).

Movement in a weightless environment obeys Newton's laws of motion. Friction forces are negligible and the angular momentum is always conserved unless acted on by an outside torque. Filmed sequences of astronauts performing a number of gymnastic moves in space were analyzed frame-by-frame. The principle of conservation for angular momentum was demonstrated as the astronauts tumbled, twisted and rotated in space. Throughout their motion and up until they entered in contact with the wall, the angular momentum was constant at 35.7 ± 1.2 kg x m^2/sec while rotating freely (NASA 1995).

5.2   Pre- and Postflight Studies

Since the legs are less used for locomotion, new sensorimotor strategies emerge in microgravity. Some of this newly developed sensorimotor program "carries over" to the postflight period, which leads to postural and gait instabilities upon return to Earth. Both U.S. astronauts and Russian cosmonauts have reported these instabilities even after short-duration space flights. Postflight, subjects experience a turning sensation while attempting to walk a straight path, encountered sudden loss of postural stability especially when rounding corners, perceived exaggerated pitch and rolling head movements while walking, and experience sudden loss of orientation in unstructured visual environments. In addition, oscillopsia and disorienting illusions of self-motion and surround-motion are observed during the head movements induced by locomotion.

In an early and intensive program, Russian investigators (Bryanov *et al.* 1976) studied locomotor behavior in 14 cosmonauts following missions lasting from 2 to 30 days in the Soyuz spacecraft (Soyuz-9, 12, 13, 14, 15, 16, and 17). Using motion picture analysis techniques, the sequential position of various body joints and limbs were recorded and analyzed to determine performance associated with walking, running, standing, long jumps, and high jumps. Distinct postflight performance decrements in gait and jumping behavior were observed with the duration of the decrements related, in most cases, to the length of the flight. Postflight gait was modified for 15 to 30 minutes after two days in space, but was affected for up to two days after flights of six to eight days. This same trend was observed for flights lasting 16 to 18 days, with performance on the Soyuz-9 (18 days) mission showing more degradation (disturbances in walking were still apparent 25 days after flight) than that observed following the sixteen-day Soyuz-14 mission (almost complete recovery two weeks postflight). Surprisingly, gait and related responses (jumping performance) following the thirty-day Soyuz-17 flight were more analogous with the postflight performances of the Soyuz-14 crew.

A typical postflight profile of the Russian cosmonauts is similar to that observed in the returning U.S. astronaut population. In walking, the cosmonauts place the legs wide apart, with the trunk held to the side of the supporting leg, and the intended path is not maintained. For greater stability, they frequently raise their arms to the side and they walk with small steps of irregular length. It is highly characteristic that in the transfer of weight during a forward step, the downward movement of the foot accelerates. At the moment of impact with the ground, the foot is "thrown" rather than being placed normally, creating the appearance of a stamping gait (Bryanov *et al.* 1976).

It is not uncommon when walking with returning crewmembers the length of the O&C Building at Kennedy Space Center, which is about 100 meters in length, to observe that they deviate to their right or left, then they realize that they had almost run into the wall, and they make a quick correction back to center. This turning sensation while attempting to walk a straight path is presumably related to the asymmetry in the re-adaptation of the vestibular system.

5.2.1   Head and Gaze Stability

Grossman *et al.* (1988) demonstrated that during walking and running in place in normal gravity, the peak velocity of head rotation in all axes is generally constrained below 100 deg/sec, and is thus below the saturation velocity (350 deg/sec) of the vestibulo-ocular reflex (Pulaski *et al.* 1981) However, the predominant frequency of head rotation during walking in place may range up to 4 Hz, and during running in place, to 8 Hz. Grossman and his colleagues (1989) have characterized gaze stability during walking and running, and have found that the angle of gaze is relatively stable. However, individuals with loss of vestibular function experience impaired visual acuity and oscillopsia during locomotion, stressing the importance of the VOR in maintaining gaze stability during locomotion (Grossman & Leigh 1990, Pozzo *et al.* 1991).

During the performance of various postural and locomotor tasks in terrestrial gravity, angular head deviation is maintained with a precision of a few degrees (Grossman *et al.* 1988, Berthoz & Pozzo 1988). Berthoz & Pozzo (1988) traced several of the figures from the classic Muybridge (1955) book showing successive photographs of human subjects engaged in a variety of different tasks. When these figures were superimposed around a common point (external auditory meatus), they noted that the head is stabilized in space within a few degrees. Berthoz & Pozzo (1988) also performed a quantitative examination of head stabilization during locomotion and found that, like the subjects photographed by Muybridge, the head did not exceed angular rotations of more than 3-6 deg in amplitude.

These ground-based results suggest that coordination of the body during locomotion is driven by the requirement to maintain head stability, and thus gaze. This concept represents a "top down" approach to the problem of gait stability. The underlying hypothesis is that gait stability is established to maintain head position in space reducing gaze error. Therefore, the maintenance of posture and gait stability is a goal-directed response designed to stabilize the head relative to the Earth's vertical ensuring gaze stability and the maintenance of visual acuity. This "top down" approach contrasts with the concept that maintenance of posture and gait following space flight is exclusively a function of in-flight changes in locomotion, the reduction of muscle tonus and a corresponding loss of muscle strength.

This novel concept was applied to data obtained from the H-reflex experiment flown on Spacelab-1 (Reschke *et al.* 1984, 1986). Linear acceleration was provided by a vertical drop of approximately 12 cm. High speed photographs (2400 frames/sec) were taken during selected drops before and immediately after flight, and the angle of the head was computed from markers placed on the head. There was approximately twice as much angular deviation of the head three hours after landing than there was before flight (8-10 deg preflight; 20 deg postflight); by the third day postflight, a strategy had developed that allowed the subject to maintain a stable head position despite the observation that orientation of the trunk and limbs continued to be more variable than that recorded preflight. Thus, under most tasks, the head seems to be stabilized in a very precise fashion suggesting that postural and gait motor control strategies are organized around achieving this goal. During movement in the microgravity environment of space flight, the requirement to stabilize the head is presumably reduced. Thus gait and postural instabilities experienced by astronauts upon return to Earth may be caused by in-flight adaptive acquisition of new "top down" motor strategies designed to maintain head and gaze stability during body movement in microgravity. Novel and potentially unstable gait strategies may be adopted postflight in an attempt to maintain head stability in the face of conflicting sensory cues during the period of sensory recalibration on re exposure to a one-g environment.

More recently, Bloomberg *et al.* (1997) have reported changes in head pitch variability, a reduction of coherence between the trunk and compensatory pitch head movements, and self reports from crewmembers indicating an increased incidence of oscillopsia (the illusion of a visual surround motion) during postflight treadmill walking. These results are reported in greater details below.

5.2.2 Dynamic Visual Acuity

In recent experiments designed to investigate the effects of space flight on head and gaze stability during locomotion, astronaut subjects were asked to walk and run on a motorized treadmill while visually fixating a stationary target positioned in the center of view (Figure 5-11, left). Tests were conducted 10 to 15 days before launch and two to four hours after landing. A video-based motion analyzing system was used to record and analyze head movements (Bloomberg *et al.* 1999). Data from 14 crewmembers collected following their long-duration (~ 6 months) stays in space showed a decrement in dynamic visual acuity while walking. For some subjects the decrement in dynamic visual acuity was greater than the mean acuity decrement seen in a population of vestibular impaired patients collected using a similar protocol. This decreased dynamic visual acuity is presumably related to the degree of oscillopsia experienced during postflight locomotion (Bloomberg & Mulavara 2003).

It is also clear from these studies that head motion displays more variability during locomotion following space flight. Analyzing each subject's amplitude of the predominant frequency for the head angular roll, pitch and yaw movement during locomotion showed that, after space flight, there was a significant change in the head roll and pitch orientations, respectively, during walking. In contrast, only smaller percentage of subjects showed a significant change in head movement magnitudes in the yaw orientation, during walking.

Comparison between responses from astronauts who had experienced more than one space flight and first-time fliers indicated that the former demonstrated less postflight alteration in the frequency spectrum of pitch head movements than the latter. Postflight behavioral differences between astronauts based on their experience level have been previously observed in tests of dynamic postural equilibrium control (Paloski *et al.* 1993). In these tests, inexperienced astronauts show greater postflight decrements in postural stability than their more experienced counterparts. Such differences may be the result of many factors. However, they could indicate that repeated exposure to space flight leads to facilitation in formulating the adaptive sensorimotor transition from a microgravity to a terrestrial environment.

The significant reduction in predominant frequency amplitude of pitch head movements observed in astronauts postflight may be caused by attempts to reduce the amount of angular head movement during locomotion, and reduce potential canal-otolith ambiguities during the critical period of terrestrial re-adaptation. This in turn, further simplifies the coordinate transformation between the head and trunk, presumably allowing an easier determination of head position relative to space. Yet, this strategy is not optimal for gaze stabilization because it results in a disruption in the regularity of the compensatory nature of pitch head movements during locomotion. This strategy also restricts behavioral options for visual scanning during locomotion. Consequently, there may be trade-offs between head movement strategies depending on the imposed constraints. Once significant re-adaptation takes place, a decrease in constraints on the degrees-of-freedom of head movement likely occurs, returning performance back to preflight levels. Interestingly, patients with vestibular deficits (Keshner 1994) and children prior to development of the mature head stabilization response (Assaiante & Amblard 1993) also show head movement restriction during locomotion.

Changes in head and torso movements during locomotion postflight, predominantly in the pitch and roll planes, are presumably due to the central reinterpretation of otolith information. These changes in coordination between the head and torso, added to the changes in the performance of the vestibulo-ocular reflex (see Chapter 6, Section 2.4) would then be at the origin of the alteration in gaze stabilization during locomotion. These results support the hypothesis that changes in head stability and coordination induced by adaptive modification in "top down" motor control schemes may indeed be a contributing factor to postflight locomotor impairment.

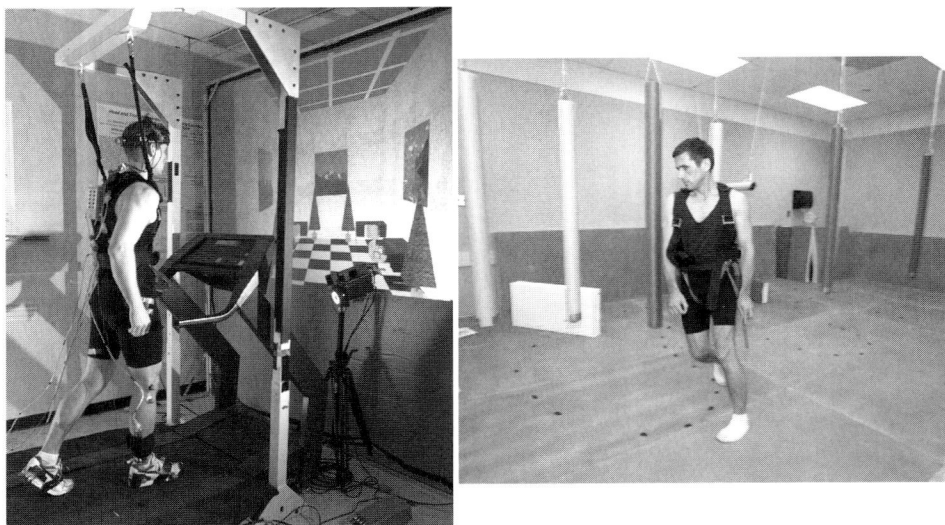

Figure 5-11. Left. While subjects walk at 6.4 km/h on a motorized treadmill, three-dimensional full-body motion data are acquired using a video-based motion analysis system; gait cycle timing is measured using foot switches placed in the shoes and dynamic visual acuity is assessed. Right. The Functional Mobility Test *provides an assessment of the functional and operational changes in locomotor function by testing subject's ability to negotiate an obstacle course placed over a medium-density foam floor. Photos courtesy of NASA.*

5.2.3   Lower Limb Kinematics

During locomotion, foot contact with the ground, weight transfers from one foot to the other and the push off with the toe from the ground are critical phases as these interactions result in forces that create vibrations, which if unattenuated, could interfere with the visual-vestibular sensory systems in the head. The musculoskeletal system controls these vibrations: muscles and joints act as filters to minimize the perturbing effects of impacts with the ground and help to maintain a stable trajectory at the head Hence, appropriate attenuation of energy transmission during locomotion, achieved by the modulation of the lower limbs' joint configuration coupled with appropriate eye-head-trunk coordination strategies, form the fundamental features of an integrated gaze stabilization system. From this point of view, the whole body is an integrated gaze stabilization system, in which several subsystems contribute, leading to accurate visual acuity during body motion. After space flight, changes have been documented in both head-trunk and lower limb patterns of coordination, which may exacerbate the on-going visual-vestibular disturbances.

McDonald *et al.* (1994, 1996) have evaluated the variability and stability of the motion observed in the hip, knee and ankle joints during treadmill walking (6.4 km/h) following space flight. The temporal characteristics of the gait patterns were remarkably robust, and there was no significant change at both toe off and heel strike postflight relative to preflight. However, increased variability was observed after space flight in hip joint at toe off and in knee joint at heel strike.

Lower limb EMG signals were collected during treadmill locomotion after short duration space flight (Layne *et al.* 1994). In general, high correlations were found between preflight and postflight activation waveforms for each muscle and each subject. However, relative activation amplitude around heel strike and toe off changed as a result of space flight. The level of muscle co-contraction, activation variability and the relationship between the phasic characteristics of the ankle musculature in preparation for toe off were also altered by space flight (Layne *et al.* 1996). During walking after long-duration space flight, astronauts also showed modified transmission characteristics of the shock wave at heel strike and increased total knee movement during the subsequent stance phase (Mulavara *et al.* 2000).

Related studies revealed disruptions in endpoint toe-trajectory control of lower limb kinematics during the swing phase of gait cycle (Courtine *et al.* 2002), increased lateral motion of the trunk during overground locomotion suggesting instability during gait (Courtine & Pozzo 2004), and impairment in the ability to coordinate effective landing strategies during jump tasks (Newman *et al.* 1997). These sensorimotor disturbances may lead to disruption in the ability to ambulate and perform functional tasks during initial reintroduction to a gravitational environment following a prolonged transit.

5.2.4 Functional Mobility Test

To further elucidate the underlying basic sensorimotor mechanisms responsible for postflight locomotor dysfunction, Bloomberg and his colleagues also used an integrative approach. They designed a *functional mobility test* (FMT) that serves as a global test of locomotor performance that relates to activities required for emergency egress after landing. In the FMT, the astronauts walk at their preferred paces through an obstacle course set up on a base of 10-cm thick medium density foam. The foam provides an unstable surface that increases the challenge of the test. The 6.0 m x 4.0 m course consists of several pylons and obstacles made of foam (Figure 5-11, right). Subjects are instructed to walk through the course as fast as possible without touching any of the objects on the course.

The dependent measure is the time to complete the FMT. Data collected on 18 crewmembers of ISS Expeditions 5-12 indicate that adaptation to space flight led to a 52% increase in time to complete the FMT one day after landing. Recovery to preflight scores took an average of two weeks after landing. Furthermore, three of 18 subjects were unable to perform the FMT up to one day after their return from space flight. These disturbances may have significant implications for performance of operational tasks immediately following landing in case of an emergency or on a planetary surface.

5.3   Walking on the Moon and Mars

Studies at NASA Langley Research Center, Hampton VA, carried out on a simulator equipped with an inclined plane (see Figure 1-06), showed that humans walking and running was approximately 40% slower under lunar gravity conditions compared with terrestrial conditions (Pestov & Gerathwohl 1975). As the rate of movement increased, the inclination of the trunk forward increased to a greater degree under lunar gravity than under terrestrial conditions (Figure 5-12). The effects of actual lunar gravity on human activities were evaluated during the Apollo missions. Interestingly, the energy expenditures of astronauts during activities on the Moon averaged 220-200 kcal/h, about the same as walking without any equipment under terrestrial conditions. A comparison of postflight medical data showed that the astronauts who did not experience lunar gravity were physically less fit than the other crewmembers. Their weight loss was considerable, orthostatic intolerance was increased, red cell mass decrease was more pronounced, work capacity was lower, and they showed greater loss in all body fluid volumes (Berry & Homick 1973).

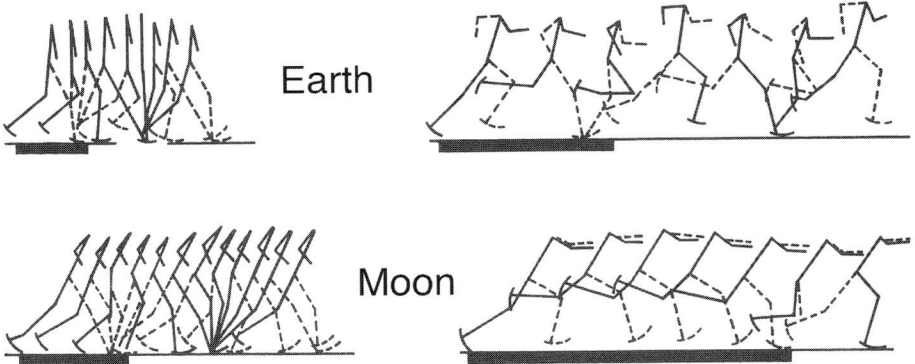

Figure 5-12. Changes in body kinematics during walking (upper diagrams) and running (lower diagrams) under lunar and terrestrial gravity levels. The heavy line shows the length of stride. Time interval between stick figures is 0.16 sec. Although more ground is covered in one single stride in lunar gravity compared to Earth gravity, locomotion is much slower. Adapted from Pestov & Gerathwohl (1975).

Despite training in ground simulations and in the one-sixth g airplane flying parabolas, falls were frequent among astronauts during extravehicular activity on the lunar surface. Eugene Cernan, Apollo-10 astronaut, on the Moon recalls "Jack (Schmitt) reached for a rock, lost his balance and toppled into a pratfall. … Jack fell again while trying to grab another Moonstone. 'I haven't learned to pick up rocks, which is a very embarrassing thing for a geologist,' he admitted" (Cernan & Davis 1999, p. 323). The high and rearward center of gravity of the Apollo suit influenced upslope walking and the stiffness of the inflated suit strongly influenced gait, making it impossible to squat to retrieve dropped objects.

Different lunar gaits were tested and adopted by the crew. These included a "loping gait" in which the astronaut alternated feet, pushed off with each step and floated forward before planting the next foot; a "skipping stride," in which he kept one foot always forward, hit with the trailing foot just a fraction of a second before the lead foot, than pushed off with each foot, launching into the next glide; as well as a "kangaroo hop," which few Apollo astronauts ever employed, except playfully, because its movements were so stilted (Hansen 2005, p. 502). Learning each gait was relatively fast: Eugene Cernan, Apollo-10 astronaut, on the Moon: "I skipped around to get my sea legs in the low gravity of this strange new world. Learning how to walk was like balancing on a bowl of Jell-O, until I figured out how to shift my weight while doing a sort of bunny hop" (Cernan & Davis 1999, p. 322).

Contributing to the problem of locomotion on the lunar surface was the ruggedness of the terrain and the lower visibility. When looking out in any direction toward the horizon, the astronauts on the Moon felt a bit disoriented. Because the Moon was such a smaller sphere than the Earth, the planetoid curved much more visibly down and away than they were accustomed to. Also, because the terrain varied a good bit relative to their ability to move over it, they had to be constantly alert. "On Earth, you only worry about one or two steps ahead," Buzz has recalled (Figure 5-13). "On the Moon, you have to keep a good eye out four or five steps ahead." (Hansen 2005, p. 502). "Exacerbating the problem was the fact that astronauts really could not see their feet very well… The fact that the cables [on the ground] got dusty almost immediately also contributed to the problem" (Hansen 2005, p. 502).

In the planned Moon missions, lunar polar terrain may be more sloped than that explored by the Apollo astronauts. The polar sun angle will be far lower (1 deg, rather than 15 deg) so astronauts will be traversing areas of deeper shadow, possibly requiring the use of lights. Options for sensory supplementation during extra-vehicular activity should therefore be investigated. The effectiveness of vibrotactile cueing systems has been demonstrated in pilots and patients. They could be easily integrated in the suit. Also, night vision sensor imagery, an artificial horizon and a navigation display could be incorporated into an add-on external head-up display.

Figure 5-13. Astronaut Buzz Aldrin descends the steps of the Lunar Module *ladder as he prepares to walk on the Moon. Astronaut Neil A. Armstrong took this photograph during the only lunar extra-vehicular activity of the Apollo-11 mission. Photo courtesy of NASA.*

Ground-based simulations indicate that both the optimal walking speed and the range of possible walking speeds on Mars will be reduced compared to Earth. It was calculated that the optimal walking speed will be reduced to 3.4 km/h (down from 5.5 km/h on Earth) and the walk-run transition on Mars will occur near the optimal walking speed on Earth. However, because of the reduced gravity, the mechanical work done per unit distance to move the center of mass on Mars will be about half than on Earth (Cavagna *et al.* 1998).

6   SUMMARY

Numerous astronauts have been systematically subjected to posture and balance measurements within as little as two to four hours after landing since the very first space missions. The measurements have been obtained using standardized equipment, like balancing rails of variable width, stabilometry, and the NeuroCom EquiTest, and standard procedures, like voluntary arm or toe rises and deep knee bends. With rare exception, they all suffer from substantial disequilibrium (ataxia), especially on tests when their eyes are closed or where the support surface or the visual surroundings are caused to sway in conjunction with changes in the subject's center of mass. These situations leave the vestibular system as the only source of accurate information about orientation.

After short-duration missions, the astronauts recover rapidly for the first eight to ten hours and then gradually return to pre-mission levels over the next four to eight days. Some performance decrements are still observable weeks later. There is an inverse relationship between the initial severity of balance problems and the number of previous space flights. This indicates that one of the best countermeasures for space travel is space travel. Surprisingly, the otolith-spinal reflex appears to be no different in postflight tests than preflight performance, even though it was so greatly attenuated in microgravity. This perhaps indicates that recovery of this capacity to one-g is so rapid the problem disappears before it can be measured.

Astronauts experience substantial awkwardness, ataxia, vertigo, and slowing of gait for one week or more postflight. This is according to both anecdotal reports and controlled tests executed on a motorized treadmill, over a maze path, and on rails. A tendency to maintain a wide stance while walking, difficulty ambulating around corners, abnormal ankle angle, postural compensation for arm movements, and a substantial attenuation of the otolith-spinal reflex, which serves to prepare the body for the impact of unexpected falls, are specific problems that have been observed. About half of these aftereffects disappear within the first two to three hours after landing following short-duration missions. These aftereffects can last for much longer after long-duration missions (Figure 5-14).

Bloomberg *et al.* (1997) have reported reduced dynamic visual acuity in postflight astronauts while they were walking on a treadmill, especially for far distances. This deficit appears to be due to gaze destabilization (oscillopsia) because of a reduced ability to engage in compensatory head pitch movements during locomotion. These visual effects have been measured after two to four hours postflight and subsequently for as many as ten days postflight. This visual disability poses a potential hazard to reading cockpit displays, especially when making head movements, because it must certainly be present during the re-entry phase of the mission.

Adaptation to space flight also led to a 50% increase in time to traverse an obstacle course on landing day, and recovery of function took an average of two weeks after return. Importantly, alterations in kinematics and dynamic visual acuity were accompanied by commensurate changes in functional mobility. Such alterations in locomotion seen after space flight raise some concern about the crew capability for unaided egress from the Space Shuttle or the Soyuz in a case of emergency. Many crewmembers experience marked vertigo when making head movements during re-entry, landing and afterwards. This vertigo could be a major obstacle to successful egress if vision were impaired, as with a smoke-filled cabin.

The most significant visual-motor problems astronauts will encounter during their stay on the Moon and Mars are likely to occur when moving about in their space suits. The suits are quite large and bulky and alter the center of gravity. They will also need to learn the "lunar bounce" form of locomotion employed by the Apollo astronauts. Another possible problem will be reduced dynamic visual acuity due to changes in gait.

Our experiences on the Moon are limited and dated. Therefore, the only way to assess the effects of lunar gravity on perceptual-motor coordination is by Earth-based simulation. Partially unloading the body by means of springs or lower body negative pressure is one way to do this. This has already been done to test the effects of lunar gravity on treadmill walking (e.g., Donelan & Kram 2003). However, these procedures have no effect on the otolith organs. The one ground-based procedure that can produce all of the effects of lunar gravity is parabolic flight maneuvers, with all of the shortcomings and difficulties previously described.

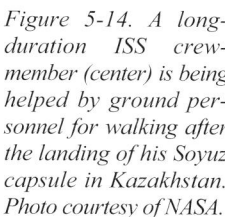 

Figure 5-14. A long-duration ISS crew-member (center) is being helped by ground personnel for walking after the landing of his Soyuz capsule in Kazakhstan. Photo courtesy of NASA.

Chapter 6

COMPENSATORY EYE MOVEMENTS

For several decades, the study of eye movements has been a source of valuable information to both scientists and clinicians. The singular value of studying eye movements stems from the fact that they are restricted to rotations in three planes and the eyeball offers very little inertia. This facilitates accurate measurement, for example using video eye recording in near infrared light, a prerequisite for quantitative analysis. Eye movements must continuously compensate for head movements so that the image of the world is held fairly steady on the retina, and thus appears clear and stationary. During head movements, the vestibular apparatus measures head velocity and relays this information to those centers controlling eye velocity and position to generate compensatory eye movements; this reflex behavior ensures that a steady image is maintained on the retina and vision is not blurred. When performed in darkness, the eye movements compensate for head movements. In the light, visual inputs serve to hold gaze steady or to shift gaze to an object of interest.

The absence of perceived gravity alters the inputs to the vestibular system and therefore could affect the compensatory eye movements. Eye movement is probably the response of the vestibular system that has been the most studied during space flight. It has been studied during voluntary (active) as well as during involuntary (passive) head movements. Indeed, one problem in studying eye movements by asking subjects to perform voluntary head movements is that the central nervous system is "aware" of the movement to be performed. A copy of the motor command (the so-called *efference copy*) is presumably sent to the eye-head coordination control system, and this helps to achieve the adequate, compensatory eye movements. For this reason, scientists also use passive rotation generated by servo-controlled rotating chair, sled or centrifuges in order to generate unpredictable inertial stimulation of the vestibular system, and to study the resulting responses.

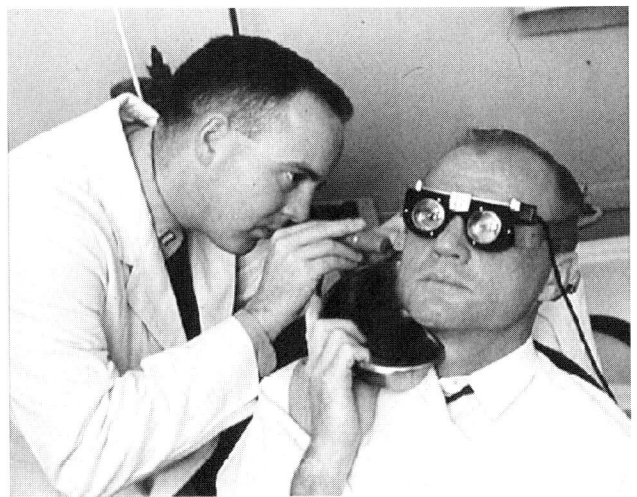

Figure 6-01. Mercury Astronaut John Glenn is wearing Fresnel glasses for examination of his eye movements during irrigation of his external auditory ear with hot water. This caloric testing is a functional investigation of the semicircular canals in the vestibular system. Photo courtesy of NASA.

G. Clément, M.F. Reschke, *Neuroscience in Space.*
DOI: 10.1007/978-0-387-78950-7_6, © Springer Science+Business Media, LLC 2008

1   CALORIC NYSTAGMUS

The most widely used clinical test of the functioning of the peripheral vestibular system is the *caloric test*. During this test, irrigation of the external ear canal with water or air above or below body temperature generates by thermal conduction a temperature gradient across the inner ear (Figure 6-01). As a result, the horizontal semicircular canal when orientated parallel to the gravitational vector is stimulated, producing characteristic rhythmic eye movements called *nystagmus*, and the subject experiences slight rotatory vertigo. For many years it was generally believed that this response was initiated by motion of the endolymph along the horizontal semicircular canal, as generated by the induced thermal gradient, and which in turn leads to a thermo convective force. This would produce a displacement of the endolymph, which stimulates the canal's sensory cells, in the same way that an actual rotation of the head. At the turn of the last century, Robert Bàràny, a Viennese otoneurologist, received the Nobel Prize for proposing this mechanism for caloric nystagmus (Bàràny 1906).

The weightlessness conditions of space flight were ideal for verifying this mechanism: if the thermo convective hypothesis was correct, no nystagmus response should have been observed in space. Caloric irrigation was first performed during the Spacelab-1 mission (Scherer *et al.* 1986). The equipment included insufflation of heated or cooled air in the ear and measurement of eye movements by EOG (Kass *et al.* 1986). Contrary to general expectation at the time (Scherer & Clarke 1985), a clear caloric nystagmus response was elicited in all test subjects (Figure 6-02).

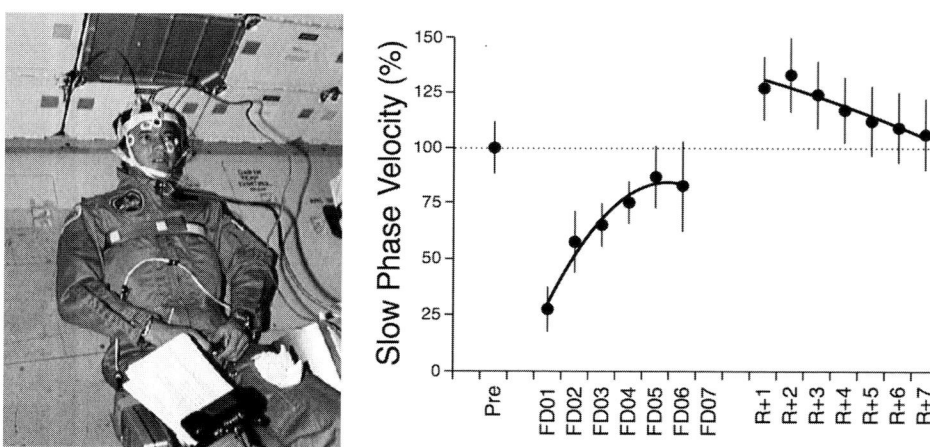

Figure 6-02. Left. During early space missions, eye movements during voluntary head rotation or caloric stimulation were recorded by electrodes placed around the subjects' eyes. Right. Results from caloric tests carried out during (flight day, FD) and after (R+1 to R+7 days) three space flights. The slow phase velocity of the caloric nystagmus was calculated for each subject and expressed in percentage relative to the preflight (Pre) values. Mean and standard error of four subjects in-flight and five subjects pre- and postflight. Adapted from Clarke et al. (1993b).

A caloric experiment was subsequently performed during the Spacelab-D1 mission to verify the Spacelab-1 findings and to investigate the thermal vestibular responses using an improved test procedure on three more subjects. Unilateral thermal stimulation was included in the stimulus profile in order to clarify whether hot and/or cold stimuli alone were able to elicit a response. The influence of controlled linear acceleration on the ongoing caloric response was also studied during sled motion. All three subjects tested exhibited a clear caloric response, with a nystagmus always directed toward the warmer earSled runs revealed a cyclical modulation of the caloric nystagmus by concomitant linear oscillation. Finally, caloric nystagmus was enhanced when subjects were released from their seat, thus reducing somatosensory inputs to the central vestibular system (von Baumgarten *et al.* 1986). Air at 15° C insufflated in the left ear also induced a caloric nystagmus in one crewmember during the German mission on board Mir in 1992. However, the horizontal slow phase velocity of caloric nystagmus in-flight was about 40% less than that measured before flight (Clarke *et al.* 1993b). A re-adaptation to normal gravity was observed over the ten-day period after landing.

Thermal convection in the endolymph is therefore not the only mechanism involved in caloric nystagmus. A number of possible alternative mechanisms have been discussed for the caloric-induced nystagmus. These include a direct thermal effect on the sensory hair cells or the afferent nerve connections from the hair cells to the CNS, or differential pressure effects due to thermal expansion of the endolymph fluid in the labyrinth (von Baumgarten *et al.* 1987).

2   VESTIBULO-OCULAR REFLEX

The vestibular system helps maintain a fixed gaze on a stationary or moving external object while we are undergoing complex head and body movements. The eye movements that compensate for head movements are driven by stimuli arising in the utricles and semicircular canals of the vestibular system. The motion of the visual image also drives eye movements that compensate for movements of the visual scene relative to the subject. If the movements of the head or the visual scene continues in one direction, slow pursuit movements alternate with rapid saccadic return movements, leading to vestibular or optokinetic nystagmus, respectively. These compensatory reflexes are very primitive and are controlled almost entirely by sub-cortical centers (except for foveal pursuit eye movements), which make interesting models for studying the role of gravity on the vestibular system.

2.1   Background

The *vestibulo-ocular reflex* (VOR) responds to transient rotation of the head in yaw, pitch, or roll with eye movements in the opposite direction. This action prevents slippage of images off the retina since the latency of the VOR is less than 16 msec vs. the 70 msec required for retinal processing. During prolonged rotation, acceleration is no longer sensed. In darkness, the initial nystagmus response to rotation produced by the VOR decays over a 30 to 45-sec period after which the eyes are stationary. Post-rotatory eye movements are also produced by the VOR in response to deceleration (Leigh & Zee 1991).

VOR can be quantitatively described by gain, phase and time constant. For sinusoidal rotation, VOR gain is calculated as the ratio of peak slow-phase eye velocity to peak head velocity. Phase describes the synchronicity of eye and head movements. For the frequency range of natural movement (0.5 to 5 Hz), eye movements should be of equal velocity and opposite in direction to head movements resulting in a gain of 1.0

and a phase of 0 deg (phase shift of 180 deg). For sustained rotation, gain is computed as the ratio of initial eye velocity to head velocity. In darkness, slow-phase nystagmus velocity decays exponentially and the VOR time constant is defined as the time required for eye velocity to decrease to 37% of its initial value. The nystagmus initially observed in sustained rotation is longer in duration than the signal provided by the vestibular nerve due to velocity storage, a mechanism that prolongs the nystagmic response (Leigh & Zee 1991, Raphan *et al.* 1979).

VOR gain varies between subjects and is affected by the distance from the subject to the target, the mental set chosen by the subject (especially in darkness), adaptation to corrective lenses, and age. The VOR time constant also exhibits inter-subject variability, is shortened by habituation, differs between horizontal and vertical components, and displays vertical directional asymmetry.

2.2 VOR Asymmetry

The horizontal VOR gain is nearly compensatory, i.e., the velocity of eye slow phase velocity is equal and opposite to the velocity of the head and their ratio corresponds to unity. Yet, the gain of the vertical VOR elicited by pitch motion in subjects lying on their side is not fully compensatory (Clément *et al.* 1992). In contrast, during pitch motion in upright subjects, vertical VOR gain is nearly compensatory (Boehmer & Henn 1983). This suggests that the vertical VOR gain is more compensatory when there is gravity input due to head tilt. At low frequencies, phases between eye velocity and head velocity are near zero during upright pitch motion, whereas phases for onside pitch motion develop progressively larger leads as the frequency decreases (Tomko *et al.* 1988). The gravity-sensitive signal that is normally present during upright pitch acts to make the phase of the vertical VOR at low frequencies more compensatory, and to enhance gain of the reflex over the same range. The role of gravity sensors in keeping low frequency gain from falling off and phase close to zero is anticipated from neurophysiological data on otolith function, since otolith gain is relatively flat down to DC an there is little phase lead or lag (Goldberg & Fernández 1982).

In both onside and upright pitch motion, the time constant for upward slow-phase eye movements is greater than for downward movements. This asymmetry in vertical VOR time constant may indicate underlying asymmetries in neural systems that control head and eye movements, such as the velocity storage mechanism. Signals originating before eye control would then cause asymmetric eye movements during pitch motion. Such neural asymmetries may be required for three reasons. First, the vertical visual field is not symmetric about the horizontal plane through the eye. More of the field is below such a plane than above it when the head is in its normal position. Therefore, a larger number of downward fast components may be required to reset the eye to its normal upright position because there is less upward than downward range of motion.

A second possible reason for more downward than upward fast components during vertical VOR is that muscular control of the head is distinctly different for upward and downward head motion. Downward head motion is assisted by gravity while upward head motion is opposed to it. This asymmetry in the force required to move the head requires an asymmetric set of dorsi- and ventri-flexor contractions, and at some point in the neural control circuitry an asymmetric control signal is needed. In support of this hypothesis, VOR adaptation has been observed during horizontal head inertia changes on Earth (Gauthier *et al.* 1986). However, in microgravity, Pozzo *et al.* (1998) did not measure noticeable changes in the asymmetry of muscular contraction between upward or downward voluntary arm movements.

A third interpretation for the upward slow phase eye velocity being greater than downward slow phase eye velocity could be explained to be a consequence of suppressing downward eye movement produced by the optical flow during forward motion such as walking (Guedry & Benson 1970).

2.3 Effects of Gravity on VOR

The observations above indicate that reflexive vertical eye movement control depends on interactions between the signals of the otolith organs and those of the vertical semicircular canals. The vertical VOR therefore differs in a fundamental way from the horizontal VOR, which normally relies solely on semicircular canal stimulation (Tomko *et al.* 1988).

Gravity effects on the vertical VOR might have been anticipated since it has long been known that otolithic stimulation results in vertical eye movements. Both vertical and torsional eye movements are produced by electrical stimulation of the utricular nerve in cats (Suzuki *et al.* 1969). Vertical eye movements are also produced during constant velocity pitch, i.e., when changing head orientation relative to gravity. If the pitch is not vertical the slope of the slow component of nystagmus is both cyclically modulated and directionally biased (Baloh *et al.* 1983). Based on the results of their pioneering electrophysiological studies of the otolith signals, Fernández & Goldberg (1976) suggested that the bias component might result from central processing of otolith data. Their studies confirmed that semicircular canals and otolith organs interact in controlling vertical eye movements.

Tilting of the head during the post-rotational period following horizontal rotation suppresses the magnitude of the slow-phase eye movements and the time course of decay decreases as if the apparent time constant of this response had been reduced. This phenomenon, termed *nystagmus dumping*, is assumed to be mediated by the otolith organs, although the exact mechanism is subject to some debate (Leigh & Zee 1991, Benson 1974).

The VOR has been shown to adapt to a number of environmental changes. Habituation to repetitive stimulation in darkness, especially constant velocity rotation or low frequency oscillation, decreases both time constant and gain, which may function to reduce motion sickness. Persistent vestibular stimulation also alters the VOR. For constant rotation in darkness, a reverse nystagmus (slow phases in the same direction as rotation) may develop after decay of the initial nystagmus. The vestibular system also adjusts the VOR in response to altered visual stimuli. For example, if the relationship between eye movements and the visual image is altered by reversing prisms or eyeglasses, then the VOR gain and phase can adapt to this new relationship. However, conflict between visual and vestibular cues may provoke motion sickness symptoms during the adaptation period (Leigh & Zee 1991, Reason & Brand 1975).

2.4 Microgravity Investigations

On the ground, head yaw motion generates only semicircular canal signals. In contrast, all but the slowest pitch and rolling movements generate both otolith and canal signals. Otolith signals would, however, not be present in microgravity. If a compensatory vertical VOR depends on convergence of semicircular canal and otolith signals, then head pitch and roll in microgravity should not lead to appropriate control of eye movements, until some form of adaptation occurs. Thus, head pitch and roll in microgravity should cause inappropriate retinal slip and visual-vestibular mismatch, while yaw will not. Such a mismatch is probably an adequate stimulus for motion sickness. Gravity sensitivity in the vertical VOR may therefore explain why, in microgravity, head movements in pitch or roll with the eyes open result in SMS more quickly and more severely than head movements in yaw (Lackner & Graybiel 1985).

2.4.1 Horizontal VOR

In-flight experiments have relied on voluntary head oscillations at frequencies ranging from 0.25 to 1 Hz (Thornton *et al.* 1989, Viéville *et al.* 1986, Watt *et al.* 1985, Benson & Viéville 1986). Passive rotation using rotating chairs has also been employed pre- and postflight (Benson & Viéville 1986). Head oscillations are performed with eyes open fixating a wall target, where gain is presumably unity, and with eyes open in darkness or eyes closed while imagining a wall-fixed target (Figure 6-02, left). Most studies have used the conventional electro-oculography technique to measure eye movements. Few studies have detected significant in- or postflight changes in horizontal VOR compared to preflight (Thornton *et al.* 1985, Watt *et al.* 1985, Benson & Viéville 1986) or the direction of the changes noted has varied between subjects (Grigoriev & Yegorov 1990).

A single subject on board STS-51G exhibited decreased horizontal VOR gain at 0.25 Hz on his first test, conducted six hours into the mission, which recovered to preflight levels by flight day 7 (Viéville *et al.* 1986). A decrease in horizontal VOR gain early in microgravity is consistent with the parabolic flight results that showed decreased horizontal VOR gain with decreasing gravity (Lackner & Graybiel 1981b). Since no phase shift accompanied the in-flight reduction in horizontal VOR, the subject on STS-51G may have suppressed vestibular input to avoid sensory conflict, possibly a learned response from his prior training as a pilot. It is also possible that the subject could not be able to accurately imagine a wall-fixed target in the absence of gravity (Viéville *et al.* 1986).

Passive horizontal VOR in humans has been studied in very rare occasions during space flight: the body restraint system during Spacelab-1 (see Figure 4-12, left), the MVI rotator during IML-1 (see Figure 4-15, left), the hand-spun rotator during SLS-1 (Figure 6-03), and the human centrifuge during Neurolab (see Figure 4-15, right). Previous experiments performed during parabolic flights by DiZio and his colleagues (1987, 1988) had demonstrated that the dominant time constant of the post-rotatory horizontal VOR was shorter during acute exposure to weightlessness, but that there was no consistent change in the magnitude of the initial peak slow-phase velocity response. The post-rotatory horizontal VOR was monitored in-flight in one crewmember on the SL-1 mission using a hand spun rotating chair, and in four crewmembers using a servo-controlled chair. Results indicated no change in gain but did suggest a shortened time constant in-flight (Oman & Balkwill 1993).

A residual shortening of the time constant was also observed in crewmembers tested during the first several days after return from week-long flights (Benson & Viéville 1986, Oman & Kulbaski 1988, Oman *et al.* 1989, 1996). The effects were thus qualitatively similar to those observed by DiZio *et al.* (1987, 1988) in parabolic flight. Responses gradually returned to preflight norms during the first postflight week. Oman

& Balkwill (1993) have speculated that as a consequence of the altered graviceptive input in weightlessness, the CNS may reduce the vestibular component driving central velocity storage in favor of visual inputs.

The horizontal VOR characteristics (gain and dominant time constant) of two rhesus monkeys tested 15 hours after returning from a two-week flight on board an unmanned Russian biosatellite were not significantly different from preflight (Cohen *et al.* 1992). This discrepancy between monkey and human data may be due to the differences between velocity storage in these two species.

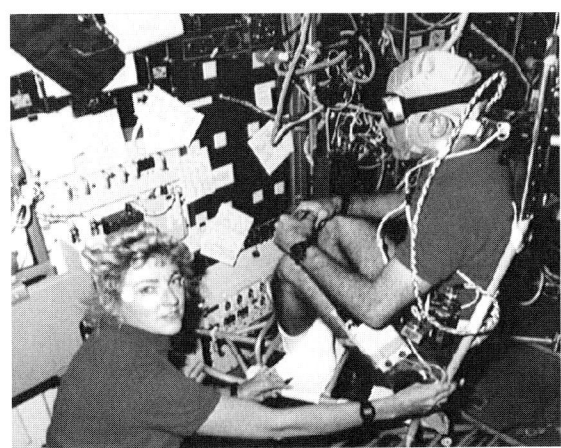

Figure 6-03. To induce the post-rotatory VOR and the eye movements associated with it, the astronauts of the SLS-1 and SLS-2 missions were rotated quickly in this hand-spun chair and then suddenly stopped. Eye movements were recorded using electro-oculography. Photo courtesy of NASA.

2.4.2 Vertical VOR

On Earth, in contrast to head movements in yaw, head oscillations in pitch produce changes in the direction of the gravity vector sensed by the otoliths. Such difference is no longer present in microgravity. Therefore, the weightless environment offers an ideal way to investigate the contribution of the otoliths to vertical VOR (Berthoz *et al.* 1986).

In-flight investigations of vertical VOR have employed voluntary (active) head oscillations at various frequencies. While pre- and postflight changes have not been observed in some instances (Watt *et al.* 1985, Berthoz *et al.* 1986), other investigations have noted alterations in vertical VOR. Two subjects tested during STS-51G showed a decrease in vertical VOR gain during 0.2 Hz oscillation for the first four flight days after which the gain began returning to preflight levels (Viéville *et al.* 1986). This decrease was accompanied by a reversal of the up-down gain asymmetry (Figure 6-04). This reversal of vertical VOR gain asymmetry during space flight has been confirmed during another space study involving two cosmonauts (Clarke *et al.* 2000).

Seven out of thirteen cosmonauts (52%) tested on the first day after space flights ranging in duration from 7 to 365 days showed no nystagmic response during active head rotation in pitch with the eyes closed, but a low amplitude compensatory eye deviation (Kornilova 1997). When measured with video infrared oculography, vertical VOR gain for oscillations ranging from 0.12 Hz to 2 Hz. was found to decrease in cosmonauts returning from space flight (Clarke *et al.* 1993a).

The single study on the passive vertical VOR in space to date was conducted during the IML-1 Spacelab mission using a servo-controlled rotating chair and monocular video recordings (see Figure 4-15, left). When tested during passive sinusoidal oscillations at 0.2 and 0.8 Hz, no systematic changes in pitch VOR phase or absolute gain across four subjects tested were noted before, during and after this eight-day mission (Clément *et al.* 1999). However, the angular deviation between the direction of the eye movement during VOR and the head rotation axis was calculated during pitch head rotation in three subjects throughout the flight. This measurement showed that the slow phases were tilted by 9.2 to 31.8 deg relative to the head z-axis during rotation at 0.2 Hz in-flight (Clément 1998). Since the tilt was always in the same direction, it is possible that a somatosensory input was introducing a perceptive bias in the oculomotor response. Of note is that pre- and postflight testing was performed with the subject in the 90-deg left-side down position. The head rotation axis was therefore aligned with gravity, and head pitch up or pitch down did not stimulate the otolith organs.

In monkeys, vertical VOR was only tested postflight, with both sinusoidal oscillations ranging from 0.025 to 0.125 Hz and with angular velocity steps in both directions about an Earth-vertical, interaural axis, i.e., in left-side down position, too. Vertical VOR gain was reduced by 6-9% postflight when measured with sinusoids (Dai *et al.* 1994). However, when measured with velocity steps, the upward VOR gain decreased and the downward VOR gain increased compared to preflight. Unfortunately, the time constant of decay of post-rotatory nystagmus was not measured in this last study.

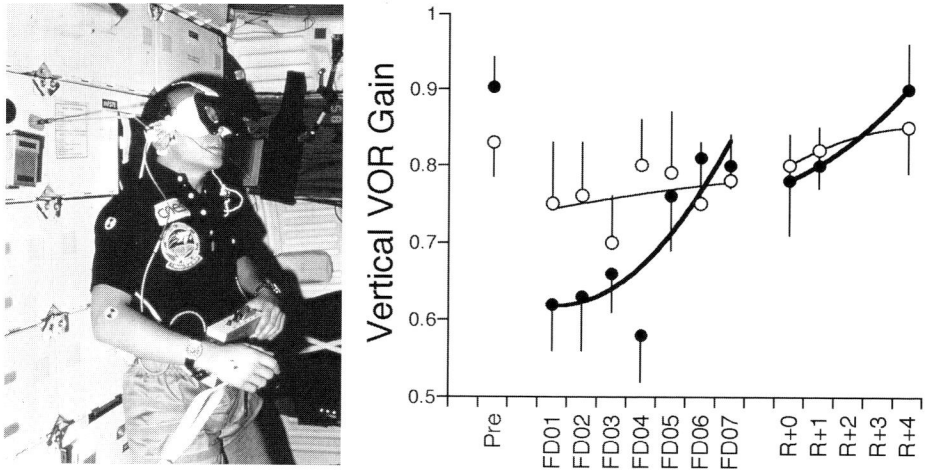

Figure 6-04. Mean and standard error of vertical VOR gain (ratio of eye velocity and head velocity) elicited by voluntary head movements in pitch at 0.2 Hz while imagining a wall-fixed target, averaged in two subjects before, during and after a seven-day space flight. The photograph on the left shows one astronaut performing this experiment during STS-51G. Photo courtesy of NASA.

The decrease in vertical VOR gain during space flight can been attributed to the absence of the otolith contribution during pitch head tilt in microgravity. Indirectly, this reduction may also be caused by the potential impairment at imagining a wall-fixed target in darkness in microgravity in the same manner as on Earth. It is well known that the VOR gain is strongly dependent on cognitive processes and that the gravitational reference is used for the internal representation of the environment (see Chapter 7, Section 4.4). In absence of this reference in microgravity, the subjects may experience difficulties in memorizing or representing the imagined target.

2.4.3 Torsional VOR

During voluntary head movements in roll, torsional VOR gain increased in two cosmonauts during a short-term mission, but decreased over a six-month stay in a further two cosmonauts (Clarke *et al.* 1993). Also, discrepancies between intended and performed head movement were observed in-flight. Although the subjects were convinced they were performing pure roll head movement, there were clear components of both roll and yaw motion, indicating an impairment in sensorimotor coordination after pro longed exposure to weightlessness. Interestingly, this combined roll-and-yaw head movement did not elicit a combination of torsional and horizontal eye nystagmus, but rather a combination of torsional and vertical nystagmus. After long-duration space flights, the gain of torsional VOR was initially reduced in the postflight phase, but was observed to return to preflight values over a two-week period (Clarke 2006).

Torsional passive VOR was also elicited in two astronauts on the seventh day of STS-42 Shuttle mission by means of angular velocity steps in roll delivered by a servo-controlled rotating chair. The torsional VOR component was found to be reduced in both astronauts in-flight, compared to preflight measurements, but a large horizontal component of nystagmus was present, suggesting a cross coupling from the roll to the yaw eye rotation axis in microgravity. Following the STS-11 Shuttle mission, three astronauts were tested using a dynamic roll stimulation about an Earth-horizontal axis (Figure 6-05, left). This 30-deg peak-to-peak roll stimulation elicited greater horizontal nystagmus in tests conducted 70-150 min after landing on three subjects compared to tests conducted preflight or later postflight (Harm *et al.* 1993).

The gain of torsional VOR was also found to be significantly reduced in two monkeys tested after the twelve-day Kosmos-2229 flight (Dai *et al.* 1994). In this experiment the passive torsional VOR was measured in two experimental paradigms. First, the monkeys were rotated at constant velocities about a vertical naso-occipital axis while prone and the chair was suddenly stopped. And second, they were sinusoidally oscillated at frequencies ranging from 0.025 to 0.125 Hz about a horizontal axis while upright. Between preflight and postflight, the gain of the torsional VOR was reduced in both modes of stimulation on average about 50% in one animal and about 15% in the other animal. This postflight decrease in roll VOR gain was still visible when monkeys were tested 11 days after landing. This postflight maintenance of reduced roll VOR gain in monkeys differs from the response behavior in humans, which clearly re-adapts to values comparable with baseline over a similar period. The reduction in the torsional VOR in monkeys postflight is unexplained, and simultaneous measurements of the other components of eye movements i.e., horizontal and vertical during the roll study are not reported in the paper. The authors evoke the possibility that their restrained monkeys experienced fewer low frequency roll head movements in space, compared to the ground conditions, resulting in depression of the canal-ocular roll reflex over the course of space flight (Dai *et al.* 1994).

The torsional VOR was also studied during passive roll tilt in tadpoles of the clawed toad *Xenopus laevis* following three 7-12 day space flights (STS-55 in 1993,

STS-84 in 1997, ISS taxi flight in 2001). The objective was to compare the responses across animals at different developmental stages. At onset of microgravity, tadpoles were at stages 25-28, 33-36, or 45. The torsional VOR of tadpoles from the groups 25-28 and 33-36 was not affected by microgravity, while the torsional VOR of tadpoles from the stage-45 group revealed a significant augmentation. The authors suggest, "after a critical status of vestibular maturation obtained during the appearance of first swimming, microgravity activates an adaptation mechanism that causes a sensitization of the vestibular system" (Horn 2006).

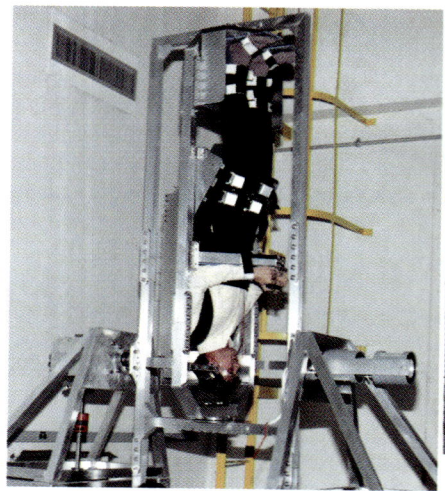 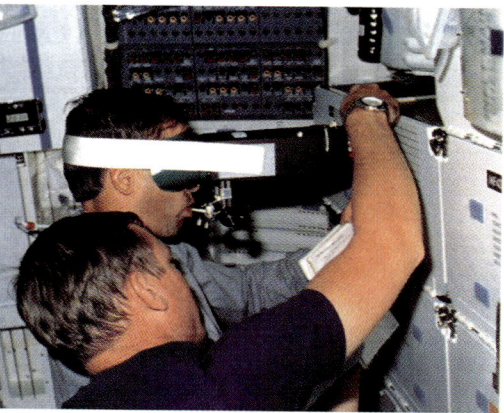

Figure 6-05. Left. The NASA JSC Pitch-and-Roll Device *provided passive pitch or roll (seen here) simulation about the subject's head y- or x-axis, respectively. The body was retrained in the upright position. A light-tight shroud covered the entire system. Self-motion perception and eye movements were recorded during sinusoidal oscillations ranging from 0.1-0.4 Hz or continuous rotation. Some Shuttle astronauts were tested pre- and postflight with this device. Right. Two astronauts during a test of ocular counter-rolling on board the Space Shuttle. The subject viewed an inversed "T" that appeared as an after-image on the retina. Measurement of eye torsion was accomplished by then matching the retinal image with a reference target. Photos courtesy of NASA.*

2.4.4 Ocular Counter-Rolling

In terrestrial conditions, *ocular counter-rolling* (OCR) is an otolith-driven orienting eye movement that is generated during head roll tilt. This reflex, presumably activated by the shear force exerted by the component of gravity along the maculae of the otolith utricular organs, tends to maintain the retinal meridian in a vertical orientation. The ocular response consists of a small, torsional conjugate eye movement opposite to the direction of the static head roll. This reflex is presumably activated by the shear force exerted by the component of gravity along the maculae of the otolith utricular organs. It is virtually absent in individuals without functioning otolith organs.

Because of its strong dependence on the otolith organs, this reflex has been used in many postflight studies to gauge the effect of microgravity exposure on otolith function. One problem is that OCR compensates for only about 10 to 20% of static head roll tilt in humans, with large inter-individual differences, and its amplitude rarely exceeds 8-10 deg (Collewijn *et al.* 1985).

In microgravity, static head tilt does not stimulate the otoliths and OCR is not produced by statically rolling the head on the neck. This result was confirmed in six astronauts on STS-24 through 26 and STS-28 (Figure 6-05, right). Unlike the preflight results (recorded in both the upright and supine positions), little or no ocular torsion was observed in-flight (Reschke *et al.* 1991).

Because of the virtual absence of this reflex in-flight during head tilt (but not during head translation), the amplitude of OCR was expected to be lower after space flight compared to preflight. However, the findings of studies using static whole body tilt are inconsistent. Some studies report decreases in astronauts' OCR after the flight relative to preflight, while others have shown postflight increases of OCR or no changes at all (Figure 6-06). When averaged across all the pre- and postflight studies on OCR, from Gemini to Shuttle missions (Graybiel *et al.* 1967, Reschke *et al.* 1985, Vogel & Kass 1986, Hofstetter-Degen *et al.* 1993, Diamond & Markham 1998, Young & Sinha 1998, Moore *et al.* 2001) the difference between pre- and postflight OCR measurements was found to be less than 0.6 deg for body tilt angles ranging from 15 to 45 deg. The inconsistency in results across studies may be due to the various experimental procedures employed, including flash afterimages, flash photography of the eyes, or video-oculography. Another problem is related to the high variance of the OCR across individuals.

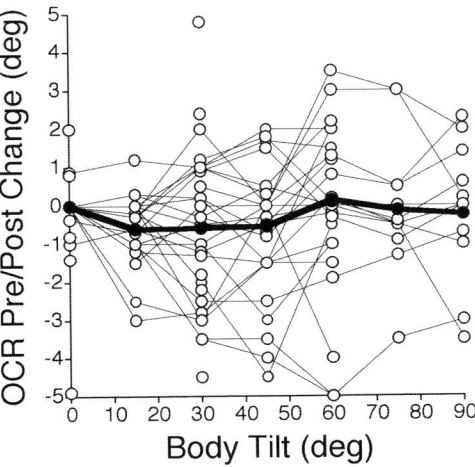

Figure 6-06. Differences between pre- and postflight ocular counter-rolling (OCR) amplitude in 18 astronauts (open symbols: individuals; closed symbols: average) during static whole body tilt. There is a trend for a small (< 1 deg) decrease in OCR amplitude postflight relative to preflight, but the results are inconsistent. The fact that OCR does not change after space flight indicates that the otolith end organs function normally after short-duration weightlessness.

In the Russian space program, all cosmonauts returning from long-duration missions (30 to 175 days) and 14 of 18 cosmonauts returning from short flights (four and seven days) exhibited changes in OCR when tilted to the right or left of vertical one to two days postflight. Sixteen of these subjects demonstrated increased OCR while only two demonstrated a decrease (one following a long-term and one after a short-term flight). These tests, however, were not conducted immediately after landing. Of the eight subjects tested on landing day following short flights, four showed no change in OCR while the other four demonstrated inappropriate torsional eye movements in the direction of the head movements. In general, responses returned to normal 2-4 days following short flights and 8-10 days after long flights (Yakovleva *et al.* 1982, Kornilova *et al.* 1987).

2.4.5   Linear Acceleration

Transient lateral translation of the head stimulates the utricles and evokes torsion of the eye in the opposite direction. This response has been compared pre- and postflight. Two subjects exposed to transient lateral accelerations three to five hours after the landing of Spacelab-1 demonstrated smaller torsional amplitudes than three of the four preflight measures. Torsional amplitude in these subjects steadily increased over most of the postflight tests. The changes noted were not statistically significant because of high variability in the preflight measurements (Arrott & Young 1986). Passive y-axis linear translation also enhanced horizontal eye movements in two astronauts on the second and third day after landing of the STS-11 Shuttle mission (Parker *et al.* 1986). Further tests were performed on a linear sled after the Spacelab-D1 and the SLS-1 and SLS-2 missions, but here too the results were not conclusive. Subjects were accelerated sinusoidally at 1.0 Hz and 0.25 Hz with a peak acceleration of 0.5 g and with a series of low acceleration steps. Horizontal eye movements during y-axis lateral acceleration were found to have a gain that was frequency dependent, but no clear changes in gain were observed postflight compared to preflight (Young *et al.* 1993).

In 1998, a human-rated centrifuge flew on the sixteen-day Neurolab mission (see Figure 4-15, right), which allowed the exposure of crewmembers to sustained linear acceleration of 0.5 g and 1 g for several minutes per trial. An ocular torsion was generated by the otoliths in response to the tilted gravito-inertial acceleration (or just the centripetal acceleration in orbit). During the Neurolab mission, a mere 10% decrease in torsion magnitude during 1-g centrifugation was measured in microgravity compared to Earth. Moreover, the amplitude of torsion was roughly proportional to the applied interaural linear acceleration, with a magnitude during 0.5-g centrifugation approximately 60% of that generated during 1-g centrifugation (Moore *et al.* 2001). The Neurolab studies also showed no reduction in postflight torsion magnitude compared to preflight values.

Before this experiment, it had been stated that weightlessness would be equivalent to a functional deafferentation of the otolith organs. The fact that otolith-driven eye movements, such as the ocular torsion, are largely unchanged during and after space flight denies such hypothesis and changes considerably our view of the vestibular function in weightlessness. This result also suggests that torsion level is primarily generated by utricular units with polarization vectors along the head interaural axis. Following this study, it has recently been shown in the cat that tilts of the body are reflected in vestibular nuclei activity in the absence of the labyrinths (Yates & Miller 1996). This may provide a mechanism for the integration of somesthetic and otolithic input for the generation of ocular torsion.

A follow-up study, which will compare the torsion eye movement in response to unilateral stimulation of the otoliths by linear acceleration, is in preparation before and after longer duration space flights. This experiment uses a centrifuge where sitting subjects are displaced minimally from the rotation axis, so that one labyrinth becomes aligned on-axis, while the second labyrinth alone is exposed to the centripetal acceleration (Clarke & Engelhorn 1998). This technique will allow the investigation of both otolith-dependent eye movement and perceptive responses during unilateral stimulation of the otolith organs.

2.4.6   Off-Vertical Axis Rotation (OVAR)

When subjects are rotated in yaw about a rotation axis that is tilted relative to the direction of gravity (see Figure 4-13, left), a stimulation referred to as *off-vertical axis rotation* (OVAR), the semicircular canals of the vestibular system will initially sense the rotation, but their activity will die out following an exponential decay (Guedry 1965). The otolith organs, however, will be stimulated continuously by a rotating gravity component, which induces a sinusoidally varying linear stimulus along the utricular macula. The frequency of these sinusoidal variations in shearing force is proportional to the rotation velocity, whereas their amplitude is proportional to the tilt angle. Compared with static head roll tilt, OVAR thus presents the advantage of generating a continuous sinusoidal modulation of ocular torsion, allowing a more robust computation of mean response across several cycles. Also OVAR at constant velocity at low angle of tilt has been shown to induce a percept of head sway around a cone, hence a sense of roll tilt, which persists for as long as the rotation continues (see Chapter 4, Section 5.4).

Two monkeys tested 15 and 18 hours after landing following a two-week space flight showed an increase in the modulation of the slow phase eye velocity of nystagmus induced by OVAR relative to preflight (Cohen *et al.* 1992), which is attributed to a change in the activity arising in the otolith organs as a result of changes in head position relative to gravity. However, these monkeys were tested with OVAR at tilt angles ranging from 20 deg to 90 deg, and the increase in modulation was only significant for tilt angles higher than 30 deg. The modulation of vergence eye movements during OVAR, which might be related to the stimulation of the otoliths by the forward and backward pitch motion, was decreased postflight in the same animals (Dai *et al.* 1996).

In humans, the modulation in horizontal eye velocity and position during OVAR at tilt angles ranging from 5 deg to 15 deg was not altered in two astronauts tested 32 hours after a seven-day space flight (Clément *et al.* 1995). More recently, we investigated eye torsion in response to OVAR at 10-deg and 20-deg tilt in seven astronauts before and immediately space flight. A distinct sinusoidal modulation of torsion eye position was generated in response to the sinusoidal changes in head position relative to gravity during OVAR at 45 deg/sec. However, there was no significant change in this torsion eye movement during OVAR immediately after landing and on subsequent postflight test days, compared with preflight values (Clément *et al.* 2007).

It is interesting to note that the motion perception of astronauts when exposed to linear translation, centrifugation, or OVAR (see Chapter 4, Section 5.4) is fundamentally different postflight compared to preflight, whereas the eye movements, in particular torsion, are not. This dissociation between otolith-driven eye movement and perception during passive vestibular stimulation after space flight suggests that eye movements and orientation perception are governed by qualitatively different neural mechanisms. Ocular torsion is primarily a response of otolith activation by low-frequency linear acceleration along the interaural axis, whereas perception of tilt is primarily governed by the integration of graviceptive cues, including somesthetic, presumably centrally processed through neural models of the physical laws of motion.

The peripheral vestibular organ would experience little or no changes after short-duration space flight, but the central processing of graviceptors inputs and the outputs of internal models for spatial orientation are likely to be affected. This dissociation would explain why otolith-driven eye movements appear relatively unaffected by microgravity, while perceptual and oculomotor responses depending on central vestibular processing can be greatly disrupted. Changes in oculomotor responses such as smooth pursuit and saccades are described in Section 4 below.

3   OPTOKINETIC SYSTEM

3.1   Background

Rotatory nystagmus produced by the VOR in darkness decays over a 30 to 45-sec period. In the light, however, eye movements are maintained by the visual drive of the optokinetic system. This *optokinetic nystagmus* (OKN) increases as the VOR response decreases thus preserving image stability. Gain for optokinetic stimulation is the ratio of eye velocity to stimulus velocity and is typically about 0.8. Vertical OKN gain is lower than horizontal OKN 54gain and exhibits directional asymmetry with most subjects displaying a higher gain for upward movement of the stimulus (Leigh & Zee 1991, Cohen *et al.* 1977).

The optokinetic system continues for some time after the visual stimulus is removed. The response, called *optokinetic after-nystagmus* (OKAN), may be controlled by the same velocity storage mechanism as the VOR apparent time constant (Raphan *et al.* 1979, Waespe & Henn 1979). OKAN is described by initial eye velocity, time constant of the decaying slow-phase velocity, cumulative slow-phase eye position (sum in degrees of slow phases), and directional symmetry. Measures of initial eye velocity and time constant show a great deal of intra-subject variability. OKAN responses are subject to habituation, usually absent in the vertical plane (although sometimes observed following upward movement of the stimulus), decreased with age, altered by changes in head position (similar to the effects of head movements on post-rotatory nystagmus), and suppressed by visual fixation of a target during optokinetic stimulation but not by visual fixation after stimulation ceases (Leigh & Zee 1991).

As for the vertical VOR, otolith stimulation is also known to affect the closely related vertical optokinetic responses. The effect of otolith inputs on vertical OKN and OKAN was demonstrated by Igarashi *et al.* (1978), who showed that selective macular ablations increased slow phase eye velocity of vertical OKN in squirrel monkeys. In addition, Clément & Lathan (1991) and Gizzi *et al.* (1994) have demonstrated that the velocity storage mechanism for vertical OKN and OKAN is different when the subject is onside than when upright, implying an otolith input to that mechanism as well.

3.2   Microgravity Investigations

There is evidence that linear acceleration can modify nystagmic responses to optokinetic stimulation. However, removal of the gravitational input is necessary to determine the exact contributions of the otolith organs to optokinetic reflexes (Clément & Berthoz 1988).

3.2.1 Optokinetic Nystagmus

Pre- vs. postflight changes in optokinetic nystagmus induced by a rotating drum were not observed in tests conducted early in the Space Shuttle program (Thornton *et al.* 1985). However, two cosmonauts were reportedly unable to track a horizontal optokinetic stimulus moving at 80 stripes/min in tests following 75 days in space, and demonstrated asymmetry at speeds of 40 and 60 stripes/min (Matsnev *et al.* 1983). Suppression of eye movements during optokinetic stimulation as well as asymmetry and alteration of normal nystagmic components were also observed immediately after flight on board Salyut-6 and 7 (Gorgiladze & Maveyev 1990). However, no in-flight changes were noted in response to vertical and horizontal optokinetic stimulation (6 deg/sec) with the exception of a decrease in the amplitude of vertical OKN early in-flight (Kornilova *et al.* 1991, 1992).

OKN data obtained during early space missions were limited to a few subjects, and were recorded using EOG, which is not a very reliable technique for measuring vertical eye movements. Also, the visual optokinetic stimulators had a small, monocular field of view. More recently, horizontal and vertical OKN data have been recorded using wider optokinetic displays (Figure 6-07, left) and video-oculography.

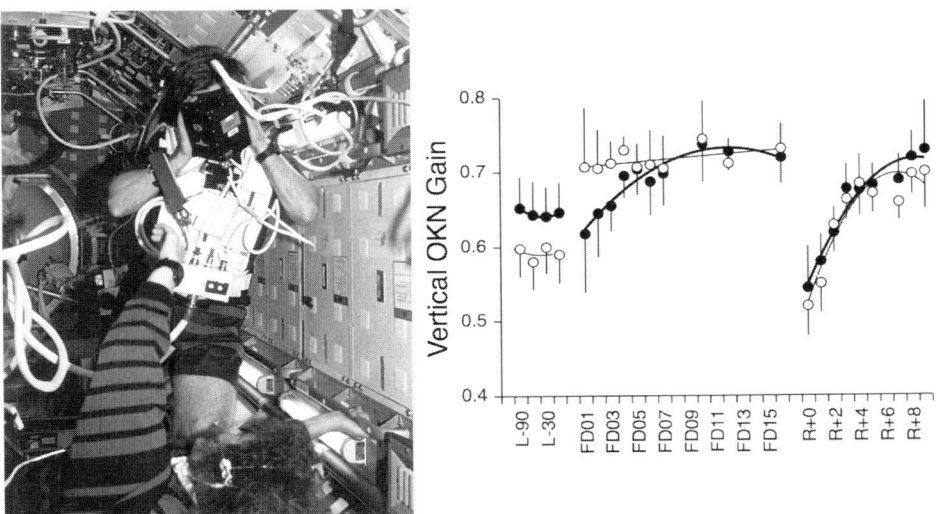

Figure 6-07. Left. A free-floating astronaut is looking into a binocular optokinetic display made of vertical stripes moving at constant speeds ranging from 18 to 54 deg/sec in the horizontal, vertical or oblique directions. In this picture the astronaut holds his head tilted over one shoulder while watching horizontal optokinetic stimulation relative to his head, as part of an experiment aimed at studying the effect of head tilt on the spatial orientation of the velocity storage mechanism. Photo courtesy of NASA. Right. Mean and standard error of the vertical optokinetic nystagmus (OKN) gain, measured as the ratio between slow phase velocity and stimulus velocity before (L-), during (FD) and after (R+) space flight in ten subjects during stimulus velocity ranging from 20 to 80 deg/sec. On Earth, the gain of the OKN with slow phases directed upward (filled symbols) is larger than the OKN with slow phases directed downwards (open symbols). Note the inversion in this asymmetry early in-flight and the trend towards symmetry after two weeks in space and immediately after return to Earth.

The mean OKN slow phase velocity measured in ten subjects during four space missions showed a slight increase in the horizontal OKN gain for both directions of stimulation throughout the repetition of the test in-flight and postflight (Clément *et al.* 1986, 1993, 2003). This increase was more likely due to a training effect, rather than the consequence of adaptive changes to microgravity and re-adaptation to Earth gravity. An increase in the vertical OKN gain was also observed with the repetition of the test. However, at the beginning of the flight, this increase was much larger for the downward slow phase velocity. The downward slow phase velocity was faster than preflight; whereas the upward slow phase velocity was basically unchanged early in flight. Consequently, the normal (terrestrial) up-down asymmetry of the vertical OKN gain, with upward optokinetic stimulation eliciting higher gain response than downward optokinetic stimulation, was reversed during the first three days of flight. No asymmetry was observed on the subsequent flight days. The preflight gains and up-down asymmetry were restored after two weeks after landing (Figure 6-07, right). A close look at the eye position during OKN also revealed that the nystagmus beating field, i.e., the mean eye position of gaze, was displaced upward early in-flight (Clément 2003).

The otolith inputs could also have a role in the vertical OKN asymmetry: in the presence of gravity, they would exert an upward drive on the eye movements (Clément *et al.* 1986), perhaps as the result of the extensor muscle tone generated by the labyrinthine reflex (see Figure 1-10). Several ground-based studies support this hypothesis. For example, in squirrel monkeys, macular and vestibular nerve lesions induce an improvement of the vertical OKN slow phase eye velocity (Igarashi *et al.* 1978), and bilateral sacculectomies affect vertical OKN asymmetry and nystagmus beating field (Igarashi *et al.* 1987). The vertical asymmetry has been shown to dominate in the optokinetic system, which involves subcortical pathways (Murashugi & Howard 1989) and relays activity to the eye velocity storage mechanism. Changes in head position relative to gravity, and thus changes in otolithic information, modify vertical OKN (Matsuo & Cohen 1984). A reversal asymmetry effect (downward slow phase eye velocity greater than upward slow phase eye velocity) has previously been observed in one subject with the head declined 30 deg below horizontal and during parabolic flight (Clément & Lathan 1991, Wei *et al.* 1997).

However, rather than a direct influence of the otoliths on the nystagmus, another interpretation has been proposed for the changes in vertical OKN asymmetry. The overall effect both during head tilt relative to gravity and space flight may be related to otolith-dependent changes in eye position, which affect slow phase velocity according to *Alexander's Law* (Lackner & DiZio 2000). Alexander's Law is based upon the observation that slow phase eye velocity increases as gaze is displaced in the direction of the nystagmus fast phase and diminishes with gaze in the opposite direction. The changes in the vertical OKN beating field seen during space flight are in agreement with this interpretation. An upward displacement of the beating field would result in both an increase in the downward slow phase velocity and a decrease in the upward slow phase velocity, which would invert the original up-down asymmetry.

Changes in the preferred vertical direction of gaze, i.e., the *straight-ahead direction*, could also be at the origin of the reversal in vertical VOR gain asymmetry in microgravity, according to Alexander's Law (see Figure 6-04, right). Unfortunately, all of the space flight experiments to date have focused on measurements of eye *velocity*, but few have investigated actual changes in eye *position*. Further studies are needed to confirm that, as suggested by the above interpretation, the unloading of the otoliths in microgravity induces an upward deviation in static eye position, which in turn could alter the symmetry of vertical nystagmus.

3.2.2 Spatial Orientation of Eye Movements

Optokinetic stimulation effectively acts as a velocity storage mechanism, for after the lights are turned out, OKAN continues in the same direction for some seconds, particularly in monkeys, with a declining slow phase velocity. The time constant of decay of OKAN is similar to the dominant time constant of the VOR (Raphan *et al.* 1979) and is a direct reflection of the time constant of the velocity storage mechanism.

Ground-based studies have shown that the direction of the OKN slow phase velocity in humans (Clément & Lathan 1991, Gizzi *et al.* 1994) and OKAN in primates (Dai *et al.* 1991) is strongly affected by head position with respect to gravity. When subjects or animals are tilted laterally, the direction of the eye rotation axis during OKN shifts gradually from the axis of the visual stimulation towards the gravitational vertical (Figure 6-08). In monkeys where the velocity storage is even more effective, the eye rotation axis during OKAN also shifts during head roll tilt and tends to align with gravity. The direction of OKN and OKAN slow phases can therefore be regarded as a representation of the spatial orientation of the velocity storage mechanism.

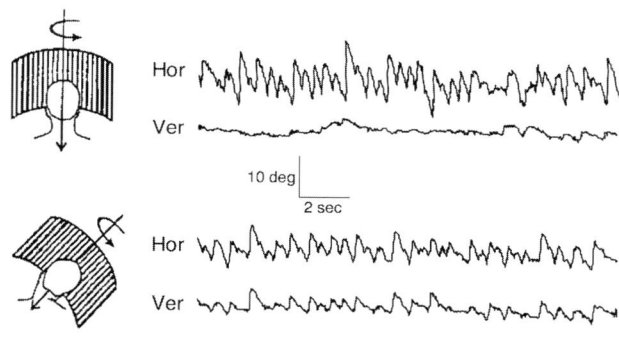

Figure 6-08. Horizontal (Hor) and vertical (Ver) components of eye movement during horizontal optokinetic stimulation in a subject with the head upright (upper traces) and roll tilted over the trunk (lower traces) on Earth. When the head is tilted, the optokinetic nystagmus is oblique, indicating that the eye rotation axis during OKN tends to align with gravity.

On Earth, the velocity storage uses all linear acceleration, including gravity, for its orientation and the orientation of the OKN and OKAN responses. It is hypothesized that this orientation disappears in microgravity and that a new reference, such as the longitudinal z-axis of the head or body, could be used (Dai *et al.* 1994). In support for this hypothesis, we showed that when astronauts are exposed to horizontal optokinetic stimulation while free-floating in space even when they tilt their head to one side, the eye rotation axis during OKN stays aligned with the head vertical axis (Figure 6-09). This result indicates that in microgravity the orientation of the velocity storage mechanism moves from an allocentric, gravity-referenced frame to an egocentric, head-referenced frame. When roll tilt positions were compared before and after flight, the vertical component of OKN, which occurred preflight during horizontal OKN in roll tilt position, did not occur early postflight. This result suggests that the gravitational orientation of the velocity storage mechanisms was lost immediately after space flight.

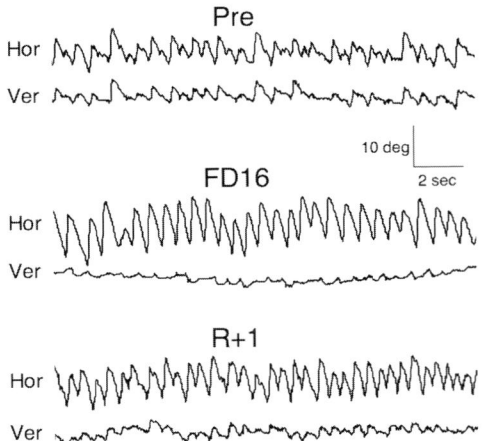

Figure 6-09. Comparison between recordings obtained during horizontal optokinetic stimulation with the head tilted in roll preflight (Pre), in-flight (FD16) and one day after landing (R+1). During the flight when the head was tilted the OKN was purely horizontal. An oblique OKN, composed of both horizontal and vertical eye movements, begins to reappear one day after return of the astronaut to Earth.

A confirmation of this finding was obtained during a follow-up study. When astronauts were exposed to a 1-g centripetal linear acceleration along the interaural y-axis by means of the Neurolab centrifuge, the eye rotation axis was tilted in the direction of the linear acceleration vector during horizontal OKN (Moore *et al.* 2005). So, in presence of a steady-state linear acceleration in microgravity, such as during centrifugation, the orientation of the velocity storage mechanism utilized an artificial-gravity referenced frame.

Another evidence from a shift of the velocity storage from gravitational spatial to an egocentric reference frame after adaptation to microgravity has been obtained in one monkey returning from an eleven-day space flight (Dai *et al.* 1994). Before flight, when the animal was tilted on its side by 90 deg, the axis of eye rotation during OKAN following horizontal (with respect to the head) optokinetic stimulation was only tilted by 5 deg with regard to the gravitational vertical. One day after landing the same test yielded an orientation angle of 28 deg with respect to the gravitational vertical, revealing a significant shift toward the head or body z-axis. At seven days after recovery, the angle of the eye rotation axis had returned to 7 deg relative to gravitational vertical, indicating that it was again closely aligned with gravity (Dai *et al.* 1994). In humans, cross-coupled OKAN from horizontal to vertical was observed after horizontal optokinetic stimulation on the first day of exposure to microgravity in one subject, but not on later days (Clément & Berthoz 1990). In another subject during a longer flight, the vertical OKAN peak slow-phase velocity was increased at the end of the flight relative to preflight measures (Clément *et al.* 1993)

Recently, ground-based studies have shown that the spatial organization of all eye orientations during visually-guided saccadic eye movements, known as the *Listing's plane*, varies systematically as a function of static and dynamic head orientation in space. Listing's plane is defined as the equatorial plane of the eye when the eye is in primary position, i.e., when the line of sight is perpendicular to the plane on which the axes of ocular torsion lie (Leigh & Zee 1999). When the subject is stationary with the head upright the rotation axes of the eye for changes in the line of sight are confined to Listing's plane (Hess & Angelaki 2003). Listing's plane has shown to be tilted in some subjects and the angle of tilt varies between subjects (Halswanter *et al.* 1994).

Using a state-of-the-art binocular video-oculographic eye movement measuring system (Figure 6-10), Clarke and his collaborators are currently investigating the potential changes in the orientation of Listing's plane during visually guided saccadic eye movements with different static head tilts and with or without visual input in a weightless environment. Their hypothesis is that in microgravity the orientation of Listing's plane is altered, probably to a small and individually variable degree. Further, with the loss of the otolith-mediated gravitational reference, it is expected that changes in the orientation of the coordinate framework of the vestibular system occur, and thus a divergence between Listing's plane and the vestibular coordinate frame should be observed. Preliminary results obtained during ISS Expeditions 9-13 indicate a backward tilt of Listing's plane in all subjects tested in microgravity. There is a return to preflight values during the first two weeks after landing (Clarke & Haslwanter 2007).

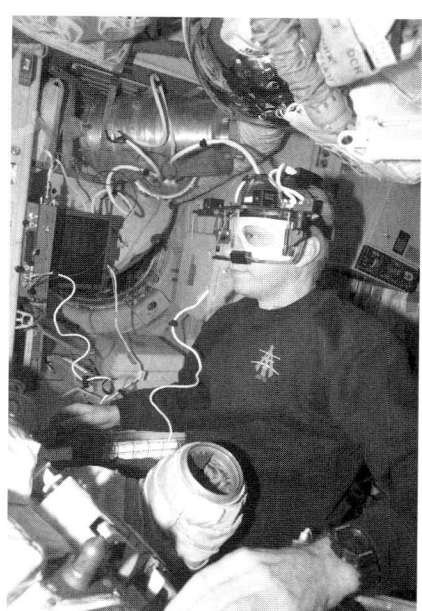

Figure 6-10. Video-oculography system developed by the German Space Agency (DLR) for the International Space Station. This system offers accurate three-dimensional eye-in-head measurements (< 0.1 deg spatial resolution, 200 Hz sampling frequency), using integrated, intelligent cameras. The customized facemask provides a comfortable fit that eliminates head slippage. Photo courtesy of ESA.

4   GAZE, SACCADES AND SMOOTH PURSUIT

The VOR and OKN are field-stabilizing reflexes. However, the eyes can also move alone to fixate on a small object of interest without moving the head, acquire visual targets, or follow a target that is moving relative to a fixed background. The effects of space flight on these gaze holding, saccades and smooth pursuit responses are described thereafter.

4.1   Gaze Holding

One measure of spatial localization shared by the astronauts and those suffering from cerebellar disorders that is easily quantified and for which a neurobiological substrate has been identified, is the control of the *angle of gaze*. In order for our visual perception of an object in the environment to be clear and spatially accurate, we must aim and hold the line of sight (gaze) on the object of interest. When this is achieved, the image of the object lies on the foveal region of the retina, i.e., the area of highest density of photoreceptors. Holding a stable gaze not only provides the best visual acuity, but also influences the perceived spatial localization of objects. If images drift away from the fovea, they will be seen less clearly and localized less accurately in space. This is the case in patients with cerebellar disease, who often complain that they cannot clearly see or localize objects that are eccentrically placed in their environment (Leigh & Zee 1999).

To maintain stable eccentric gaze, the CNS must be capable of performing appropriate neural integration of a velocity command signal to generate and maintain an eye movement of position. The position command is generated presumably through a neural integrator lying in the brainstem and the cerebellum, since lesions here cause impaired gaze holding (Zee *et al.* 1981). Lesions of the neural integrator usually cause it to become leaky, so that the eye drifts back to the central position with a negative exponential waveform, leading to *gaze-evoked nystagmus*.

A common clinical finding in patients with cerebellar disease who show deficient gaze-holding is *rebound nystagmus*, i.e., a transient nystagmus following a prolonged attempt at eccentric gaze (with slow phases toward the direction of prior gaze) that occurs after the patient returns the eye to central position. One possibility is that the generation of rebound nystagmus depends upon the ability to internally monitor eye movement signals (i.e., through efference copy or extra-ocular muscles proprioception) and activate compensatory eye drifts. One prediction of such a mechanism is that the slow phases of rebound nystagmus might show increasing-velocity waveforms.

Data recently obtained after space flight support this hypothesis. Figure 6-11A shows representative records from a cosmonaut three days after a flight of approximately 96 days. It is evident that when the cosmonaut looks either right or left (up and down directions on the figure), he develops centripetal drifts of the eye with corrective quick phases, called a *gaze-evoked nystagmus*. This nystagmus is not present when the eyes were held in the central position prior to the rightward shift in gaze (see beginning of record) and the overall gaze-evoked nystagmus was reduced five days later regardless of the shift in gaze. Also, after returning towards the center position following the rightward shift in gaze, the eye drifts back to the right, with centripetal quick phases, typical of a *rebound nystagmus*.

Downbeat nystagmus during lateral gaze has also been encountered in astronauts and cosmonauts postflight (Viéville *et al.* 1986, Kornilova *et al.* 1983). These data suggest that a sustained change in graviceptor inputs may cause the gaze-holding mechanism (neural integrator) for horizontal eye movement to become leaky, and the gaze-holding mechanism for vertical eye movements to develop an imbalance (causing upward eye drifts). The rebound phenomenon, being a product of a more generalized motor outflow control that is related to adaptive strategies, is presumably not confined to eye movements alone. A better understanding of this phenomenon may lead to development useful on-orbit and postflight adaptation, as well as the evaluation and development of new countermeasures for astronauts and rehabilitative strategies for patients with cerebellar disease.

4.2   Orienting Gaze

Typically an orienting gaze movement initiated to bring a selected part of the visual world onto the fovea consists of an eye movement saccade and a head movement followed by a reflexive compensatory eye movement driven by VOR. An experiment was specifically designed to investigate the astronauts' ability to perform saccades, smooth pursuit and VOR, and was repeatedly used on board Space Shuttle missions between 1991 and 1996 as part of the NASA *Extended Duration Orbiter Medical Project* (EDOMP). This equipment included portable devices to assist in recording head and eye movements, and a special goggle that instantaneously occluded the subject's vision (Figure 6-11, left).

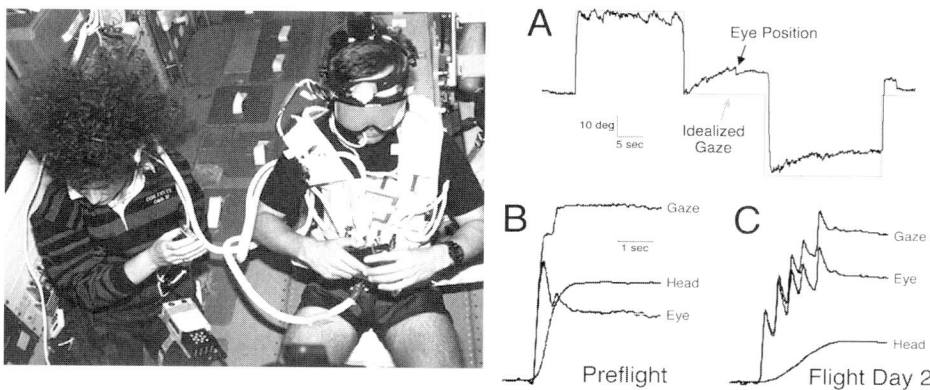

Figure 6-11. The photograph on the left shows an operator and a subject wearing an electronic occluding goggle during gaze tests on board the LMS mission. A. Eccentric gaze holding in one astronaut three days following a 96-day space flight. The subject was instructed to look at imagined targets in darkness located successively at 30 deg to the right, center, and 30 deg to the left. There was a clear shift of gaze from the centered and eccentric positions. B. In the usual sequence of acquiring an eccentric visual target, a saccade directs the eye towards the target when the angular displacement exceeds either the physical or physiological limits of eye rotation. The head being a larger mechanical object with greater inertia compared with the eyeball typically moves after the eye has moved in the orbit. The head movement excites the semicircular canals and produces an eye movement through the VOR that is opposite in direction and velocity to that of the head. The compensatory VOR returns the eye to the primary straight-ahead position in the skull's orbit, exchanging the head's final angular position for the initial eye saccade. C. During the flight, there was a consistent trend for the head movement to the target to be delayed. When targets appeared in the vertical plane, the subject for these trials used mainly the eyes to attempt acquisition of the target by means of multiple saccades prior to reaching final gaze position.

During the *target acquisition* tasks, the subjects were required, using a time optimal strategy, to look from a central fixation point to a specified target as quickly and accurately as possible using both the head and eyes to acquire the target. Targets were presented on a cruciform display that was fixed to the Space Shuttle's mid-deck lockers. During separated tests, *pursuit tracking* studies were designed to measure the effectiveness of both smooth pursuit eye movements and combined eye-head tracking in acquiring and holding gaze on a moving target. Finally, *gaze stabilization* studies tried to characterize the VOR while the subject consciously attempted "visual" fixation at a just-viewed and memorized wall-fixed target.

During these studies, data collection took place before, during and after the flight. Interestingly, at the end of the mission, the trials for gaze stabilization began at the Space Shuttle's re-entry interface and continued non-stop until 5 min had elapsed or the Shuttle had landed. Following Shuttle wheels stop the trials, patterned after those accomplished during re-entry, were performed for another 5 min. The re-entry and wheels stop protocols were difficult because the head movements were performed inside of the astronaut's helmet. As a consequence, only a few subjects were tested with this protocol. Results from these series of tests showed that there is degradation in the astronauts' ability to acquire targets with the head and eyes, and to pursue moving targets, even when the location of these targets is known, and the acquisition or pursuit process has been practiced and rehearsed. The timing and accuracy of target capture is particularly degraded when the object to be acquired is outside of the central field of view (i.e., offset from center by 60 deg) and is located in the vertical plane thus requiring a pitch head movement for target acquisition.

4.2.1   Target Acquisition

The panels in Figure 6-11B and C illustrate the acquisition of a target beyond the effective oculomotor range in the vertical plane. It is interesting to note that the subject for these trials used the eyes to attempt acquisition of the target. This can be seen clearly in the preflight trial. The eye moves prior to the head and gaze is established with the eye's position. Once the head begins to move, the VOR is established and the reflex pulls gaze off of the target. Both the head and a corrective eye saccade are then used to maintain gaze (Figure 6-11B).

During space flight a different strategy is developed. The eye is still used to establish gaze, but the head movement is greatly reduced in both velocity and displacement. Of particular interest in this example is the number of saccades made by the eyes and the velocity of these saccades (Figure 6-11C). They do not represent a typical VOR response. Rather they show a considerably higher gain than normal. The responses early after flight show most of the strategy components developed during the flight (i.e., attainment of target with eyes, low head velocity, and multiple saccades). A return to preflight levels is observed by the fourth day postflight.

4.2.2   Gaze Stabilization

Another approach to characterization of the ocular stabilization of a stationary target was investigated using a gaze stabilization paradigm, which was composed of the following steps:

   a.   The subject first visually fixates a wall-fixed target with head in a central position.
   b.   When the goggles become opaque and vision is occluded, the subject rotates the head while maintaining ocular fixation on the just-seen wall-fixed target.
   c.   When the goggles become clear the subject refixates on the target (if necessary) with the eyes only.
   d.   Rotate the head back to center, keeping eyes on the target.

Landing day data indicates that the CNS has developed strategies to compensate for vestibular control of target capture and pursuit tracking. Typically, the compensatory response is to limit head movements and attempt to capture the target with the eyes only. When this is not possible, a smaller than normal head movement is initiated too late in the sequence to provide the necessary accuracy and speed for target acquisition. Together the head and eye movements are inadequate and several additional shifts of the eye are necessary to place the target properly on the retina. This can result in significant delays (up to 1.5 sec) before the target is acquired. During the phase of re-entry where the change in gravitational forces was the greatest, there did not appear to be an adequate VOR for the head movement in both the horizontal and vertical planes (Figure 6-12). It is at this stage of flight that small head movements frequently evoke sensations of either self- or surround-motion. One probable explanation for compromised VOR function and subsequent gaze drift centers on the idea that the altered eye movements are compensatory for the false perception of self- or surround-motion.

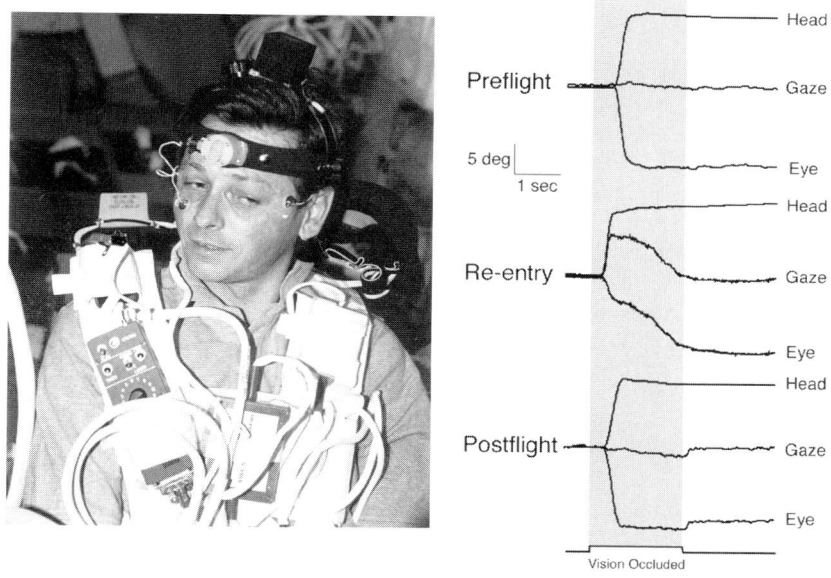

Figure 6-12. Right. "One-shot" gaze stabilization paradigm. Postflight performance required a large saccadic eye movement to bring the eye back on target once vision was restored for the first trial or two. The saccadic correction is illustrated when comparing the preflight response (upper panel) with the postflight response (lower panel). Subsequent postflight trials showed an immediate trend toward preflight baseline values, usually returning to normal within four gaze stabilization trials. During re-entry (middle panel) the eyes slowly drifted to the original position. This pattern was seen for both horizontal and vertical head movements. Left. Eye and head movements were recorded on a ingenious portable electronic device called SuperPocket *that was developed jointly by CNES and NASA. Eye movements were recorded using standard electro-oculography. Signals were amplified with a gain of 4000 and recorded on tape during flight or digitized with a sampling rate of 500 Hz directly using a computer system during the preflight and postflight data collection systems. Active head movements were measured using a tri-axial rate sensor bundle integrated on the head cap. Photo courtesy of NASA.*

4.2.3 Smooth Pursuit

During pursuit tracking it is typical to see saccadic activity wherein subjects use saccades to either anticipate or catch-up the moving target. Exposure to space flight has a tendency to modify this saccadic behavior. Early in microgravity, during tracking of sinusoidal movement of a point stimulus, the eye movement decreases in amplitude, resulting in an undershooting, and corrective saccades appear The effects of microgravity on the pursuit function were most pronounced early in-flight (FD3), after long exposure to weightlessness (FD50, 116 and 164), as well as after landing (Figure 6-13). (André-Deshays *et al.* 1993). Pursuit improved following in-flight execution of active head movements, indicating that the deficiencies in pursuit function noted in microgravity may be of central origin (Kornilova *et al.* 1993).

The velocity of eye saccades, whether elicited during pursuit or simply acquiring a visual target, is also reduced in microgravity (Uri *et al.* 1989, Israel *et al.* 1993). It is unclear what mechanism is responsible for this decreased peak saccadic velocity during flight unless the change is related to the control of retinal slip. It is beneficial for visual performance to maintain the spatial representation of the target on the same side of the fovea (as opposed to racing across the fovea), and hence in the same cerebral hemisphere that initiated the primary saccade.

Overall, corrective saccades appear to be used to maintaining gaze on target, reducing retinal slip, and assisting the astronauts in maintaining clear vision throughout the different phases of the space flight (Somers *et al.* 2002).

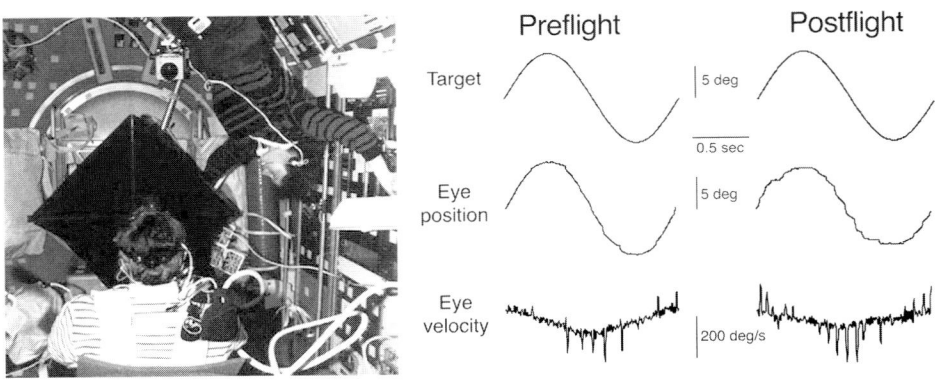

Figure 6-13. Left. Cruciform target used for displaying smooth pursuit stimulus during space flight. Right. In-flight and postflight, subjects have difficulties following a sinusoidally moving visual dot; the eyes try to catch-up the visual target with fast saccades rather than smooth pursuit. Photo courtesy of NASA.

5   SUMMARY

The sensorimotor function of the vestibular system of the inner ear in space flight is by far the most carefully studied of all. This is especially true of the gravity-sensing otolith organs and their relationship to eye movements. In particular, prolonged microgravity during orbital flight is a unique way to modify the otolith inputs and to determine the extent of their contribution to the vertical vestibulo-ocular reflex (VOR) and optokinetic nystagmus (OKN).

Visual acuity can be significantly degraded with a retinal slip of only a few degrees per second; therefore, eye movements are really important for keeping the images of objects stationary on our retinas. Impairment of this ability can lead to disorientation and reduced performance in sensorimotor tasks like piloting a spacecraft.

The following effects, or lack of effects, of microgravity on reflex eye movements have been documented:

a. The vestibular semicircular canal function is unaltered because the horizontal VOR during head yaw movements and the thresholds for angular acceleration in all directions are unaffected by microgravity.

b. As a probable result of the offloading of the saccules, the normal asymmetry of the vertical VOR, which is greater for downward head movements than upward movements, during head pitch is inverted in microgravity.

c. The absence of gravity stimulation of the otoliths, primarily the utricles, reduces the torsional VOR gain during head roll in microgravity. This deficit appears to recover after several months of exposure, which is a longer period for adaptation than the few days often assumed and subjectively reported by most astronauts and cosmonauts. It is unclear if this change is based on a re-weighting of neck-proprioceptive afferents, enhancement of efferent copies, or both.

d. During the first days in microgravity, the asymmetry of vertical OKN (greater gain for upward visual flow than downward visual flow) is inverted. A return to symmetry of the eye movements is then observed. These changes in asymmetry of vertical OKN and VOR may be related to otolith-dependent changes in vertical eye position that in turn affect slow phase eye velocity.

e. The orientation properties of the OKN are attributable to a velocity storage component generated in the vestibular nuclei and controlled through the nodulus and uvula of the vestibulo-cerebellum. This spatial orientation of the velocity storage is utilized to orient eye and body movements with the estimated tilt of the resultant gravito-inertial forces in order to maintain gaze and postural stability. This function is preserved in microgravity when a sustained linear acceleration is generated. Otherwise, the velocity storage tends to orient its axis toward the body longitudinal axis.

e. Some studies have shown increased latencies and decreased peak velocities of saccadic eye movements, while others have found just the opposite. It is possible that these conflicting results are related to when during the mission the measures were obtained.

f. There is a serious disruption of pursuit movements, especially in the vertical plane.

Planetary exploration missions will include several transitions between different gravito-inertial force environments. These changes will eventually affect the reflex eye movements. The question of whether astronauts will be able to perform adequately in a weightless or an artificial gravity environment during a mission to Mars if they have learned certain sensorimotor skills, like piloting, in the normal gravity of Earth, is certainly valid. More generally, the question of whether a person can have two different sets of reflexes, between which they are able to switch rapidly based on the environment in which they are immersed, arises. Are there procedures that could help to transfer (or to inhibit) training from one situation to another? Determination of the dual-adaptive capabilities of reflex eye movements in such circumstances is vitally important so that it can be determined to what extent the sensorimotor skills acquired in one gravity environment will transfer to others (Shelhamer 2007).

Chapter 7

SPATIAL ORIENTATION

The up-down spatial reference has simplified tasks that are dependent upon spatial orientation throughout everyone's life (Figure 7-01). When a space traveler enters weightlessness, this reference immediately disappears. Without this ever-present up-down perception vector, even conducting a visual search for an object within an orbiting spacecraft is quite difficult. As an astronaut floats about in weightlessness, her body orientation changes such that an object last seen above her head and to the right could soon after appear below her feet and to the left. The senses that usually maintain subconscious awareness of whole-body position relative to the immediate spatial environment no longer perform the task. The visual system is not capable of detecting the direction of gravity on Earth. However, in this new environment, it must now perform the unaccustomed task of maintaining orientation awareness. A disrupted orientational awareness can contribute to space motion sickness by generating the requirement for additional "hunting" head movements and by the fact that continual conflict in the spatial orientation system provokes motion sickness.

Analysis of crewmember perception reporting of their body position and body motion during their normal activities on board the spacecraft is one way to obtain an appreciation of the adaptation of spatial orientation to weightlessness. This can also be accomplished by analyzing data obtained during controlled stimulation on experimental rotating or translating devices.

Figure 7-01. Which way is up? Which way is down? Inside a spacecraft in orbit, there is no physical sensation to let you know when you are upside down. Photo courtesy of NASA.

G. Clément, M.F. Reschke, *Neuroscience in Space.*
DOI: 10.1007/978-0-387-78950-7_7, © Springer Science+Business Media, LLC 2008

1   INTRODUCTION

The ability to undertake goal-directed action is required for human survival. Perception must permit answers to basic questions including: What am I in? What is the environment layout? And what is my orientation with respect to that environment? (Warren 1990). Spatial orientation is the relationship between a body-oriented coordinate system and external references frames. It results from the integration of auditory, visual, vestibular, tactile, and proprioceptive signals, and from a comparison between them and the motor-command or efference-copy signals arising in the brain. Traditional research on exteroception, such as vision and hearing, focuses on perception of the environment apart from the passive observer. In contrast, research on perception and control of self-orientation and self-motion addresses interactions between action and perception (Shaw *et al.* 1990). Self-orientation and self-motion perception are required for and modified by goal-directed action.

1.1   Maps and Internal Models

The CNS must develop, maintain and modify as needed, neural models of the self and the environment. These models may represent three-dimensional Cartesian coordinates for both the self (intrinsic coordinates) and the environment (extrinsic coordinates). Extrinsic coordinate neural models derive from the observer's ability to detect up-down vector signals produced by gravity and by the polarity of the visual scene. Horizontal coordinates, i.e., front-back, side-side, are incompletely specified by the up-down vector. Additional complexity is introduced because extrinsic coordinate models derive from multimodal processes. For example, detection of gravity is mediated by graviceptors at several locations in the body, including the vestibular apparatus, somatic receptors, and visceral receptors (Mittelstaedt & Fricke 1988).

Intrinsic coordinate models must be more complex because they may be eye-centric, head-centric, or torso-centric (Howard 1982, p. 309). Intrinsic coordinate models also should differ from those for extrinsic coordinates in that x-, y- and z-axis vectors are all non-arbitrary and physiologically specified (Cornilleau-Peres & Droulez 1990).

Effective action in the normal environment requires mapping of relationships between models for intrinsic coordinates relative to the model for extrinsic coordinates. The resulting maps may be used in at least two ways: (a) through perception of body orientation; and (b) through setting of initial conditions for central motor control command systems. Eye-head movements during visual target acquisition, limb movements during reaching for targets, and locomotion toward goals all require motor control. While earlier studies suggested a common (shared) central motor command system, more recent research suggests parallel command pathways, at least for limb and visual target acquisition control (Findlay & Walker 1999).

Recent advances in neuroscience suggest that central neural processing involves activity in multiple, parallel pathways, also known as *distributed functions* and *distributed networks*. Based on these advances and the evidence for parallel motor control systems, multiple, parallel maps relating intrinsic and extrinsic coordinate neural models have been postulated (Calvin 1995). These parallel maps may be associated with different processes including perception of whole-body motion, limb target acquisition and head-eye target acquisition. For effective reaching or moving toward a target, the map that provides initial conditions for the limb motor control system would require weighting of the intrinsic body z-axis. For effective looking for a target, the map that provides initial conditions for head-eye motor control would require weighting of the intrinsic head and retinal meridian z-axes.

1.2 Adaptive Effects

Alteration of sensory processing (e.g., labyrinthectomy) or rearrangement of environmental features (e.g., prolonged exposure to microgravity) requires adaptation for effective motor control. One aspect of adaptation may involve re-mapping of intrinsic and extrinsic coordinate relationships. In the normal adapted state, parallel maps are likely to be congruent. During adaptation, these maps may differ and adaptation may be complete when the parallel maps are once again congruent.

Perceptual and oculomotor response discrepancies observed during adaptation to stimulus rearrangements support the above postulates. Except for the case of ocular torsion and perceived tilt (Wolfe & Held 1979), perceptual and oculomotor responses are normally approximately congruent (Guedry 1974). However, response incongruence has been noted during adaptation to unilateral loss of vestibular function when the spinning sensation gradually subsides while peripheral asymmetry, as revealed by eye-movement records, remains (Perterka & Benolken 1992). Similar response incongruence has been observed following exposure to stimulus rearrangements including the inertial-visual stimulus rearrangement produced by microgravity (Figure 7-02).

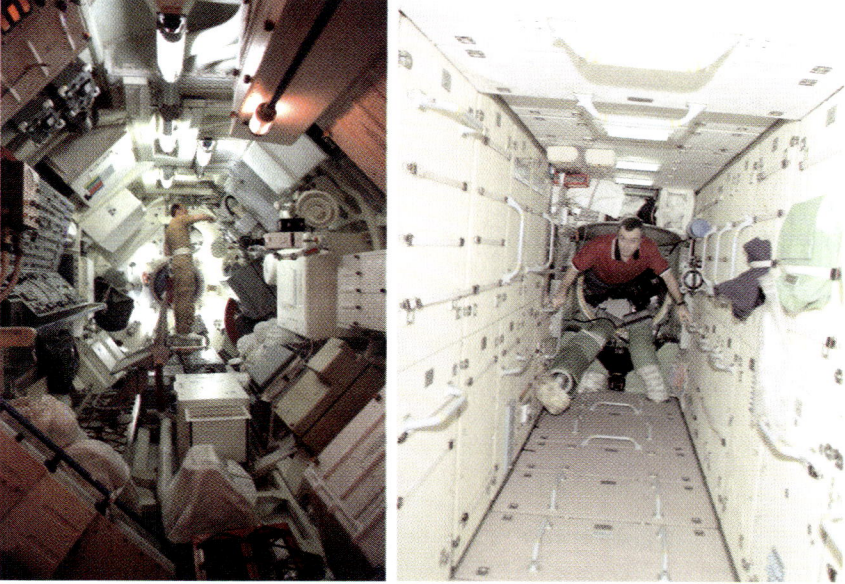

Figure 7-02. Inside of the Skylab (left) and ISS (right) space stations. Note the absence of distinct "floor", "ceiling" and "walls." Light sources, however, are coming in from one side only, and during their daily activities, the crewmembers are often oriented such as the light is coming from above. Photo courtesy of NASA.

Perhaps the most dramatic case of perceptual-oculomotor response incongruence was reported by Oman *et al.* (1980). After one to three hours of wearing goggles that produced a left-right reversal of the visual field, subjects exposed to a moving striped display reported illusory self-rotation in the same direction as the seen stripe motion. However, no subject showed evidence of reversal of the vestibular-ocular reflex slow-phase component. More recently, Oman & Balkwill (1993) reported that a 90-deg forward head pitch following a sudden stop from 120 deg/sec yaw rotation in microgravity resulted in "almost instantaneous" termination of perceived self-rotation. However, the duration of the post-rotatory nystagmus during this procedure was as long (no dumping) as that observed pre- and postflight when the head was held erect.

These and related observations led Perterka & Benolken (1992) to suggest that the neural mechanisms underlying central compensation may not be fully shared by vestibular reflex and self-motion perception systems. A re-mapping of relationships between intrinsic and extrinsic coordinate neural models appears to be a variation of their hypothesis.

If the fully adapted state is characterized by congruence among parallel maps, one implication is that different re-mapping processes may occur across different time intervals during adaptation. These re-mapping processes would be a form of sensorimotor learning. Of the conditions that facilitate sensorimotor learning, active, voluntary motion is among the most important (Welch 1986). The rate of re-mapping would then be dependent upon the classes of voluntary actions performed. If the observer were to engage only in head-eye target acquisition behaviors one might expect that the map serving the head-eye motor control system would be altered sooner than would the map serving limb motor control.

Self-orientation and self-motion perception derives from a multimodal sensory process that integrates information from the eyes, vestibular apparatus and somatosensory receptors. Perhaps due to these underlying multimodal processes, self-orientation perception is not "referred" (in the sense that a tactile stimulus is *referred* to a location on the body surface, or that visual stimuli are *referred* to the eyes) to any single receptor or body location (Parker 1991). For example, self-orientation with respect to a gravitationally defined vertical can be reported employing numerous procedures such as setting a luminous line, positioning a limb in darkness, or verbally reporting perceived head position in darkness.

Reviews of ground-based studies on spatial orientation have been undertaken by several authors, including Howard & Templeton (1966), Guedry (1974), Howard (1982), and Howard (1986). The following findings are particularly relevant to the question on the effects of gravity on reference frames:

a.   Observers are able to report perceived orientation with respect to extrinsic reference vectors (axes) defined by gravity, visual scene polarity and tactile polarity, and to intrinsic reference vectors such as the eye, head or torso z-axis (Lackner and Graybiel 1983).

b.   Reports can be obtained verbally as well as by movements of the eyes, movements of the limbs, manipulation of a tactile stimulus (rod) and movement of a visual line, and report accuracy can be judged with respect to the reference vectors (Wood *et al.* 1998).

c.   Numerous studies (e.g., rod-and-frame studies, tilted room experiments) show that reports indicate a compromise when visual and gravitational reference vectors are not parallel (Young 1984).

d.   Studies show that discrepancies between gravity and internal z-axis vectors may also influence reports. For example, reported tilt of a truly vertical line in the direction opposite to the head tilt implies that the subjective visual vertical is tilted in the same direction as the head tilt. This A- (Aubert) effect predominates when body tilt is large (> 60 deg) and can be understood by relating extrinsic gravity and intrinsic body z-axis vectors (Mittelstaedt 1983).

e.   Observers are able to estimate accurately rotational displacement solely on the basis of semicircular canal cues, within known limits of angularvelocity and amplitude (Guedry 1974). However, observers cannot estimate velocity, only position can be judged. Consequently, when whole-body rotation is used in space, an imperfect compensation indicates potential changes in extrinsic or intrinsic reference vectors, or changes in the inputs weighting due to microgravity adaptation.

2   ANECDOCTAL REPORTS FROM CREWMEMBERS

Illusory perceptions of self-orientation with respect to the vehicle result when the CNS "wrongly" interprets static visual cues. On Earth, the force of gravity and visual cues naturally provide an important reference for spatial orientation. People stand with their feet on the floor, trees grow up, and the horizon is horizontal. During free-floating in microgravity, the familiar otolith, tactile and somatosensory cues are largely absent, and the body can be in an unfamiliar orientation with respect to the spacecraft and the visual environment.

It is interesting to note that those crewmembers who remained seated in the relatively small Soyuz, Mercury, Gemini, and Soyuz capsules rarely encountered spatial orientation problems. However crews of the larger spacecraft reported occasional disorientation, particularly when they left their seats, and worked in unpracticed, visually unfamiliar orientations. The problem occurred both inside the spacecraft and outside, as when performing a spacewalk, as well as during re-entry and immediately after landing.

Perception of self-orientation and self-motion with respect to the environment may be dependent on the ability to detect and integrate information regarding four determinants:

a.   The direction of the gravito-inertial acceleration vector.

b.   The polarity of the visual scene.

c.   The internal vectors aligned with the observer's head or trunk z-axis (idiotropic vector).

d.   The expectations of the observer based on his/her own actions.

In ground-based conditions, there are large individual differences based on the weighting attributed to each of these determinants, presumably due to past experience. Similarly, during space flight, although episodes of visual disorientation are observed by many crewmembers, some seem more affected than other. In some individuals static visual cues become increasingly dominant in establishing spatial orientation in microgravity. These astronauts report that they become primarily "visual creatures" in microgravity (Figure 7-03), and they frequently experience visual reorientation illusions (see Section 2.1.2 below). Other astronauts appear to ignore visual polarity information and indicate that "whatever position I'm in defines up" (Harm & Parker 1993). One

Skylab crewmember clearly described this sensation as follows: "You tend to feel that 'up' is over your head, 'down' is below your feet, 'right' is [on your right side] and 'left' [is on your left] and you take this world around you with you wherever you go" (Cooper 1976, p. 146). Clearly, these subjects are more "body oriented" and align their spatial vertical to be along their longitudinal (idiotropic) body axis. Such individuals exhibit less spatial disorientation episodes aloft, even in the absence of visual cues for vertical orientation.

In the following section, we review the illusions the most commonly experienced by astronauts and cosmonauts with the eyes open or closed, both spontaneously and during voluntary (active) body motion.

Figure 7-03. Left. A crewmember inside one of the cargo delivery modules of the ISS. Note that he tends to align his longitudinal body axis with the vertical axis of the equipment racks. Right. Astronauts during EVA occasionally feel uncomfortable when working upside-down in the Shuttle payload bay, or when their feet face the Earth with nothing in between. Photos courtesy of NASA.

2.1   Spontaneous Illusions

2.1.1   Inversion Illusion

The second human in space, cosmonaut Titov, reported that he experienced a sensation of flying in an inverted position that lasted for about one and half minute. Other cosmonauts have often described an inversion illusion, i.e., the sensation of "hanging upside down", which usually occurs immediately after onset of weightlessness and persists for periods ranging from minutes to hours (Yuganov & Kopanev 1975, Yakovleva *et al.* 1982). This illusion that both the spacecraft and its occupants are flying upside-down continues after the eyes are closed. Given the fact that space travelers are actually in free-fall, it makes sense that sensation of fullness in the head resulting from to the headward fluid shift and the gravitational unloading of the saccular otoliths in weightlessness would promote such a subjective sense of inversion, regardless of the subject's actual orientation in the spacecraft. Being firmly restrained within a seat is also thought to contribute to the illusion of being inverted. Mittelstaedt (1986) has proposed that the inversion illusion may be understood using a model that includes an internal, *idiotropic* orientation vector (Mittelstaedt & Glasauer 1993).

It is interesting to note that astronauts did not perceived an inversion illusion when exposed to a -0.22 g linear acceleration along the longitudinal body axis (force vector from toe to head) during off-center axis rotation on the MVI rotator (see Figure 4-15, left). What looks like a negative result is in fact a significant finding (Benson *et al.* 1997). This absence of inversion illusion in space during a sustained -0.22 g stimulus to the otolith system indicates that the CNS did not "grasp" the 0.22-g otolith stimulus to acquire a vertical reference. Note that the gravity on the Moon's surface is less that 0.22 g!

A variant of the inversion illusion is the sensation that the body is tilted relative to the local vertical. For example, two astronauts reported the sensation of being at a 10 to 30 deg pitch up angle when they were parallel to the middeck walls, with one of the astronauts reporting that this sensation persisted throughout the entire mission. Many astronauts also report the perception of a 10 to 30-deg head-down tilt while lying in bed for the first one or two nights following return to Earth. Astronaut Tom Jones recalled, "For nearly three days after landing I couldn't stand erect without using my eyes to determine the up direction; without them I was in imminent danger of keeling over. If I accidentally dropped something, I could pick it up only by bracing myself against a wall or table; bending over for it would quickly cause me to topple. My inner ears didn't start working well again until about 72 hours after touchdown. […] I was desperate for some rest, but I kept dreaming I was weightless. My legs felt particularly light: I was certain that if I let go of my pillow, I would instantly float to the ceiling and be stuck there for the rest of the night. I found myself holding desperately onto my anchor with both arms, and it worked—I managed to stay firmly atop the mattress all night" (Jones 2006, p. 154 & 212).

2.1.2 Visual Reorientation Illusion

Clearly, an available and important reference frame is the inside of the vehicle itself. Astronauts naturally try to recognize the surfaces around them, so as to remain spatially oriented with respect to the familiar "floor" and "ceiling", fore and aft, and port- or starboard side directions. Visual recognition of various objects and surfaces in the spacecraft interior takes place on the basis of a lifetime of previous experience in visual orientation in one-g and some a priori knowledge, or visual memories, of how various objects and surfaces are arranged with respect to each other (Oman 1989).

Part of the difficulty of the people who predominantly rely on visual cues for spatial orientation is due to the natural tendency to assume that the surface seen beneath our feet is the floor. When working "upside down" in the spacecraft, the walls, ceiling and floors then frequently exchange subjective identities. Visual reorientation illusion episodes seem then to occur with some consistency in certain situations. For example, when viewing another crewmember floating upside down in the spacecraft, people may often suddenly feel upside down themselves, because of the subconscious assumption carried over from life on Earth that people are normally upright. Another example is when an observer views another person floating with feet toward the true ceiling, while he is himself in a similar orientation, he may suddenly feel that the true floor has subjectively become a generic "ceiling" and so he no longer feels upside down (see

Figure 7-01). This visual reorientation illusion typically reverses when the people in view reassume a "normal" feet towards the rue floor orientation. The observer then experiences a visual reorientation episode and suddenly feels "upside down" (Oman 2003).

Visual reorientation illusions also typically occur during transition from one module to another, especially when modules do not have a continuous floor to ceiling relationship, and are connected by a tunnel with no well-defined floor, walls or ceiling. When traveling through the tunnel, crewmembers may perceive that they are going "up" into a well. Encountering another crewmember coming the other way can make them suddenly feel as if they are upside down, descending headfirst. Then, they could move into a given module sideways or upside-down and not recognize it: "It is one of the biggest mysteries in the world when you go in there [the lunar module] to find something" (Cooper 1976, p. 234). "I floated through the tunnel, popped the lunar module hatch and dove inside, like Alice going through the mirror, right into a completely different world. I was disoriented because the floor was now above my head, so I rolled into a weightless ball, flipped and let my brain adjust to the new environment until my equilibrium returned" (Cernan & Davis 1999, p. 211).

Unlike the inversion illusion that is limited to the first hours after entering in weightlessness, visual reorientation illusions occur repeatedly throughout a mission. Also, whereas inversion illusions are difficult to reverse and continue when the eyes are closed, visual reorientation illusions are easily reversed and typically depend on the visual context. Ed Lu, an ISS Expedition-7 crewmember, experienced these illusions even after months in space, "It is interesting how the same module can look like a completely different place when viewed upside down. Every once in a while here, I spend a few hours upside down just for fun, treating the ceiling like the floor. The familiar modules don't look so familiar anymore. This is true even though I am completely used to flipping my body around to whatever orientation needed to work on things."

Several crewmembers have reported that life on the ISS is replete with visual reorientation illusions (Petitt 2003). This is not surprising given the architectural design of the ISS modules. The aisle cross section is a symmetric square, and scientific instruments and stowage bins are located on the ceiling, floor and walls. Consequently, there is not a clear identity of each of the interior surfaces. One crewmember of the Skylab missions, which included a similar architectural design (Connors *et al.* 1985), had warned, "There's been some thoughts about mounting some furniture on the floor, some on the walls, some on the ceiling, but this doesn't work out. You tend to orient yourself when you're in a room, even though you are in zero-g, and when you orient yourself, you should find everything is the same. You don't like something up, something under…" (Cooper 1976, p. 321).

2.1.3 Falling Sensation

On Earth, sensation of falling is experienced during actual falling. However, during actual free-fall experienced in parabolic flight, individuals do not generally experience sensations of falling; they feel stationary (Lackner 1992a, 1992b). Nevertheless, sensation of falling has been occasionally reported during orbital flight. Astronaut Mike Mullane wrote, "In my upper-cockpit perch, I had no sense of [free] fall but in the windowless middeck I had experienced brief moments in which the sensation had been overwhelmingly powerful. The day before, I had been seized with an illusion that the middeck cockpit floor was steeply tilted and if I didn't grasp something I would slide down it. Try as I might I could not convince myself that I would not fall. I actually seized the canvas loop of a foot restraint to keep down from sliding off my imaginary cliff. The sensation was so distracting I finally abandoned the middeck and floated upstairs. The view of the Earth's horizon immediately eradicated any sense of the fall" (Mullane 2006, p. 334). These observations suggest that there may be a strong cognitive component to the elicitation of the sensation of falling. Visual motion and quick loss of support and knowledge that one is falling may trigger this sensation under terrestrial conditions.

On Earth, it is well known that height vertigo, i.e., sensations of fright and instability usually accompanied by increased body sway, occurs when the viewing perspective is elevated and there are no nearby objects visible. The most expressive state of height vertigo can lead to *acrophobia*, from a Greek word meaning 'summit', i.e., an extreme or irrational fear of heights. It belongs to a category of specific phobias, called *space and motion discomfort*, that share both similar etiology and options for treatment.

Astronauts have occasionally reported fear of height during extra-vehicular activities, depending on the perspectives generated. For example, Bernard Harris, an astronaut who performed several EVAs during the STS-63 Shuttle mission reported, "As I was getting ready to step out of the spaceship, it felt like gravity was going to grab hold of me and pull me down toward Earth. Your natural response is to hesitate and grab on harder. I felt myself hanging on to the handrail and saying: "No, you're not going to fall toward the Earth, this is the same thing you've been seeing for the last five days" (Figure 7-03, right).

Episodes of spatial disorientation are also commonly during EVA. "Looking around, I discovered I had run to the end on my fifty-foot tether, the taut steel wire snaking off to my left into the darkness. *Isn't this the right way down?* In the blackness surrounding my helmet lights, I couldn't see the rest of the truss or even the Orbiter, but the strand of undulating wire gave me a clue. 'Down' must be along the tether, since it was secured to the port side of the Orbiter's payload bay. That makes sense. I paused for a minute, making sure, before swinging my body back into the familiar NBL [Neutral Buoyancy Laboratory] orientation. I must have twisted around while working in the trash bag, gotten disoriented, and then just taken off in the wrong direction. Soon the radiator appeared in my helmet lamps, and I was back in business" (Jones 2006, p. 299).

"Although I faced the sky as I worked, my brain was convinced that I was standing upright, putting the rest of the station on its side, the way the mock-ups lay in the NBL. It proved very hard to defeat that underwater indoctrination" (Jones 2006, p. 312).

2.1.4 Oscillopsia

During re-entry and immediately after landing, astronauts frequently report that voluntary head motions produce illusions of motion of the visual surroundings (Reschke & Parker 1987). These illusions are likely related to disturbances of the gaze control system. An apparent oscillation of visual targets was reported by some Skylab astronauts (Graybiel *et al.* 1977) (Figure 7-04), but was attributed to drowsiness. Young and his colleagues (1984) systematically investigated this phenomenon during the Spacelab-1 mission and concluded that the apparent displacement of visual targets during head movement immediately after landing in their subjects was due to oscillopsia. Oscillopsia is readily understood in terms of miscalibrated compensatory eye movements, most likely an inappropriate VOR gain, during head movement in an unusual force environment. The appearance of oscillopsia postflight indicates that the compensatory eye movements had adjusted to the microgravity condition and were no longer appropriate to normal gravity. This suggests that a carefully designed experiment might reveal oscillopsia during the period immediately following orbital insertion, before astronauts have adapted to weightlessness. Such experiment, to our knowledge, has never been performed.

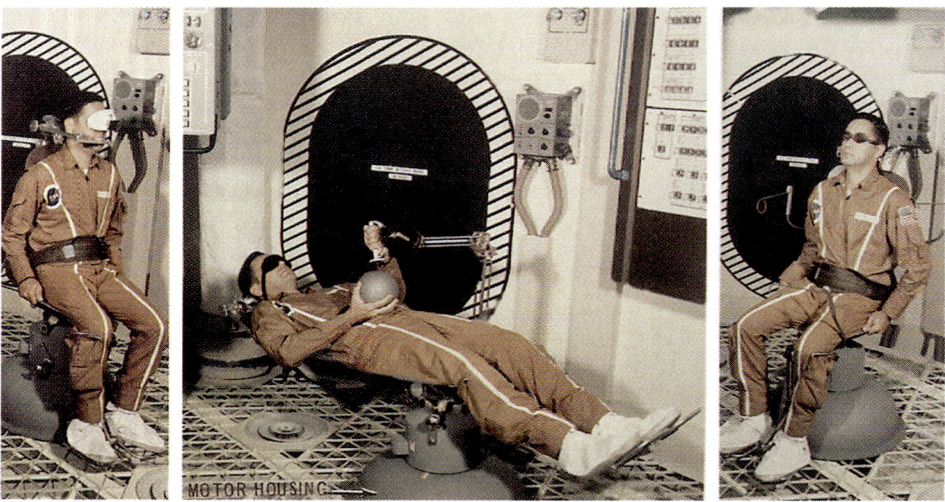

Figure 7-04. This set of photographs details the Human Vestibular Function *(M-131) experiment on board Skylab. Left: An astronaut is wearing a goggle displaying a line of light that he could adjust along his perceived "horizon" or "vertical." Middle: In the* Spatial Localization *test, the astronaut holds in his left hand a pointer mounted at the end of an articulated arm and in his right hand a hollow steel ball. With eyes covered, the subject slides the pointer over the ball to indicate the orientation of various features in the laboratory relative to his body axis. Right: The susceptibility to space motion sickness was evaluated by counting the number of active head movements required for eliciting symptoms during passive body yaw rotation in yaw. Photos courtesy of NASA.*

2.2   Illusions Generated by Motion

The vestibular system operates as a silent partner with the other senses, improving the efficiency of control of goal-directed head and body movement relative to the Earth. Vestibular sensations do not reach conscious awareness as we skillfully move about. Vestibular sensations achieve conscious awareness only when they are "disorderly" in relation to concomitant information from other senses that participate in the voluntary control of head and body motion. The "dizziness" that accompanies vestibular disorders is usually poorly described because the perceptual event is characterized by the confusion and disturbance that comes from mixed signals among the various senses involved in the control of motion.

Expected voluntary turns involve sensory feedforward against which feedback is compared. The same vestibular feedback message can be generated actively or passively but the effects of active turns and passive turns are very different. Many astronauts have reported that the effects of active and passive motion are still different in orbit and active motion generates the most disturbances and confusion (Oman 1982), hence the importance of restraining head motion for avoiding SMS symptoms.

Even though individual experiences with self- and surround-motion vary, three types or categories of disturbances are commonly reported by crewmembers:

a.   *Input-output gain disturbances*, where the perceived self- or surround-motion appears exaggerated either in rate, amplitude, or position following the physical head or body movement.

b.   *Temporal disturbances*, where the perception of self- or surround-motion lags the head or body movement and/or persists after the real physical motion has stopped.

c.   *Path disturbances*, where angular head and body movements elicit perceptions of linear or combined linear and angular self- or surround-motion.

Some examples of astronaut descriptions of perceptual experiences associated with each of the categories are presented below. Several general points or comments should be made concerning these reports of perceptual disturbances. First, the intensity or compellingness of perceptual disturbances during different phases of the mission, from strongest to weakest, generally occurs in the following order: (1) during re-entry; (2) immediately after wheels stop; (3) postflight for several days (decaying slowly) with final duration dependent on length of time in zero-g; (4) late on-orbit; and (5) early on-orbit.

Second, it should be noted that a given head or body movement usually induces perceptual disturbances in more than one category. For example, one crewmember reported that sinusoidal roll head movements made late in-flight resulted in self-translation (path), that the translation was greater than the roll input (gain), that there was a delay between the roll input and the self-translation and that self-translation persisted after the roll input ceased (temporal).

Finally, except where otherwise stated, the examples of perceptual disturbances presented in each category are derived from an investigation in which Shuttle crewmembers made low amplitude (± 20 deg), slow sinusoidal (± 0.25 Hz) head movements in pitch, roll and yaw; the head movement protocol was performed early and late on-orbit, during re-entry and immediately after wheels stop (Harm *et al.* 1993).

2.2.1   Input-Output Gain Disturbances

Gain disturbances occur in-flight, during re-entry and for as long as several days postflight. One astronaut reported that head or body movements made in any axis during re-entry and immediately postflight were perceived as being five to ten times greater than the actual physical movement. "High velocity without displacement" is a fairly common description of this exaggeration in perceived self-motion. One astronaut described this sensation as follows: "... I didn't feel like I was going anywhere, but I was moving like a bandit." Another crewmember reported that when driving his car the day after returning from space flight, he perceived his speed to be 45 mph when the actual speed was 35 mph.

Pitch Head Movements

Several crewmembers have reported the perception of whole body "tumbling" in the direction of pitch head movements made early on-orbit. The tumbling sensation intensified later in-flight and was increased in magnitude with larger amplitude head movements. Another crewmember reported perceived surround translation of 75-100 mm during the early portion of re-entry and 150-200 mm during the later portion of re-entry while performing pitch head movements. Immediately after wheels stop, another crewmember stated, "I'm accelerating faster than I thought I would be; I tried to move it [head] slower but it's still faster than I would want it to be."

Roll Head Movements

Several crewmembers also reported that roll head movements performed on-orbit and immediately postflight resulted in perceived whole body roll which, in most cases, was reported to be in the opposite direction of the head roll, i.e., a right head roll results in leftward whole body roll "as if the torso and lower limbs are coming up to meet the head." Occasionally, however, crewmembers report that the perceived whole body rotation is in the same direction as the head roll, "as if the body is following the head." One crewmember reported that perceived self-translation was greater than the actual roll input and another crewmember described the perceived head roll as exaggerated (in rate and displacement amplitude) compared to the real roll head movement made in-flight. During the re-entry phase of the mission, one crewmember reported that a 20-deg roll head movement resulted in perceived surround roll motion of 70-80 deg. Finally, at wheels stop one crewmember said, "I'm really accelerating to the left and right; my head is picking up speed as it goes through the center point."

Yaw Head Movements

Two crewmembers reported that yaw head movements made late in-flight felt exaggerated in displacement amplitude; one astronaut stated that a 20-deg head movement felt like an 80-deg head movement. At wheels stop, another crewmember reported that the 20-deg head movement was perceived as 0.6-0.9 m of self-translation.

2.2.2   Temporal Disturbances

Perceived self or surround-motion lags the "real" head or body movement by 0.5 to 1 second and can persist for two seconds or more after the real physical motion has stopped. The lag and persistence time is a function of the rate and amplitude of the physical movement. Temporal disturbances have been reported during all phases of the mission and during the first one or two days postflight. One astronaut reported that when he rolled over in bed the first night postflight, he grabbed the edge of the bed because the sensation of continued roll after real roll stopped made him feel that he might roll off the bed.

Pitch Head Movements

One astronaut who performed the head movement protocol daily for the first three days of flight reported that perceived self-motion lagged the real pitch head movement and persisted after the head movement stopped. Lag and persistence times were approximately equal and both increased over the first three days of flight from barely perceptible to 0.5 seconds. He also reported that the lag and persistence was most dramatic during the 1.3-g phase of re-entry. By increasing the frequency of the sinusoidal head movements, this subject was able to almost completely reverse the phase relation between the real (input) motion and the perceived (output) motion.

Roll Head Movements

One crewmember noted a delay in the onset of perceived self-motion and a persistence of self-motion following cessation of the real head movement late in-flight. Another crewmember reported that he made a 45-deg right roll movement from the waist during re-entry to pick something up off the deck. When he stopped the body motion, he felt that he continued to roll right; upon his return to upright, he still perceived right roll for several seconds. A third astronaut who also made a 45-deg head roll movement reported that the perceived motion persisted after the head movement stopped and then damped out. He stated that "it's almost like a second order equation because if you make a big head movement, you get a big response, then damping out. Almost like overshoot, comes back and damps out."

Yaw Head Movements

Perceptions of lag and persistence of self-motion for ± 20 deg yaw head movement at 0.25 Hz were reported by only one crewmember and then only for yaw head movements made immediately after wheels stop.

2.2.3 Path Disturbances

Pitch Head Movements

One U.S. astronaut reported the perception of a 50-60 deg pitch forward (at 5 deg/sec) attitude immediately upon achieving weightlessness (at main engine cut-off) that persisted for approximately 30 seconds; similar sensations were also associated with the offset of the Shuttle Orbital Maneuvering System burns throughout the flight. Late in-flight, one astronaut reported that sinusoidal pitch head movements were perceived as linear self-motion along the body x-axis, i.e., fore and aft self-translation. Another crewmember reported that making pitch head movements late in-flight while fixating on a far target resulted in a combination of self-pitch and self-translation along the body x-axis, whereas the same head movements performed with his eyes closed and while fixating on a near target resulted in a combination of self-pitch and self-translation along the body z-axis. When pitch head movements were performed immediately after wheels stop, three crewmembers reported perceived surround-translation along the body x-axis; the surround-translation was described as being 180 deg out of phase with the head movement by two subjects and in-phase by the third. One crewmember also reported that pitch head movements immediately after wheels stop produced perceived self-translation along the x-axis when performed with eyes open and surround-translation along the x-axis with eyes closed and fixating on the imagined target position.

Roll Head Movements

Two crewmembers performing roll head movements early in-flight reported y-axis translation 180 deg out of phase with the head movements; one perceived this motion as self-translation, the other as surround-translation. Roll head movements made late in-flight, during re-entry and at wheels stop resulted in similar reports of perceived self- or surround-motion. One astronaut stated that rapid roll head movements made during re-entry resulted in perceived self-translation along the y-axis that was 180 deg out of phase with the head movement, whereas slower head movements produced y-axis self-translation in-phase with the head movement. Another astronaut reported that roll head movements made during re-entry produced a combination of slight y-axis self-translation and surround roll motion 180 deg out of phase with the head roll.

Yaw Head Movements

Yaw head movements made by one crewmember late in-flight, during re-entry and immediately after wheels stop resulted in perceived y-axis self-translation in-phase with the yaw head movement. Another astronaut reported that yaw head movements made during re-entry and immediately after wheels stop produced perceived self-roll motion in-phase with the yaw head movement.

3   SPACE EXPERIMENTS ON SPATIAL ORIENTATION

3.1   Subjective Horizontal or Vertical

One of the earliest experiments to address spatial orientation during prolonged exposure to microgravity was completed by Graybiel *et al.* (1967) during the Gemini-5 and 8 flights. Astronauts were asked to set a dim line of light, located inside of a otherwise dark goggle, parallel with an external horizontal reference located on the capsule's instrument panel. The astronauts performed the task perfectly (except for one crewmember's systematic errors) while on orbit. The inference drawn was that relatively light touch, pressure and kinesthetic receptor cues available in microgravity supported orientation responses as well as the more plentiful cues, including those from the otolith receptors, available on Earth.

Alternatively, it can be argued that the lack of change in spatial orientation in the Gemini astronauts was due to experimental design. The astronauts trained on the ground and then performed the experiments while aloft while seated and restrained in the capsule's couches. On orbit the subjects remained in exactly the same position relative to the spacecraft and the external target while doing the experiment than on Earth. Therefore, the astronauts may have just adopted their body z-axis as the reference vector when performing the task.

A variant of this experiment was performed on the Skylab missions (Graybiel *et al.* 1977). In this experiment, the task was to again set a dim line of light inside a goggle similar to those used on the Gemini flights to an intrinsic axis (the longitudinal body axis with the center of the line referenced to the "straight ahead"), and then (after inspection of the visual surround) to the longitudinal axis ("floor") of the Skylab workshop. Following these measurements a second task, designed to involve proprioception in the judgment of the longitudinal axis, was performed. The astronauts grasped a metal sphere in one hand, to position a magnetic pointer (rod) on the sphere, and made judgments analogous to those made with the line of light (Figure 7-04). Preflight, these tests were repeatedly done on the Skylab one-g trainer. It is not surprising that the astronauts had difficulty understanding the task in zero-g, and as a consequence, it is also not surprising that there was little difference between the terrestrial and microgravity responses (with perhaps more scatter of data while aloft). Again it is likely that even though they were asked to make judgments with respect to extrinsic Skylab axes, actual judgments, both on the ground and in space, were made with respect to the intrinsic body z-axis.

On the basis of ground-based experiments that have shown differences in how subjects set a luminous line to either the vertical or horizontal, and depending on factors such as changing the gravito-inertial force (Schöne 1964, Mittelstaedt & Glasauer 1993), Gurfinkel *et al.* (1993b) suggested that microgravity might also evoke changes in the perception of the subjective vertical or horizontal. Wishing to remove estimates of verticality from the realm of subjective verbal reporting, they required their cosmonaut-subjects to trace ellipses using the hand and finger such that their long axis was oriented either parallel or perpendicular to the longitudinal body axis. During the experiment the subjects were always in the "standing" position with their feet fixed to the deck of the Russian station with Velcro. Aside from the proprioceptive input from the active arm movements the only other sensory information that might indicate verticality was the tactile input from the feet. The results showed that there was no difference between the ground-based estimates of verticality and those recorded in-flight. This result is not surprising, however. Again, the training and baseline data collections were also initiated in the same standing conditions in the mock-up, and the flight conditions were not substantially different from the early experiments by Graybiel *et al.* (1967, 1977). The subjects remained in exactly the same position relative to the spacecraft while aloft and therefore the task was referred to an egocentric (i.e., intrinsic) orientation.

Another attempt to explore spatial orientation in microgravity was conducted by a joint Austrian and Russian team on board the Mir space station. One cosmonaut-subject lying supine on the floor of Mir was required to track with his outstretched arm the horizontal (side-to-side) displacement of a visual target. Then, he was asked to repeat this arm movement with the eyes closed, first with the head aligned with the trunk, and then with the head tilted in roll to his left shoulder. Infrared LEDs and scanner cameras were used to record head and arm movements. Preflight, there was a deviation of less than 10 deg of the line between the head-aligned and head-tilted conditions. After two days in microgravity this deviation was approximately 20 deg in the head-tilted condition, and after the fifth day in-flight the line was approximately perpendicular to the head z-axis in both head conditions. Immediately after the flight the deviation was still close to 20 deg with the head tilted (Figure 7-05). This result suggests that the reference used for this spatial orientation task in microgravity and immediately after return to Earth had shifted toward the head vertical axis.

Figure 7-05. Trajectory of the arm in the frontal plane during memorized arm movements toward a horizontal moving target with eyes closed by one cosmonaut before, during and after a 7-day space flight. The subject with the trunk fixed to the floor of Mir had the head upright (HU), i.e. aligned with the trunk, or the head tilted to the left (HT). Preflight, the arm movement was close to the horizontal with both the head upright and tilted. During head tilt in-flight, however, the trajectory of the arm was significantly tilted from the horizontal, and on flight day five (FD5) it was perpendicular to the head axis, suggesting that the egocentric reference used for this orientation task was the head vertical axis. Adapted from Kozlovskaya et al. *(1994).*

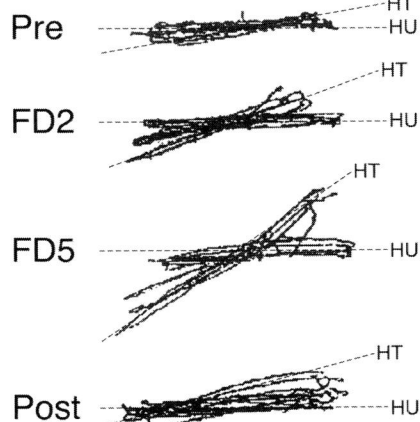

3.2   Visual Cues

Visual influence on apparent self-motion is often subsumed under the term *vection* (Dichgans & Brandt 1978). Individuals exposed to constant-velocity visual motion in a large rotating drum soon feel themselves rotating at constant velocity in the direction opposite the physical rotation of the drum and see the drum as stationary. When vection is highly compelling, it can elicit disorientation and motion sickness. Peripheral visual field stimulation is especially effective in eliciting vection, but even small central fields can have an effect.

Illusions of self-motion have been reported by astronauts exposed to a visual flow when seated in the vehicle or when passively moved by the robot arm during EVA. Shuttle Astronaut Tom Jones experienced vection in several occasions: "Luminous crystals drifted forward past the cockpit, glittering against the surrounding blackness. We were engulfed in a cloud of John Glenn's famous "fireflies." I had the momentary impression that we were flying *backward* through a soup of glowing snowflakes. Although I knew we were still racing forward and down at 17,500 miles per hour, the optical illusion was a powerful one. It reminded me instantly of how an adjacent car drifting slowly forward at a stoplight could cause me to jump on my brakes to arrest my backward 'motion'." (Jones 2006, p. 194).

"Whenever I asked Marsha to shift the robot arm slightly, I noticed another startling illusion. With each movement I requested, the arm appeared to remain stationary, while *I* drifted in the opposite direction. Once I recognized the optical illusion, I was able to enjoy it, smiling with delight at how free fall and that infinite backdrop above me could so completely fool my perceptions." (Jones 2006, p. 313).

Moving visual scenes that produce compelling illusions of self-motion become even stronger in space, since visual cues are unhindered by constraints from the otoliths, which in microgravity do not confirm or deny body tilt. This has been confirmed with experiments where crewmembers observing a rotating visual field felt a larger sense of body rotation in space than on Earth (Young & Shelhamer 1990). Bungee cords or bite boards were used to restrain the subjects in some of these experiments (Figure 7-06). When subjects were strongly tied to a seemingly solid anchor by the localized tactile cues from the bite board or bungee cord they indicated a strong inhibition of vection. After flight, all subjects showed an increase in postural instability and a strong tendency to sway when the visual field rotated. Interindividual differences in perceptual styles were presumably related to the different contributions of sensory and body-centered vectors in the normal and adaptive state of spatial orientation (Young *et al.* 1993, 1996).

3.3   Otolith Cues

On the Earth's surface, two major sources of linear acceleration are normally encountered. One is related to the Earth's gravity: the gravitational force pulls the body toward the center of the Earth and the body opposes this force to maintain an upright standing posture. The other sources of linear acceleration arise in the side-to-side, up-down, or front-back translations of the head, which commonly occur during walking or running, and from the centrifugal force sensed when turning or going around corners. On Earth, the otolith organs sense the vector sum of all linear accelerations acting on the head including gravity. This information is used by the brain to control our posture and eye movements during everyday activities such as walking and driving an automobile.

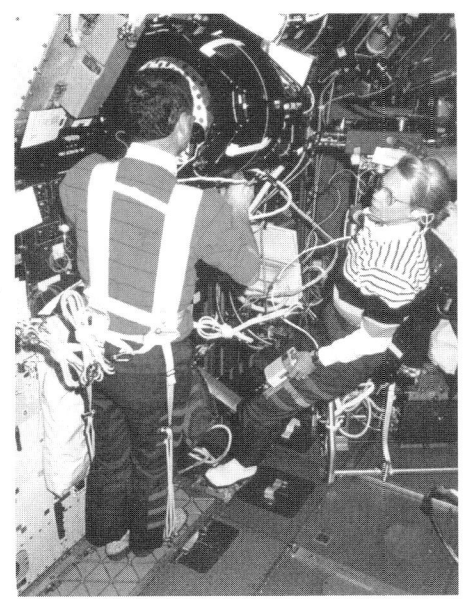

Figure 7-06. On board Spacelab, an astronaut-subject looks into a rotating dome with dot patterns and indicates with a joystick his perceived amplitude of self-motion in a direction opposite to that of the dome. In space, subjects reported stronger visual effects and illusions of self-rotation than they did on the ground. The role of tactile pressure on the feet was also investigated by means of bungee cords between the subject's hip and the module's floor. Photo courtesy of NASA.

3.3.1 Passive Head Movements

The rotating chair experiment on board Skylab had demonstrated that the crewmembers' ability to detect small changes in angular acceleration in yaw was not altered in microgravity (Graybiel *et al.* 1977). During the First International Microgravity Laboratory flight, the installation a servo-controlled rotator provided an opportunity to gather more detailed subjective reports during passive angular motion than obtained during the Skylab experiment (Reschke 1988). The IML-1 rotator could be configured in yaw, pitch or roll. In the pitch configuration, subjects were lying on their left side (see Figure 4-15, left), and in the roll orientation, they were lying on their back. Both pitch and roll configurations also placed the subjects' heads at a slight distance from the axis of rotation, thus centripetal acceleration was also generated during rotation.

Nevertheless, it was interesting to note that on orbit, during ramp and sinusoidal pitch stimuli, the perceived motion was reported to be "truly in a pitch plane and not motion in a horizontal plane" as experienced in preflight and postflight tests (Benson *et al.* 1997). One subject did note however, that after the initial exposure to pitch stimulation on orbit that subsequent exposures felt as though he was on his side, "just like on Earth." Also, during the pitch ramp stimuli, three of the four subjects tested on-orbit indicated that they felt they were tumbling "head-over-heals" rather than rotating in pitch whilst lying on their side. However, only one of the four subjects continued to report throughout the flight that the plane of motion was truly vertical pitch. Interestingly, there were multiple reports relating to yaw rotation (about the z-axis) that seemed to indicate that the subjects were confused about the axis of stimulation. One subject reported that z-axis rotation was a roll motion, and reports from another two subjects suggested similar misperception ("this could be pitch or yaw, I don't have a clue"). Therefore, it seems that without a constant gravitational vector and without vision, the subject's reference to the environment is established with intrinsic coordinates. If there is no sensory map available to establish the intrinsic coordinates, the axis of rotation or orientation relative to extrinsic coordinates cannot be determined.

Clément *et al.* (1987) also observed on FD7 of a Shuttle flight that when a free-floating subject was spun for a distance around each of the three orthogonal axes by his observer (center of spin re: hip) at a frequency equal to approximately 0.25 Hz, the perceived angular displacement around the body z-axis in microgravity was very accurate up to 360 deg. However, during both the roll and pitch movements, perception of the angle of rotation showed an overshoot (see Figure 4-14, left). At least two conclusions concerning this data set are possible. First, as expected, yaw would not be affected because in orbital flight as on the Earth the otoliths do not play a large role in yaw responses. Second, the overshoot in roll and pitch may be a product of a canal-otolith interaction where missing otolith contributions result in an altered response.

We have already reported above (see Chapter 4, Section 5.4) that astronauts often perceived translational self-motion during passive roll stimulation in darkness one to three hours after Shuttle landing (Parker *et al.* 1985). However, when exposed to sustained linear acceleration of 0.5 g and 1 g along the body y- or z-axis on a flight centrifuge, they described a sense of tilt that increased through the flight, but no sense of translation (Clément *et al.* 2001). This sense of tilt was absent during sinusoidal oscillation with peak acceleration lower than 0.2 g (Merfeld 1996).

3.3.2   Active Head Movements

Almost all crewmembers experience perceptions of self- or surround-motion associated with head movements or combined head and body movements while on-orbit, during re-entry and after landing. On-orbit disturbances appear to increase in intensity as a function of flight day: the longer the flight, the more illusory phenomena are reported. Unexpected illusory translation of either the subject or the visual surround was reported by two astronauts who performed voluntary pitch or roll head movements during re-entry and after the Shuttle had stopped on the runway (Reschke & Parker 1987). These observations, as well as similar ones reported by Young *et al.* (1984), support the hypothesis that signals from receptors that respond to linear acceleration are reinterpreted during adaptation to weightlessness (Parker *et al.* 1985). Although the data is currently limited to flights of two weeks or less, the intensity and duration of re-entry and post-landing perceptions of self- or surround-motion also appears to be a function of mission length, and perhaps of prior space flight experience or the volume of the spacecraft (e.g., Shuttle middeck vs. Spacelab or ISS).

3.3.3   Resolution of Discrepant Observations: Refinement of OTTR

The concept of *otolith tilt-translation reinterpretation* (OTTR) has been noted frequently in previous sections of this book. Although considerable evidence supports the OTTR concept, some experimental results do not. The following refinement of the original, quite simple OTTR model, which attempts to reconcile discrepancies between reported observations, derives from previous work (Parker 2003).

As noted in Chapter 4, Section 5.4.1, OTTR was based originally on the reports of surround or self-translation during head roll immediately postflight (Reschke & Parker 1987). In contrast to predictions from OTTR, in-flight centrifugation at a constant angular velocity evoked perceived tilt, not perceived translation, from the astronaut subjects as noted in Chapter 4, Section 5.4.2 (see Clément *et al.* 2001).

Subjects exposed to 0.5-g or 1-g constant linear acceleration on flight days seven and twelve invariably reported self-tilt and never reported self-translation.

As a starting point, it is assumed that the discrepant results are both valid. Initial formations of OTTR were derived from perceptual reports. It is well known that perceptual reports can be influenced by subjects' and experimenters' expectations. That "expectations" might account for the discrepancies seems unlikely in part because the original self-translation reports were unanticipated by both subjects and experimenters. Possibly more critical are the characteristics of the subject population. Most astronauts who have participated in space neuroscience experiments are independent thinkers inclined to carefully evaluate and question experimenters' assertions rather than accept them unquestioningly, to "perform their own experiments" rather than docilely follow the experimenter's protocol, and to attempt to report their experiences as clearly as possible.

The perceptual reports gathered by Reschke *et al.* (1991) and Clément *et al.* (2001) might both be considered as illusions in the sense that perceived orientation and motion don't match the stimuli that evoke those perceptions. Perceived self-tilt during centrifugation has been addressed in numerous studies and is sometimes called the *somatogravic illusion* even though the tilt perception is consistent with rotation of the gravito-inertial acceleration vector (see Chapter 4, Section 5.4.2). This illusion has operational significance when pilots of high-performance aircraft who accelerate forward, perceive their aircraft as pitched upward, and compensate by pitching the aircraft down, which may lead to a "spatial disorientation accident." In this case, illusory pitch is evoked by translational acceleration. In contrast to the somatogravic illusion, perceived self-translation elicited by head roll (tilt) with respect to gravity was not reported prior to Reschke's *et al.* observations with weightlessness-adapted astronauts.

Resolution of the discrepancies between the observations reported by Reschke *et al.* (1991) as opposed to those from Clément *et al.* (2001) must take into account differences between the stimuli to which the subjects were exposed. In the initial parallel swing studies reported by Parker *et al.* (1987), roll frequency was 0.26 Hz. For later observations during re-entry, subjects were asked to tilt their heads from upright either left or right at about 0.25 Hz. With both procedures, otolith stimulation was at a quite high frequency. For the Neurolab observations reported by Clément *et al.* (2001), the centrifuge was accelerated at 26 deg/sec^2 to constant velocities of 254 deg/sec or 180 deg/sec, which were sustained for 60 sec. The Neurolab studies performed by Clément *et al.* (2001) were thus done with a much lower frequency stimulus than those reported by Reschke *et al.* (1990).

As noted in Chapter 3, signals generated by otolith and other inertial sensors are fundamentally ambiguous due to the equivalence of gravity and linear acceleration noted by Einstein. It is unsurprising that prolonged exposure to weightless space flight would lead to disturbance of perceived self-tilt and self-translation, as suggested by Melvill Jones during the Skylab Symposium (Johnston and Dietlein 1977). The original OTTR hypothesis suggested that during adaptation to weightlessness, the brain reinterprets all graviceptor signals to indicate translation (see Chapter 3, Section 1.5.6). Benson (1974) previously proposed that the temporal dynamics of graviceptor signal changes might contribute to resolution of graviceptor ambiguity. Specifically, short-duration signal changes would tend to be perceived as linear self-motion (translation) or linear visual surround motion. Long-duration signal changes would tend to be perceived as altered orientation of the receptor (tilt) with respect to gravity.

The apparent discrepancy between the findings reported by Clément and his colleagues and by Reschke and his colleagues may be explained by considering temporal parsing of otolith signals. When the tilt-translation reinterpretation hypothesis was initially proposed, Alan Benson (personal communication) suggested that the self-translation illusion evoked by head roll postflight might be understood in terms of a shift of a "crossover" frequency rather than a complete reinterpretation. In the natural environment, translational accelerations are brief and therefore of relatively high frequency. In contrast, otolith signals produced by head tilt with respect to gravity can be of very low frequency, essentially down to DC. It is therefore reasonable to suggest that neural mechanisms develop to parse otolith signals as indicating self-translation or self-tilt based on signal frequency. Benson suggested that the initial OTTR proposal of otolith signal reinterpretation was better understood as a shift in the crossover frequency between the self-translation and self-tilt interpretations.

Benson's crossover suggestion is consistent with results from recent experiments. Paige and Seidman (1999) exposed primates to frequency modulated linear acceleration along the head interaural y-axis. Their results suggest that the CNS resolves "tilt-translation" ambiguity by applying high pass filtering to yield a "translation" response (indicated by horizontal eye movements) and low-pass filtering for the tilt-appropriate eye movement compensation (ocular counter-rolling). Paige and Seidman reported that the crossover frequency was approximately 0.4 Hz in primates (Figure 7-07). Wood's (2002) data, which were based on off-vertical-axis rotation in human observers, suggest a crossover around 0.3 Hz. The OTTR hypothesis can be refined to suggest that this crossover should be shifted toward a lower frequency immediately postflight. (This possibility will be examined by proposed experiments described below and in Chapter 4, Section 5.4.1). Consequently, the apparently discrepant results reported by Clément *et al.* and Reschke *et al.* may be attributed to differences in the stimulus frequencies used in the two sets of studies.

As noted in Chapter 3, Section 1.5.6, observer's roll and pitch motions from an upright position on Earth elicit both graviceptor and angular motion receptor signal changes. In his refinement of OTTR, Merfeld (2003) hypothesized that on Earth the brain may resolve tilt-translation ambiguity by integrating signals from the semicircular canals and the otolith receptors. Merfeld proposed *rotation otolith tilt-translation reinterpretation* (ROTTR) to account for postflight self-translation reports. He suggested that during orbital flight, the normal integration between semicircular canal and graviceptor signals is disrupted due to the absence of gravity. Consequently, the ability of the nervous system to use rotational cues to help accurately estimate the relative orientation of gravity (tilt) may be altered during adaptation to microgravity. If the tilt interpretation of otolith signal is no longer associated with canal signals, the interpretation of otolith signals by themselves during in-flight centrifugation as indicating tilt may be facilitated, as reported by Clément *et al.* (2001). Normally, facilitation by the semicircular canals of otolith signals' interpretation as indicating self-tilt would be frequency dependent. The semicircular canals can be modeled as angular accelerometers whose output is reduced above 0.05 Hz. Consequently, the tilt response to centrifugation might be observed at lower frequency both in- and postflight than prior to flight. Alternatively, the gain of the tilt perception function, analogous to the curve labeled 'torsional position gain' in Figure 7-07, may increase.

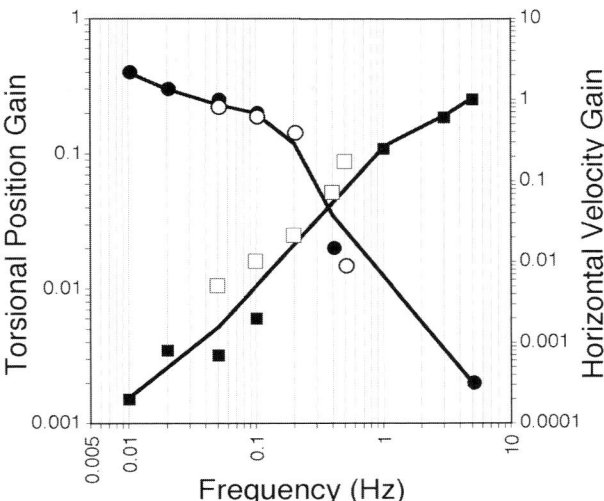

Figure 7-07. Gain for the torsional eye position (dots symbols) and the horizontal slow phase eye velocity (squares symbols) generated by interaural linear acceleration (filled symbols; data from Paige & Seidman 1999) or off-vertical axis rotation (open symbols; data from Wood 2002) plotted together across a broad frequency band. The CNS apparently resolves tilt-translation ambiguity on the basis of linear acceleration stimulus frequency. Higher frequency oscillation, which occurs in the natural world, is interpreted as head translation. Horizontal eye movement, which facilitates maintenance of gaze direction during translation, is evoked by higher frequency stimuli. Low frequency and constant stimuli are interpreted as changes in head orientation with respect to gravity (tilt) and evoke ocular torsion (counter-rolling). Crossover frequency is between 0.3 and 0.4 Hz.

Further research is likely to facilitate resolution of discrepant findings presented above. For example, investigations by Angeleki *et al.* (1999) of neural correlates of eye movement responses evoked by various combination of tilt and translation stimuli may clarify questions related to frequency parsing of otolith signals and semicircular canal contributions. Any resolution will also have to take into account additional findings such as reduced ability to compensate for roll tilt postflight reported by Merfeld (1996) (see Chapter 4, Section 5.4.1).

As noted in Chapter 4, Section 5.4.1, at the time we are writing this book, we are preparing an experiment that will measure perceived tilt in astronauts returning from space flight when they are exposed to ambiguous inertial motion cues. Subjects will be exposed to concomitant tilts and translations on a motion-base platform, which are phased such that the resultant linear acceleration remains aligned with the subject's longitudinal axis. During this *z-axis aligned gravito-inertial force* (ZAG) paradigm, the chair tilts (± 20 deg) within an enclosure that simultaneously translates so that the resultant GIA vector remains aligned with the longitudinal body axis. This condition will provide a space flight analog in that tilt signals from the canals conflict with otolith signals that do not indicate tilt. During such ZAG stimulation preflight, subjects have a tilt sensation of "swinging on a giant swing" (Golding *et al.* 2003). It is expected that, in agreement with OTTR, the crewmembers should experience a clear perception of translational motion postflight.

A third possibility to resolve discrepant findings is a variation of the well known "measuring instrument problem." It is possible that repeated centrifugation to evaluation tilt versus translation response may inhibit the "normal" adaptation process and lead to persistence of a 1-g adapted state.

It is important to refine our understanding of tilt-translation changes in microgravity because artificial gravity produced by centrifugation has been proposed as a countermeasure for the likely deleterious effects of long-duration exposure to microgravity during a mission to Mars and because possible countermeasures for space motion sickness include a preflight adaptation training procedure based in part on OTTR (Chapter 8, Section 2.2.2).

3.4   Proprioceptive Cues

In the same vein, there have been numerous reports of loss of orientation occurring when the only information available was related to proprioceptive cues. For example, on one space flight a crewmember reported that if he stopped moving his limbs he did not know their position in relation to his body unless he looked directly at them. Even once he established the location of his limbs relative to his body and the surrounding craft, he found it was difficult to initiate movement of his limbs. He reported that the affected limbs felt as though they had a "conformal coating" covering their surface.

In addition, movement of the limbs in the absence of vision often resulted in erroneous spatial orientation information. With eyes shut and feet in foot-loops attached to the deck of the shuttle, this crewmember reported that he felt he could oscillate the Shuttle about his apparent stationary body by slowly swaying to and fro in the foot restraints. Upon opening his eyes, he found that he was floating about 18 cm above the restraints, and had not actually moved around his ankle joints, but was flexing his feet. Other astronauts have reported similar perceptions of pitching and rolling the Shuttle while making voluntary pitch and roll head movements with their feet in the foot loops when their eyes were closed (Harm & Parker 1993). This suggests that tactile information is not required for orientation during flight, or is weighted low and will not be used if other information is available. These observations are supported by a considerable body of anecdotal reports (Harm & Parker 1993) suggesting that tactile input from Shuttle foot loops eliminates or greatly attenuates illusory self-motion associated with making voluntary head movements, particularly with their eyes closed.

Another experiment by Clément *et al.* (1987) examined adaptive changes in perception of orientation, including perception of subjective body orientation, during exposure to microgravity. Using two subjects who flew on the June 1985 Shuttle Discovery flight, it was observed that, when the subjects' feet were held in foot restraints, perception of subjective body orientation depended greatly on visual cues. Subjects tended to pitch forward while rotating around the ankles with respect to the spacecraft floor when vision was absent. Tilt angle was also strongly influenced with stabilized vision (restricted to foveal vision), and as might be expected, least affected with normal vision (see Figure 4-06).

It is interesting that proprioceptive input (rotation around the ankle joint) had little or no effect on correcting tilt angle in this experiment, even during the non-visual conditions. This may be related to the lack of proprioceptive input that has been observed when the limbs are held in a static position. For example, without a gravitational vertical, without visual or auditory cues, without input to the semicircular canals or responses to linear acceleration, and without touch or pressure cues, astronauts are left with only intrinsic vectors and static proprioceptive cues to determine orientation. It was noted as early as the Gemini flights that unless a limb is moved awareness of limb position may be lost. During one Apollo mission a crewmember awoke to see four floating luminous watch dials. After first determining that the dials were not part of the instrument display, he realized that the dials were those of the watches worn on each of the two astronauts' wrists. The crewmember did not know which watches belonged to him until he moved his arms (Schmitt & Reid 1985).

3.5 Tactile Cues

Gurfinkel *et al.* (1993a, 1993b) investigated the role of egocentric references for human spatial orientation as part of the joint French-Russian space flights. In their investigation, the perception of the orientation of complex tactile characters (letters and numbers) were applied to different areas of the skin while either proprioceptive information was varied or parts of the body were moved relative to the spacecraft. It was concluded that neither transition from one-g to zero-g, nor changing limb position, affected the subjects' ability to correctly identify the "skin writing." It was concluded that the egocentric body scheme can't be easily modified. An alternative explanation could be that there was no a priori reason to believe that the correct interpretation of tactile stimuli applied to either the trunk or different limbs would be affected by gravity. There is no evidence that changes in orientation relative to gravity either enhances or modifies our ability to correctly interpret "skin writing." Nor is it correct to believe, as the authors have suggested, that the perception of complex tactile stimuli is dependent upon an egocentric reference system.

In an awareness of position experiment performed during the First Spacelab Life Sciences flight (SLS-1), Watt attempted to separate two factors that may contribute to the apparent alteration of limb and body position in space that have been anecdotally reported on previous space flight missions (Young *et al.* 1993). He required subjects to point at five remembered target positions on a screen marked with a grid pattern with either the eyes closed for the duration of the test, or closed only during the pointing (see Figure 4-05). Reasoning was that alterations in proprioceptive accuracy would influence both tests, but that alterations in the extrinsic coordinate neural model would have been observed during the continuous pointing task. Compared to preflight, in-flight performance was poor (bias toward pointing low), but better when the eyes were closed only during the pointing task. Recovery to preflight levels required several days.

Overall these data suggest that, in the absence of vision, the maintenance of a stable extrinsic neural spatial map is dependent on the presence of a one-g gravitational force vector, and that proprioception during arm movement is normal in microgravity.

4 COGNITION IN SPACE

Spatial disorientation is not only due to sensory conflicts but can also be due to a conflict between defective or biased sensory inputs and the internal representations of the body and three-dimensional space. In particular, some forms of spatial disorientation are not due to peripheral deficits but to higher level mechanisms involving the cerebral cortex and the hippocampus, which contribute to the construction of the coherence of perception and the solution of perceptual ambiguities. Recent clinical studies have shown that a number of spatial orientation disorders involve the hippocampus and the parieto-hippocampal-prefontal networks for spatial memory during navigation, "spatial neglect" and the general problem of the perception of the subjective mid-sagittal plane of the body, and visuo-spatial anxiety. In the previous sections, we reviewed spatial disorientation resulting from illusory movements of the body or the environment. In this section, we will review the disorders resulting from conscious perceptions of the disorientation between the body and space, i.e., from cognition. The word 'cognition' is often used in computer science-related fields to denote the level of activities that require "understanding" of what is going on, rather than merely signal-level reaction.

An accurate representation of the environment is crucial for successful interaction with the objects in that environment. It is clear that humans have mental representations of their spatial environment and that these representations are useful, if not essential, in a wide variety of cognitive tasks such as identification of objects and landmarks, guiding actions, navigation, and in directing spatial awareness and attention. *Spatial awareness* is, very simply, an organized awareness of the objects in the space around us. Spatial awareness requires that we have a model of the three-dimensional space around us, which is also based on the integration of sensory information and experience.

Perhaps the most comprehensive examination of spatial awareness in microgravity was completed by Friederici & Levelt (1987). These investigators designed a series of experiments that systematically varied retinal information, visual background information, proprioceptive information, and gravity. Verbal descriptions of visually presented arrays were solicited under different head positions (straight vs. tilt) both on the ground and in-flight. The results clearly showed that different coordinate systems are used under the two different gravity conditions. Under terrestrial conditions, it is the gravitational vertical that is chosen for primary reference, and in microgravity, retinal information is primary. Concerning sensory weighting, the data support the hypothesis that unambiguous spatial assignment is achieved by cognitive weighting of the different perceptual cues, and that possible conflicts between different sensory input are solved by giving dominate weight to only one sensory input. Their data may support the finding by Clément *et al.* (1987) of on-orbit postural tilt when vision was restricted even though there was considerable proprioceptive input from the ankle rotation that would provide tilt cues.

In interviews with astronauts, many of them report that after looking out the windows for even short periods of time, the flight deck appears to be "upside-down" when they look back into the cabin. This reorientation visual illusion might also be due to our natural tendency to perceive the Earth as "down" (Figure 7-08). Consequently, when looking at the Earth out of a window "above" their head, some crewmembers may feel that they are just standing on their head. Experience and cognition are presumably at the origin of some of these illusions of self-orientation.

ISS Expedition-7 crewmember Ed Lu noted, "There really isn't an up or down anywhere here, but there is a direction we think of as the floor and a direction we think of as the ceiling in each module. Most of the labeling on panels and equipment is written so that it is right side up assuming this orientation, and also most of the lights are on the 'ceiling' so they cast light 'downwards'. To add to the effect, there is a simulator back on Earth we spent a lot of time in where we got used to one direction as the floor and the opposite direction as the ceiling. So up here, when Yuri and I say downwards or upwards, we mean the equivalent directions as in the training module on Earth."

4.1   Behavior and Performance

On board an orbiting vehicle or during a planetary mission, each crewmember's ability for normal cognitive functioning is essential. Such tasks as docking with a resupply spacecraft or conducting extra-vehicular activity require adequate cognitive performance. There are, however, many factors that are known to occur in space travel that may negatively affect cognitive function. Among these are fatigue, sleep loss, emotional stress, depression, mood disturbance, over-extended tasking, excessive noise, spatial disorientation, and impaired sensorimotor input, particularly time distortion, and visual dysfunction (Robbins 1988). Additionally, there are risks of illness, exposure to environmental toxins, and possible effects of medications, any of which could significantly impact cognitive ability.

Several reports by both U.S. and Russian investigators have shown that visual tracking abilities and reaction time increased in astronauts and cosmonauts. In several occasions, difficult or failed attempts at docking have been attributed to impaired psycho-motor ability. In one of these incidents, a cosmonaut "required twice as much time for ship orientation movements in the early orbits as he had on the ground and on later orbits" (Ivanov *et al.* 1972, p. 9).

Furthermore, anecdotal information suggests that, "while overall performance has been remarkably good, decrements have been evidenced in experimental errors, lost data, equipment mishandling, and a variety of behavioral disturbances..." (Robbins 1988, p. 72). This information led both the Robbins (1988) and Goldberg (1987) committees to report that the influence of the various aspects of the space environment on cognitive functions warrant investigation.

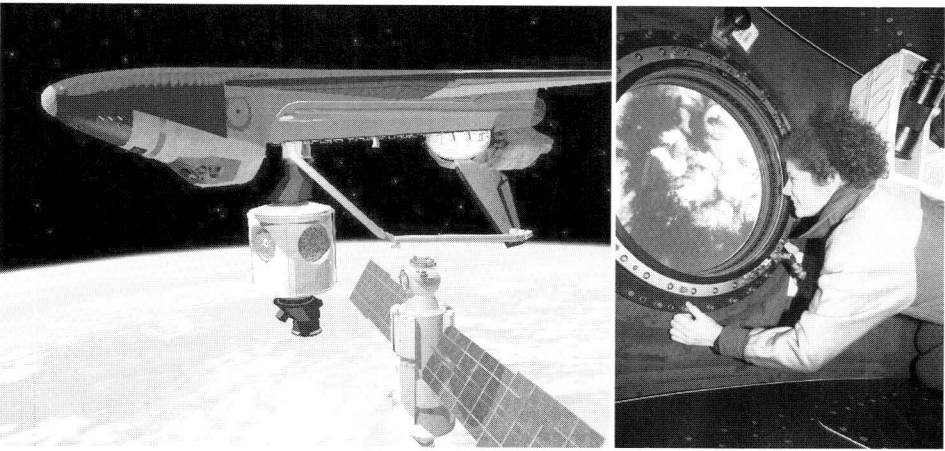

Figure 7-08. When astronauts in orbit look out the window and see the Earth "below", it generally gives then the illusion of looking down, and consequently, that they are upside-down relative to the Earth. Photo courtesy of NASA.

4.2 Navigation

Vertebrate brains form and maintain multiple neural maps of the spatial environment that provide distinctive, topographical representations of different sensory and motor systems. For example, visual space is mapped onto the retina in a two-dimensional coordinate plan. This plan is then remapped to several locations in the central nervous system. Likewise, there is a map relating the localization of sounds in space and one that corresponds to oculomotor activity. An analogous multi-sensory space map has been demonstrated in the mammalian hippocampus, which has the important function of providing short-term memory for an animal's location in a specific spatial venue. This neural map is particularly focused on body position and makes use of proprioceptive as well as visual cues. It is used to resume the location at a previous site, a process called *navigation*.

This system of maps must have appropriate information regarding the location of the head in the gravitational field. So it follows that the vestibular system must play a key role in the organization of these maps. Only recently has this been demonstrated by experiments carried out in space. During an experiment performed on board Neurolab, rats ran a track called the Escher staircase, which guided the rats along a path such that they returned to their starting location after having made only three 90-deg right turns (Knierim *et al.* 2000). On Earth, rats could not run this track. But in space, it provided a unique way to study the "place cells" in the hippocampus that encode a cognitive map of the environment. The rats had multi-electrode recording arrays chronically implanted next to their hippocampal place cells. Early in the mission, the 20-40 hippocampal cells had abnormal firing patterns and could not establish clear links with places. But by the ninth day of flight, the neurons were able to form these links (Figure 7-09).

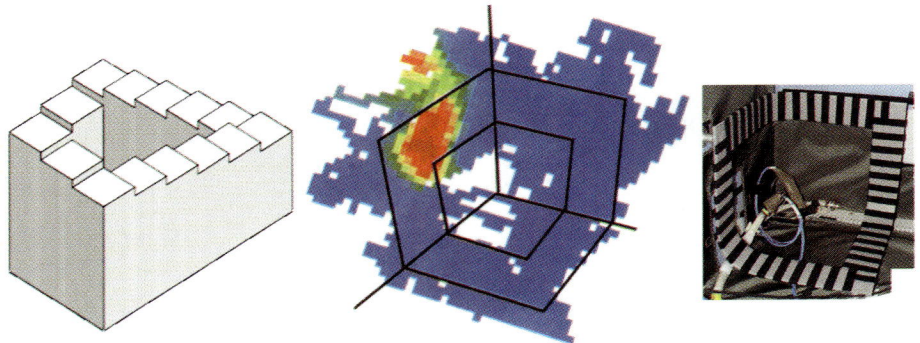

Figure 7-09. During the Neurolab flight, rats were running along a track (right) in the shape of M.C. Escher's impossible staircase (left). The animal could return to their starting location after having made only three 90-deg yaw turns instead of the usual four. The firing patterns of their "place cells" in the hippocampus (middle) was abnormal during the first testing on FD4, but returned to normal, i.e., showing no precession, during later tests on FD9. Adapted from Knierim et al. *(2000).*

The gradual adaptation of the place cells may reflect the mechanism by which rats and astronauts use visual cues to override the disorientation that weightlessness can produce. It could partly explain the inversion illusion and the navigation difficulties experienced by some astronauts when they arrive in space. A weightless environment presents a true three-dimensional setting where Newton's laws of motion prevail over Earth-based intuition. We normally think in terms of two dimensions when we move from place to place. However, in weightlessness, one might decide the best way to reach the end of a module is to go across the ceiling and then follow the opposite wall. Terrestrial constraints on orientation can be violated in a weightless environment, but our prior cognitive experience with the one-g environment may limit the perceptual patterns that are experienced (Lackner 1992a, 1992b). In other words, a cognitive map of terrestrial possibilities influences the perceived orientation.

Anecdotal reports from astronauts indicate an inability to maintain an external spatial map in the absence of both vision and gravity. During long-duration missions, astronauts report that they need at least one month to become "natural and instinctive." Furthermore, mental survey knowledge is not completely developed even after three months in some cases. In one case, for example, an ISS astronaut was filming the Earth out the porthole in the laboratory module. He frequently had to look back into the cabin and to use the on-board computer to complete the task. He said that he never "knew" where the computer was located; he always had to do a visual search to find it even though his orientation within the cabin (and hence his position with respect to the computer) remained essentially constant throughout the filming (Pettit 2003).

Because visual objects can be located in vastly more places in microgravity, it is reasonable to expect that resting focus might be altered by microgravity in spite of the absence of systematic studies on this topic. On Earth, visual objects ordinarily rest on floors and tabletops, hang on walls, or are suspended from ceilings. However, objects can be suspended anywhere in the volumes enclosed by the spacecraft in microgravity. Some astronauts have noted that one must overcome the gaze fixation habits that are appropriate on Earth to successfully search for floating objects in microgravity. The nearest we can come to the visual spatial environment on Earth could be the placement of objects in the refrigerator. Frequently the orientation of an object changes what we expect to see. Imagine a pen floating in mid-air at eye level. If the pen is viewed end-on it becomes a dot rather than a familiar writing instrument. Crewmembers on board the first *International Microgravity Laboratory* (IML-1) mission lost a huge IMAX camera within the Spacelab for a considerable amount of time. While interesting, this problem has never been systematically investigated during a space flight.

Another problem is that each module of the ISS provides a local visual frame of reference for those working inside. Once the ISS construction will be complete, the modules will eventually be connected at 90-deg angles, so not all the local frames of reference will be co-aligned. It might sometimes be difficult to remain oriented, particularly when changing modules. Even after living on board for several months, it could be difficult to visualize the three-dimensional spatial relationships among the modules, and move though the modules instinctively without using memorized landmarks. Crewmembers will not only need to learn routes, but also develop three-dimensional "survey" knowledge of the station. Disorientation and navigation difficulties could be an operational concern in case an emergency evacuation is required in the event of a sudden depressurization or fire.

An interesting investigation was performed by Bloomberg *et al.* (1999), in which the ability for crewmembers to repeat a previously seen trajectory without vision was examined. When attempting to walk a triangular path after flight, blindfolded subjects showed both under- and over-estimations of the distances walked, but a correct estimation of the angle turned (Glasauer *et al.* 1995). These results suggest a difficulty for reconstructing motion cues from the otoliths, but not from the semicircular canals. However, the changes found could also be related to the lower walking velocity during postflight testing. These results imply that mechanisms like computing self-displacement and updating spatial information (navigation) are disturbed by space flight and have to be re-acquired after return to Earth.

4.3   Mental Rotation

A large number of astronauts report that when they look outside, the Space Shuttle seems upside-down. They also report that, after several days in zero-g, they can mentally rotate whatever is in their field of view to make it rapidly become a floor or a wall. For example, Ed Lu, an ISS Expedition-7 crewmember, recalls: "In space you need to remember that you aren't limited like you are on the ground to having your feet on the floor – they can just as easily be on the wall or on the ceiling. I find that when I am working a tight space, I don't really think about any particular direction as up or down, but when out in an open space like in the middle of a module I do. If for instance I am up on the ceiling, by concentrating I can make myself think of the ceiling as the floor." These astronauts adapt to microgravity by increasing the weight of the visual polarity vector, as indicated in these reports. In these visual-dependent subjects, mental rotation ability may be important for mitigation of space motion sickness and effective performance in microgravity.

There is evidence that the way we recognize many objects is tied to the gravitational and retinal orientation in which we normally see them (Howard 1982). Familiar objects, such as chairs, tables, trees, people, and houses, are normally seen in one orientation with respect to gravity and the normally-vertical retinal meridian, and with fixed orientations with respect to omnipresent surfaces, such as walls, floors, and ceilings. For example, a chair is always seen resting on the floor, never on the walls or the ceiling. The major axis of symmetry of these objects is usually aligned with gravity. They also have asymmetries (also called *polarities*) that give them an intrinsic "top" and "bottom" by association. Simple objects can be coded in memory in orientation-free form, but visual features that are more spatially complex are represented in memory in more retinal orientation-specific ways. Examples of the latter are the letters of the alphabet or the subtle features of a human face. The inside surfaces of a spacecraft can be considered as complex visual "objects" (Oman 1989).

Recognizing orientation-specific features of an object in an abnormal orientation requires that the brain perform a mental rotation of the object image. When the rotated images matches a template stored in visual memory, the angle through which the image has been rotated probably determines the perceived orientation of the object. The rotation of mental images has been studied most intensely by means of what is known as the *Shepard task*. This task involves two stimuli that differ by a rotation and possibly by a reflection; the subject decides whether one version of the stimuli is a reflected version of the other. The most common strategy for accomplishing this task is mental rotation. The experiments by Shepard & Metzler (1971) elegantly demonstrated that the time required to identify a rotated object was proportional to the angle through which it has been rotated (Figure 7-10).

On Earth, gravity provides a convenient "down" cue. Large body rotations normally occur only in a horizontal plane (yaw). In space, the gravitational down cue is absent. When astronauts roll or pitch upside down, they must recognize where things are around them by a process of mental rotation that involves three dimensions, rather than just one. To investigate adaptation to mental rotation of the visual environment, Clément *et al.* (1987) asked the astronauts over a period of days to report the critical tilt angle relative to the Shuttle interior from which it was no longer possible to mentally reconstruct the original environment. On Earth, most subjects cannot recognize a familiar room when they are tilted by more than 60 deg with respect to gravity (Lackner

& Graybiel 1993). It was found that the subjects in microgravity got better and better in mentally rotating the Shuttle middeck. After the third day in-flight most astronauts could mentally rotate the environment even from the complete "upside down" orientation.

However, the preferred orientation remained the upright orientation. One Skylab crewmember observed, "Being upside down in the wardroom made it look like a different room that the one we were used to. After rotating back to approximately 45 deg or so of the attitude that we normally called 'up', the attitude in which we had trained, there was a sharp transition in the mind from a room which was sort of familiar to one which was intimately familiar… We observed this phenomenon throughout the whole flight." Another noted, "At certain times of the day the view outside the wardroom window was upside down. It was convenient just to flip upside down and look out […] You would finish your observation and start to come back and rejoin the world inside, when you were about 45 deg away from the [visual] vertical, all of a sudden your new frame of reference would lock in and everything was right side up again" (Cited by Oman 1989, p. 33).

Ed Lu, ISS Expedition-7 crewmember, also noted, "It is interesting how the same module can look like a completely different place when viewed upside down. Every once in a while here, I spend a few hours upside down just for fun, treating the ceiling like the floor. The familiar modules don't look so familiar anymore. This is true even though I am completely used to flipping my body around to whatever orientation needed to work on things."

After a few days in orbit, when astronauts readily transition between Earth referenced down and cabin-referenced down, they presumably become adept at performing mental rotation as mission progresses. Matsakis *et al.* (1992) studied the speed of mental rotation while subjects were exposed to microgravity during a Russian (Soviet) space mission. Cosmonauts were trained to perform a mental rotation using the Shepard & Metzler (1971) procedure prior to flight (Figure 7-10). These cosmonauts exhibited significantly improved mental rotation performance during the mission relative to that observed preflight. This difference seems to be particularly marked for stimuli calling for mental rotation in roll or pitch (an actual body rotation around both of these axes would induce different responses from the otolith organs in weightlessness compared to Earth).

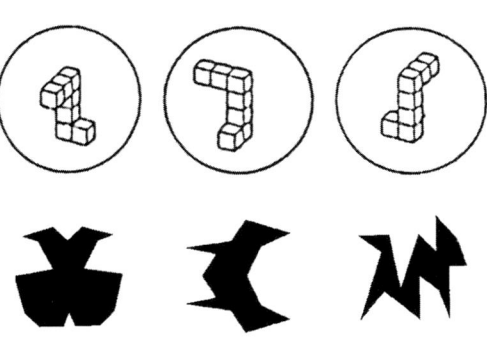

Figure 7-10. Above. Examples of shapes used for a mental rotation test. When the shape on the left is presented with the shape on the right (a 180-deg rotation), the time taken to decide that both shapes are the same is about 6 sec. When the shape on the left is presented with the shape in the middle (a 90-deg rotation), the response time is only 3 sec. The speed of mental rotation in this test is about 30 deg/sec. Bottom. These figures have a vertical symmetry axis (left), a horizontal symmetry axis (center) or no symmetry axis at all (right).

However, a later study in which the repertoire of objects was different between all experimental sessions to avoid a learning effect, showed no significant differences in rotation time in space versus ground data (Léone 1998). So, the results are inconclusive at this point and further studies are needed to confirm whether mental rotation is facilitated or not in microgravity.

Nevertheless, anecdotal reports from astronauts also suggest mental rotation of the visual environment improves over the course of the mission (Harm *et al.* 1993). Many crewmembers report that early in-flight when they want to move from their current location to another location, they don't know whether the desired location is to their left, right, above or below them if they are in an orientation other than "feet to the deck." For example, one Shuttle astronaut said, "I usually carry around with me a sense of the Orbiter vertical no matter what my orientation is. There were times, like one day just for grins I decided to turn it upside down and use the Orbiter ceiling as the floor. So you go downstairs to the flight deck and you can do that and after awhile it does flip. It's kind of fun because everybody is upside-down. It takes awhile to convert the flip in some orientations. Sometimes it flips quickly, sometimes it doesn't." The greater the misalignment between their intrinsic coordinates and the extrinsic coordinates of the visual environment, the longer it takes to determine the correct direction to move to get to the desired location. Most crewmembers report their ability to determine the correct direction to move in, regardless of their orientation, improves over time.

To understand the changes in the ability to do mental rotation in space, the following considerations may be important. Weightlessness provides a release of the gravity-dependent constraint on mental rotation. This facilitates the processing of visual images in any orientation with respect to the body axis. In addition, when astronauts and cosmonauts are in microgravity, they can be in orientations and move in ways that are not possible on Earth. Mental rotation is important for efficient goal-directed locomotion. One must orient in order to locomote efficiently on Earth as well as in microgravity. The Shepard & Metzler procedure may be viewed as "passive," though it seems unlikely that this passive procedure would invoke the same neural response as would active locomotion. Learning complex procedures often seems to elicit improved performance on the simpler, single-component procedures. Learning to perform the complex mental rotations associated with locomotion (that may start from unusual orientations) may facilitate performance on a passive mental rotation task (Parker & Harm 1993).

Other experiments have investigated whether it was easier to detect the presence of a symmetry axis in absence of gravity (Léone *et al.* 1995). For example, it is well known that on Earth, a vertical axis of symmetry is faster to identify than a horizontal and an oblique axis of symmetry (Figure 7-10). A change in the position of the head relative to the trunk on Earth influences symmetry detection. One experiment performed in space on five astronauts indicated that both vertical and horizontal axes of symmetry were equally faster to identify (Léone 1998).

In an experiment with a tactile matrix in which tactile stimuli were presented in different body parts, the interpretation of the tactile shapes depends upon the relative orientation of body parts (Gurfinkel *et al.* 1993b). This experiment showed that: (a) the types of representations of tactile shapes are the same as on the ground; and (b) the latency for recognition is the same, and therefore there is no suggestion for an added processing time like the one which one observes in mental rotation (this would have added seconds to the latency in the present experiment). It is remarkable that very little errors are made in space in the tactile matrix recognition task. This could be due to the fact that there are two possible reference frames, one intrinsic (egocentric) and one extrinsic (allocentric), and that in space there is only one that is used by the brain. This may simplify the task and reduce the number of errors.

4.4   Mental Representation

The force of gravity provides a constant reference for orientation under terrestrial conditions, thus determining unequivocally the direction of up and down (Howard & Templeton 1966, Howard 1982). Our natural environment also embodies an orientation polarization. The sky is above our head and the ground in under our feet; rooms have floors and ceilings. Only certain body orientations and configurations are possible within this environment. Locomotion can only take place on the ground or the floor of a room, not the walls or ceiling. As a result, only certain perspectives of the outside world or the inside of a room are naturally possible. For example, one cannot view the floor of the room or the walls from the perspective of being physically located at the ceiling. Similarly, one cannot (without being artificially suspended) see one's feet spatially separated from the floor when no other part of one's body is in contact with the floor of the room. Our cognitive representation of space is therefore based on the assumption that we are living on the surface of a three-dimensional world.

It has been argued that the Earth's gravitational field is one of the most fundamental constraints for the choice of reference frames for the development and the use of cognitive representations of space. For example, a subject looking at a diamond-shaped figure (in retinal coordinates) perceives a square-shaped figure when he and the figure are both tilted by 45 deg relative to gravity (Figure 7-11). This result indicates that an object's form perception generally depends more on the orientation of this object in world (spatial) coordinates than on its orientation in retinal coordinates. In other words, gravity is critical for the extraction of an object's reference frame (Rock 1956).

Figure 7-11. The appearance of geometric figures changes with the tilt of the observer's relative to gravity. In the square/diamond illusion, a 45-deg tilted square is perceived as a diamond when the head is in the upright position, but as a square when the head is tilted by 45 deg relative to gravity. Adapted from Rock (1973).

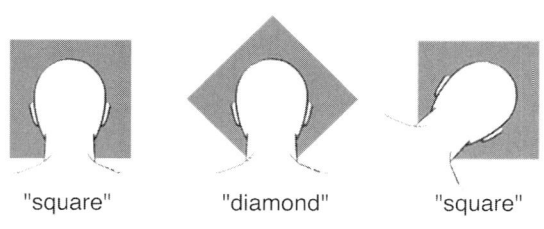

"square"   "diamond"   "square"

When we execute movements, we must know the relationship between our body parts and the external space. Representation of body is called *body scheme* or *body image* (Ramachandran & Blakeslee 1998). The body scheme is an abstract internal representation of spatial and physical-mechanical properties of one's body based on some combination of past and current information, which is supposed to be described in the egocentric reference frame (Parsons 1990). Recent studies have revealed that body scheme is not simply a representation of joint angles, but rather a complex integration of proprioceptive, kinesthetic, vestibular, visual inputs, and motor feedback (e.g., the sense of effort). In the last 15 years, a number of neuropsychological experimental studies performed with adequate technology have shown that vestibular stimulation affects a variety of disorders concerning the body scheme in patients, such as unilateral neglect, hemiplegia, motor neglect, and inattention (Vallar 1998). Optokinetic stimulation, which may be considered functionally equivalent to vestibular stimulation with respect to its perceptual and motor effects (Dichgans & Brandt 1978), has also a positive effect on some of these syndromes (Rossetti & Rode 2002). Therefore, it is reasonable to expect that weightlessness, by altering the functioning of the vestibular and proprioceptive receptors, could be responsible for perceptual distortions of the body scheme and the environment. It is also well known that mental images containing depth and spatial information are generated from sensory information and from information held in long-term memory (McCabe *et al.* 2001). Although the mechanisms by which a mental representation is forming are still not understood, there is evidence that mental representation is not a static entity; it evolves with experience. Exposure and adaptation to a novel gravitational environment may therefore potentially alter mental representation of space.

True, the effects of gravity on these cognitive processes could be more simply and cheaply investigated by tilting an individual relative to gravity, whether it is an actual tilt using a tilt table or a virtual tilt using a centrifuge. However, one problem with these ground-based studies is that tilting the observer's relative to gravity on Earth creates a conflict between perceived gravitational (extrinsic) vertical and retinal- or body-defined (intrinsic) vertical, but does not suppress the gravitational information. By contrast, the loss of the gravitational reference in space flight provides a unique opportunity to differentiate the contribution of intrinsic and extrinsic factors to the orientation system in astronauts.

Measuring the changes in the mental representation of an object throughout a space mission is a simple way to assess how the gravitational reference frame is taken into account for spatial awareness and spatial orientation. Results of space studies suggest that the absence of the gravitational reference system, which determines on Earth the vertical direction, influences the mental representation of the vertical dimension of objects and volumes. These studies, including copy drawing and handwriting tests, as well as depth and distance perception, are detailed thereafter.

4.4.1 Drawing and Handwriting Tests

The vestibular, proprioceptive, tactile, and visual senses result in eye-hand coordination, body perception, motor planning, activity level, attention span, and emotional stability. Handwriting is a very complex skill that requires many of these systems to work well together. Handwriting skills are prerequisite for cognitive, perceptual and motor development. People with difficulties in sensory processing, spatial awareness or perceptual skills usually have handwriting problems (Burton *et al.* 1990).

When we draw, we represent images with perspective to express our view of a three-dimensional world in a flat two-dimensional projection. An interpretation of one person's drawing can then be used to assess his/her active perception of the world and the construction of this equivalent self contained two-dimensional world. In fact, drawing or copying tests are commonly used in neuropsychology to investigate visuo-spatial deficits. For example, the *Clock Drawing Test* (CDT) is used to assess visuo-spatial neglect in stroke patients (Agrell & Dehlin 1998). Visuo-spatial neglect is the inability of a brain damaged patient to react to visual stimuli on the opposite side to the lesion. The test system comprises a graphics tablet connected to a portable computer.

Test overlays are placed on the tablet and are completed by the patient. The patient is asked to copy a clock-face, mark the hours, and then draw the hands to indicate a particular time (e.g., 10 minutes past 11). Although different scoring templates for the CDT exist, most often code for features such relative size, spacing and placement of numbers or hands, disorganization, preservation, completeness, and other potential errors, which are hypothesized to indicate cognitive and visuo-spatial impairment (Figure 7-12A).

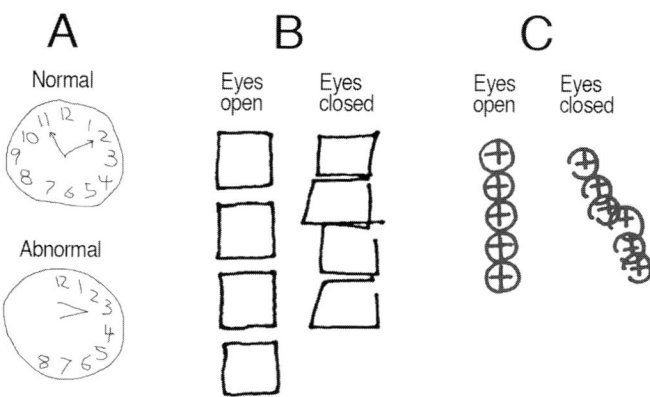

Figure 7-12. A. Clock Drawing Test *(CDT). While neurologically normal individuals are usually able to draw reasonable copies of a clock, those with unilateral neglect will usually fail to draw the left face of the clock accurately by leaving out the numbers on the left side of the picture. B.* Square Drawing Test *(SDT). Drawings of squares look normal in vestibular patients with the eyes open, but are shorter in the vertical plane with the eyes closed. Adapted from Sekitani* et al. *(1976). C. Spontaneous objects drawings in patients with vestibular syndrome. With the eyes closed, the objects and the space between them are "squished" in the vertical direction and deviated toward the side of the lesion. Adapted from Fukuda (1983).*

The *Square Drawing Test* (SDT) devised by Sekitani (1976), which is a modification of Fukuda's vertical writing test (Fukuda 1983), is utilized in Japan to evaluate vestibular dysfunction. In this test, patients are asked to draw a series of squares vertically with the eyes closed (Figure 7-12B). A correlation was found between parameters of the SDT, in particular the distance between the squares and the total length, and the performance of the vestibular system evaluated by caloric nystagmus (Miura & Sekitani 1993), demonstrating that SDT is a feasible method to evaluate and follow recovery in patients with vestibular dysfunction.

Lathan *et al.* (2000) once asked two crewmembers to draw the well-known Necker's cube on a notebook with the eyes closed, and have compared these drawings between preflight and in-flight The Necker's figure is the simplest representation of a three-dimensional object in a two-axis coordinate system (Necker 1932). Comparison between the length of line between the cubes drawn on the ground and the cubes drawn in space revealed a 9% decrease in size in the vertical dimension (i.e., the height) of the cubes drawn in weightlessness (Figure 7-13).

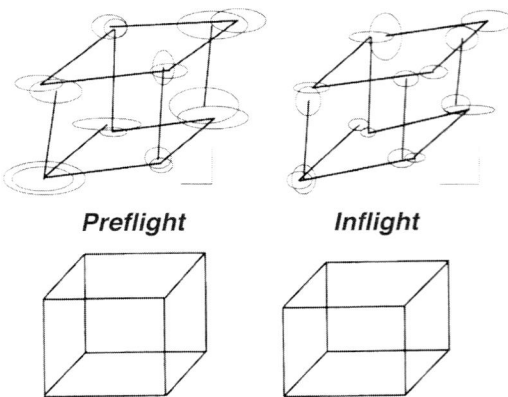

Preflight   **Inflight**

Figure 7-13. Top: Mean and standard error of each endpoint of Necker's cubes drawn by an astronaut preflight and in-flight. Bottom: Normalized mean preflight and in-flight cubes. Horizontal and oblique lines were unchanged, but the vertical lines were shorter in microgravity than on Earth (Clément 2005).

As a consequence of its gravitational basis, the upright direction has always been the most, salient, constant, and unique direction of our world. The fact that the up-down dimension is the primary dimension altered when gravity is no longer sensed is therefore not surprising. There is a natural correspondence between the forward-backward direction in the horizontal plane, which is orthogonal to the gravitationally-defined upright, and the up-down direction aligned with that upright, respectively (Shepard & Hurwitz 1984). The fact that the length of the vertical lines only is significantly different in weightlessness suggests that it is the mental representation of the height of the cube that has changed, and that adaptive changes in proprioception are presumably not responsible for this effect. It is well known that when a person judges the orientation of an object, he or she relates the direction of some recognizable axis in the object, an intrinsic axis, to some axis of reference outside the object, an extrinsic axis (Howard 1982). Gombrich (1969) has argued that the act of drawing involves first making an internal representation of the thing to be copied. In the case of a three-dimensional cube, the internal representation is presumably normalized with respect to gravity, i.e., the vertical axis of the cube is defined with reference to gravity. Consequently, although the drawing and writings are made on a horizontal surface, when the gravitational reference is absent, the mental representation of the vertical axis of these objects and words is altered.

Similar results have been found in another study involving two astronauts. The trajectory of hand-drawn ellipses in the frontal plane in the air with the eyes closed revealed a 10-13% decrease in the vertical length of the ellipses relative to preflight, whereas the horizontal length of the ellipses were basically unchanged (Gurfinkel *et al.* 1993a). This result supports our hypothesis that the mental representation of the vertical dimension of objects or volumes (in this case an ellipsoid) is altered during exposure to weightlessness.

Handwriting tests are also utilized for the assessment of vestibular disorders (Figure 7-12C). Japanese investigators have long used vertical writing tests to diagnose patients with impairment in motor function from those with vestibular disorders (Fukuda 1959, Sekitani *et al.* 1976). Compared to patient with impaired motor function where the size of written characters is irregular and their spacing uneven, the writing of patients with vestibular disorders is regular, but there is a reduced space between vertical characters with the eyes closed compared with the eyes open (Figure 7-14).

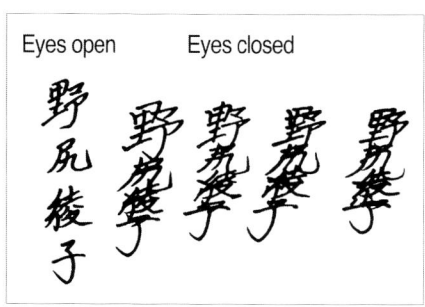

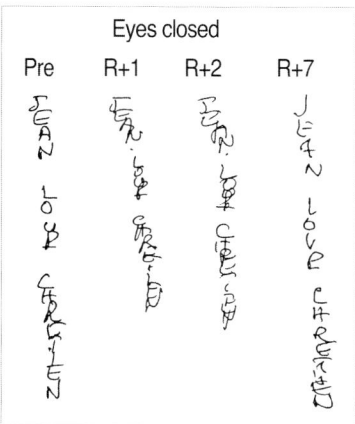

Figure 7-14. Left panel. Vertical writing test in a vestibular patient. With the eyes closed, the total length of the written words and the spacing between the vertical characters are shorter than with the eyes open. Adapted from Fukuda (1983). Right panel. Vertical writing test in an astronaut returning from a 28-day space mission (Pre: before flight, R+1: one day after landing, etc.). The total length of the words and the distance between the vertical letters also changed as a result of adaptation to space flight (Clément 2005).

When an astronaut was asked to write his name with the eyes closed during a Shuttle mission, the length of the words written "vertically" was smaller in-flight compared to preflight (Clément *et al.* 1987). In another astronaut, the reduction in the vertical length of words was observed during several days after returning from a 28-day space mission on board Mir (Figure 7-14). It is interesting to note that in both experiments, the size of the letters did not change in-flight or postflight, but the vertical distance between them and the total length decreased in-flight and early postflight compared to preflight. Since the notebook was always in the same horizontal position (velcroed to the subject's knees in-flight), it is the mental representation of the vertical layout of letters that in fact was altered when the gravitational frame of reference was removed.

Why do the astronaut draw and write objects and words shorter in the "vertical" dimension in weightlessness? It is well known that the vestibular system plays a role in the building up of the mental representation of space. Some patients with damages to the brainstem component of the vestibular disorders report a sense of heaviness and an illusory or hallucinatory perception of their body being smaller in size than the normal (Tran Ba Huy & Toupet 2001) (Figure 7-15). Like the hemineglect patients who do not draw the part of the clock that corresponds to the lesionned side, does a patient with an otolithic disorder draw and write shorter objects in the vertical dimension? Preliminary results of studies performed by our group tend to support this hypothesis.

It is interesting to note that on Earth it takes the same time to write a letter or a word at different sizes, which implies proportional changes in velocity. In other words, the time of the task is held constant while one adjusts the velocity (Michel & Schotti 1975). Also in typing, lifting a weight and grasping objects there is constancy in the shape of the velocity profile for a given work at different sizes (Jeannerod 1988). Thus, a strong tendency exists to keep the execution time of these complex trajectories independent of the movement size (Viviani & McCollum 1983). Gravity is presumably taken into account in the internal dynamic representation of the intended movement (Kingma *et al.* 1999, Kandel *et al.* 2000). However, its exact role is still unknown. In an upcoming space experiment, ISS crewmembers will write and draw memorized two- and three-dimensional objects of different sizes on a digitizing tablet. The analysis of the writings and drawings, looking both at amplitude and velocity profiles, will allow to further assess the changes in perceptual-motor alterations during space flight. Complementary experiments will use psychophysical measures to examine depth and distance perception without requiring a perceptual-motor task to further delineate the effects of exposure to microgravity on cognitive processes.

Figure 7-15. Some patients with vestibular disorders complain that their body feels shorter or larger than normal, that the floor is slanted and the walls of building are tilted or have skewed shapes. Such changes in their body scheme and erroneous mental representation of three-dimensional space could be related to the part of the central vestibular system that processes otolith input.

4.4.2 Depth Perception

An object can look shorter than it actually is because its angle relative to the viewer is different. A free-floating astronaut sees the objects restrained to the floor under a different perspective, so this may give the illusion that the objects are shorter. This effect might also be related to a change in the perception of the distance of the objects.

There is a difference between *distance* perception and *depth* perception and it is important to distinguish between the two. *Distance perception* is the ability to see and recognize distances between people or objects in any and all directions relative to the viewer's eye. It is the ability to view objects near to far, and at varying angles, and to be able to accurately and quickly estimate: (a) distance from a person's eye to a particular object; and (b) distances between specific objects no matter what the directions and distances whether outward from viewer's eye, or left-to-right distances between objects. By contrast, *depth perception* has very specific and limited meaning. This is the distance straight ahead of the viewer's eye, toward or into an object or the surface of an object. By definition, depth is looking straight into a hole or tube and estimating forward distances (Walk & Gibson 1961).

Depth perception is based on geometrical principles familiar to an observer. As an example, accommodation is the change of the focal length of the eye lens to place a focused image on the fovea of the retina, which provides a signal to the brain that can be processed to sense depth. For distances less than about 10 meters, the muscular action of convergence provides unambiguous depth information. Because our two eyes are separated by about 10 centimeters, they form slightly different images on each retina. This separation produces binocular disparity, or stereoscopic vision. As a result, nearby objects (< 20 m) appear at slightly different positions against a more distant background through each eye. Objects are also are seen from a slightly different angle by each eye. When we move our head back and forth, this image shift or parallax can be increased. Motion parallax greatly extends depth perception, which gives the brain a much larger model world in which to place ourselves.

Fusion of the two different images seen by each retina into one three-dimensional world image occurs in the CNS. Any real object reproduced in a two-dimensional image is ambiguous. The image on the retina is not the real object and does not have "depth," just like a photograph. Fortunately, many aspects of nature change with distance in predictable ways. Such changes give depth cues for the objects they affect. Monocular cues for depth include linear or *geometrical* perspective in which objects appear smaller when they are farther away, overlapping, shading and shadows, aerial perspective, which entails the loss of contrast and "blueness" with increasing distance due to scattering of light in the atmosphere, pattern changes, and color itself. Artists make use of these cues to give apparent depth to a flat picture.

The visual system also partially relies on past experience to judge the shape and depth of objects. That is, top-down processing is used. All depth cues, both binocular and monocular, are learned through experience in one way or another. Babies must learn how to look at things. Even though binocular disparity may be hardwired into our genetics, some learning of how to use it must take place. Our past experience can lead to some funny misinterpretations. The interior of a facemask appears to project outward, because faces always have the nose pointing toward us. Culturally mediated experiences may result in individual differences in depth perception, which has been proposed by several studies. On the one hand, constant exposure to photographs and pictures of representational drawings would contribute to the development of secondary depth cues. On the other hand, people who live in environments largely free of right angular corners and parallel lines would not experience depth based on linear perspective (Gregory 1965).

Space experiments have begun to investigate the role of depth cues in absence of a gravitational frame of reference. Howard *et al.* (1990) had shown that the illusory concavity or convexity people normally perceive when interpreting shading on a truly flat surface depends on a "light comes from above" assumption, where "above" depends on the relative orientation of the dark-to-light shading gradient to head orientation, and to gravity. During the Neurolab mission, crewmembers were presented with convex or concave shaded figures (Figure 7-16). After several days in space, they could not use so reliably that light information for depth, because they had been exposed to situations where the light source could come from any direction while they were free-floating in the cabin (Oman 2003).

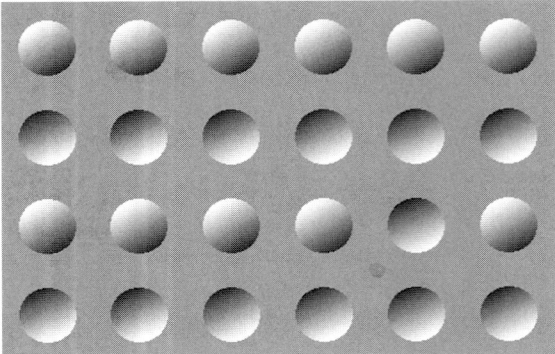

Figure 7-16. On Earth, these shaded figures provide unambiguous depth perception cues and we see them convex or concave because we are used to having light sources above us, projecting downward. In microgravity, the strength of this illusion was less than on Earth because astronauts are used to see light coming from any direction when free-floating.

Other experiments have used classic geometrical illusions of size, which on Earth generate inaccurate judgments because they provide misleading depth cues (Liphsits *et al.* 2005). For example, the size distortion provided in the illusion shown in Figure 7-17 relies heavily on perspective cues for depth. The oblique lines of the figure generate linear perspective cues; the observer is using the converging lines as depth cues, and the size distortion of the horizontal segments is being created by the application of a size-constancy adjustment (Gregory 1965). This is simply due to the fact that when objects of known distance subtend a smaller and smaller angle, they are interpreted as being further away. This is because objects of the same size but in varying distances cast different retinal image sizes. This size-constancy rule gives us indirect cues about the distance of objects of known absolute or relative sizes. The mechanisms for this constant perception are built into internal models within the CNS. They are rapidly learned or they may even be inherited; a baby a few weeks old reacts as if his perceptions already have constancy of shape and size.

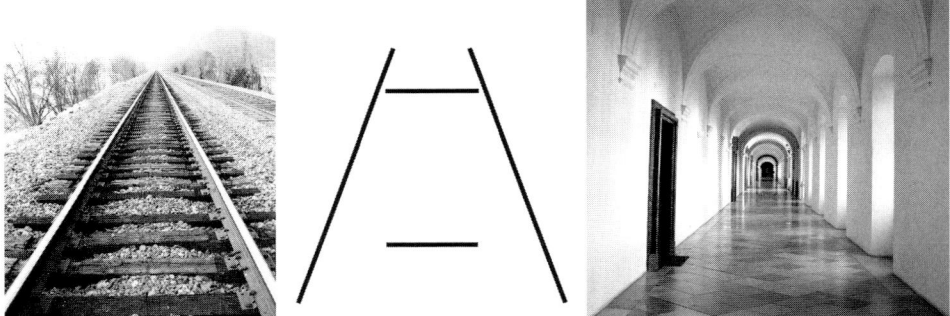

Figure 7-17. Left and center panels. The Ponzo illusion is one of a set of geometric illusions that produce misjudgment of relative line length due to depth clues. The perception of the size of the two horizontal bars depends on the interpretation of depth clues from the oblique lines. According to the size-constancy rule, we expect the lower bar to look larger than the upper bar because we think it is closer. Right. In linear perspective, all lines that are parallel with the viewer's line of sight appear to converge toward a distant point (the vanishing point) located on the horizon line, which corresponds to the viewer's eye level.

These visual illusions are not just relevant to visual processes, but also informative about the nature of human spatial orientation. Ground-based experiments have shown that visual illusions based on the arrangement of horizontal and vertical lines were influenced by an actual or artificial tilt of the study observer relative to the gravitational reference (Howard 1982). It therefore seems logical and plausible to assume that microgravity, a condition in which the gravitational reference is absent, may influence the strength of these visual illusions. Lipshits *et al.* (2001) performed an experiment on two cosmonauts on board the Mir space station where they had to adjust the length of a vertical line to match the length of a horizontal reference. On Earth, in an inverted T figure with equal length of the lines, the vertical line is perceived to be up to 25% longer than the horizontal one. Unfortunately, one of the cosmonauts tested did not have the illusion preflight. The other cosmonaut responses showed no changes in-flight relative to preflight. However, the experimental results were also confounded by several procedural flaws: (a) the cosmonauts were extensively trained on the illusions preflight; (b) they were not free-floating in weightlessness, but instead their body was firmly restrained at the hip, shoulder and feet; and (c) they were in a seat that was fixed to the space station 'floor' at the exact same location as during training. These conditions conveyed strong tactile and cognitive orientation cues. Cognitive information about the subject's orientation with respect to the space station and extensive training in that same orientation may allow the subject to use in orbit the same reference frame as on Earth even when the gravitational reference is removed.

Much of the work to be performed in weightlessness is not carried out with the body fixed in such a rigid fashion, most notably during extra-vehicular activity. Experiments designed to test the hypothesis that gravity is the unique or dominant reference frame for spatial orientation, movement and cognition must therefore be performed while free-floating. In fact, recent data suggest that some geometric illusions are decreased during free-floating in short-term microgravity during parabolic flight (Villard *et al.* 2005), but not when subjects are strapped in the airplane seats. In those illusions that were significantly affected by free-floating, the average decrease in appearance was about 40%. In other words, about 5 out of 11 participants did not experience these illusions in microgravity compared with normal gravity. It is interesting to compare this number with the 46% of astronauts who appear to ignore visual polarity information while in microgravity on board spacecraft and report that wherever their feet are is "down" (Harm *et al.* 1999). These astronauts appear to compensate for the absence of gravity primarily by increasing the weighting assigned to the vertical body axis orientation vector. If the same mechanism is taking place during short-term microgravity, as in parabolic flight, then the loss in the quality of image processing that inevitably comes with a coordinate transformation from the extrinsic gravitational reference system to an internal body reference system could be responsible for the alteration in the perspective cues.

These results confirm the assertion that these illusions are not just due to optical effects, but also depend on the integration of other sensory signals, including otolithic and somatosensory cues (Palmer 1999). Those illusions based on perspective depth cues are particularly relevant to space flight environment. Indeed, in the absence of atmosphere and with different lighting conditions affecting color and contrast, as in space flight, linear perspective is presumably the most reliable of cues for depth perception.

Linear perspective uses converging lines and vanishing points to determine how much an object's apparent size changes with space. It is based on the principle that there is a theoretical horizon line representing the point of view of the observer, and that the angle of converging lines toward a vanishing point, generally in the straight-ahead direction, provides distance information (Figure 7-17). In the absence of a gravitational reference, such as in microgravity, it is more difficult to define a horizontal line. Also, previous studies have shown significant deviations in the vertical position of the eye in microgravity due to the stimulation of the otolith organs by changes in the amplitude of the gravito-inertial forces (Clément *et al.* 1989), which could alter the direction of the "straight ahead." Consequently, because the rules of geometric perspectives are less accurately defined in microgravity, the subjects would rely less on the perspective cues for depth perception. Accordingly, we predict that the magnitude of the reversed-T, Muller-Lyer, Ponzo, and Herring illusions substantially decreases in microgravity.

Figure 7-18. Left panel. People, vehicle and trees look small when seen at a distance from above (as seen here from the second floor of the Eiffel Tower in Paris). However, when seen at the same distance in the horizontal direction, they don't (right panels). Our scale for distance in the vertical plane is different from that in the horizontal plane.

4.4.3 Distance Perception

Space studies performed on astronauts did not reveal any significant alteration in visual functions, including perception of stereoscopic depth, convergence, and accommodation (see Chapter 4, Section 1). However, several other important factors for depth perception, such as superposition, atmospheric (aerial) perspective, parallax of movement, and reference horizon do not occur when astronauts are looking outside. The objects seen inside a space station stay within distances of several centimeters to a few meters, whereas the objects outside (the Earth or the stars) are very far away. There is not an intermediate distance range, i.e., comprised from a few meters to a few kilometers.

The perception of objects distance within this intermediate range may therefore be altered after a long stay on board a space vehicle. It is well documented that divers or submariners who spend a long time in enclosed chambers have trouble evaluating distance when they get out. For this reason, submarine crewmembers are not allowed to drive immediately after returning from long duties in the confined space of a submarine (Ferris 1973). It is therefore possible that the perception of distances of objects is altered in the intermediate distance range in astronauts and cosmonauts.

Distances between objects and the observer are altered when there are no objects with familiar size (such as trees, people, vehicles, etc.) in the background. A filled distance appears greater than an empty distance. There are documented cases of people in the desert significantly underestimating the distance to the nearby mountains and failing to reach them before running out of water.

Even when objects with familiar size are present, our perception of distances is different when we look in the vertical direction. For example, when we look down from the top of a 100-m tall building, the people and the vehicles below look noticeably small (Figure 7-18). But when we look 100 m "down" the street, we don't comment on how small the people and vehicles look. The reason is we have learned the "rules" for scaling people at a distance, but not from a height. In absence of a vertical gravitational reference, the perception of distance might be distorted in the same way as when we look down or up.

In fact, many crewmembers have reported impairment in evaluating distances both during orbital flights and on the Moon surface. The debriefings of the Apollo astronauts are rich in description of the difficulties that they had to evaluate distances while on the lunar surface. For example, Eugene Cernan, during the descent of the LM module to the Moon, noticed, "Every normal frame of reference had disappeared, and beyond the thin window, the strange sunlight was richer, the shadows long and deep, the lack of color absolutely forbidding." (Cernan & Davis 1999, p. 318).

"Everything looked a lot smaller and closer together in the air than it turned out to be on the ground. When we were on the ground, things that were far away looked a lot closer than they really were. The thing that confused me was that we were so close to the Surveyor crater. I didn't realize we were as close to it as we were." –Pete Conrad, Apollo-12 (Godwin 1999).

"In appearances, it took us a long time to convince ourselves that some of the craters which looked so close were really much farther away." –Alan Bean, Apollo-12 (Godwin 1999).

"When we were at the ALSEP site, it looked as if we were about 450 feet west and 50 feet north of the position of the LM. It was a pretty good level site. Later when I got back to the LM and looked back, I noticed it didn't look as if the site were that far away. This was the continual problem we had, trying to judge distances." –Alan Bean, Appollo-12 (Godwin 1999).

"We were really having trouble, on that terrain, figuring out where the heck we were… It was frustrating. We wasted time. And that continued. That's what slowed us down the whole rest of this thing, trying to be a little more precise about where the heck we were." –Ed Mitchell, Apollo-14 (Pyle 2005, p. 115).

After landing, astronaut Dave Scott, Apollo 15, opened the hatch on the top of the LM and stood up for a 360 deg view survey. "At this time I was not sure where we were located… because there was nothing in the immediate vicinity which was recognizable." (Pyle 2005, p. 128).

"Navigating was difficult because we were dependent on visual landmarks from photographic maps that had been taken from orbit. Things on the ground tended to look very different", Apollo-16 Astronaut John Young recalled (Pyle 2005, p. 142). During this mission, the astronauts continued on to a sampling stop at Plum crater. But it was the wrong place. Duke said, "As any moonwalker can tell you, all lunar craters tend to look alike, and with no visual cues for distance, such as trees or houses, it's very hard to tell how big things on the Moon are. Man, it all looks the same, doesn't it?" "Sure does", replied Young (Pyle 2005, p. 143).

It is unclear if these illusions are direct effects of reduced gravity or due to other factors of the space environment, such as high contrast effects, confinement to cramped quarters, and the absence of known landmarks, such as trees, vehicles, buildings, in the crewmember's intermediate space. Nevertheless, these errors in visual perception and misperceptions of size, distance and shape represent potentially serious problems. For example if a crew member does not accurately gauge the distance of a target, such as a docking port or an approaching vehicle, then the speed of this target can be misevaluated. In fact, it is believed that such an error was a contributing cause to the collision between a Russian Progress supply spacecraft[9] and the Russian Mir space station in June 1997 (Linenger 2001). Also, disturbances in the mental representation of objects and the surround may influence the ability of astronauts to accurately perform perceptual-motor and perceptual-cognitive tasks such as those involved in robotic control (Figure 7-19).

A series of experiments has recently been designed to allow further identification of depth and distance perception during long-duration spaceflight (Clément 2007). This joint NASA-ESA research effort includes motor tests complemented by psychophysics measurements, designed to investigate the mental representation of spatial cues and distinguish the effects of cognitive versus perceptual-motor changes due to microgravity exposure. Identifying lasting abnormalities in the perception of distance will establish the scientific and technical foundation for development of preflight and in-flight training and rehabilitative schemes, enhancing astronaut performance of perceptual-motor and perceptual-cognitive tasks.

Figure 7-19. This photograph of the orbiting Space Shuttle is ambiguous. Are the payload bays convex or concave? Is the Shuttle flying upside-down or not? Photo courtesy of NASA.

---

[9] Because the Mir radar was turned off and the Progress was not visible out of the Mir's windows for laser range measurements at appropriate times, the Mir commander's sole source of range rate information was the changing angular size and position of Mir on a TV monitor from the vantage point of the Progress vehicle (Burrough 1998).

5   SUMMARY

Many reports suggest that in the absence of a gravitational reference axis, astronauts initially exhibit increased reliance on visual reference axes derived from extrinsic coordinates, especially from the visual scene. In microgravity, astronauts must rely much more on vision to maintain their spatial orientation because the otoliths can no longer signal the "down" direction. During prolonged microgravity exposure, however, reliance seems to shift toward intrinsic reference vectors.

The erroneous illusions of attitude or self-motion during head movements performed during and after return to Earth gravity are presumably due to a reinterpretation of vestibular signals. The simple *otolith tilt-translation reinterpretation* hypothesis has been refined based on recent data from ground-based research and space flight. These data suggest that the CNS resolves "tilt-translation" ambiguity based on the frequency content of the linear acceleration detected by the otolith organs, with low frequency indicating "tilt" and high frequency indicating "translation". A crossover exists where the otolith signals are then ambiguous. Exposure to microgravity presumably results in a shift of this crossover frequency, which could then contribute to spatial disorientation episodes and SMS.

Although investigations of higher cognitive processes, such as navigation and mental rotation are limited, the astronauts frequently report that the spacecraft interiors look longer and higher than they actually are, and a reduction in the perceived height of three-dimensional objects is observed in-flight compared with preflight, suggesting an alteration in the mental representation of three-dimensional cues in microgravity.

What we mentally reconstruct are the *constant* physical properties of objects; and so we should, because those are the things that we need to know. To do this we must not only recognize the outlines of individual objects in the images on our retina, but also look for subtle cues in the images about their distances. Only when we know both the true geometrical shapes in the images and the distances of the objects portrayed there can we derive the constancy of our perceptions. Perception is a model of the brain, a hypothesis about the world that presupposes the physical laws of movement. These laws change in weightlessness and, therefore, one could expect changes in the mental representation of objects' shape and distance during space flight.

The rare investigations carried out in space so far have not demonstrated drastic changes, probably because the CNS continues to use an internal model of gravity, at least for a short while. How the cognitive processes of spatial orientation will differ from the terrestrial norm after a long absence of a gravitational reference? It can be speculated that the way of processing three dimensions will be more developed. Creativity will certainly be more three-dimensional and definitely thinking will be out of the gravitational box. Like the way culture and language influences our ability to creatively think, being free from gravity will entice thoughts never before possible for the human mind, and thus give opportunities for new art and scientific discoveries (Pettit 2003).

There is an applied aspect of this research to the space exploration program. If changes in depth perception were to occur to astronauts during long-duration exposure to microgravity, this could result in spatial disorientation episodes, errors in object (e.g., an approaching vehicle) distance perception, and difficulties in navigating through or envisioning the inner structures of a complex habitat (e.g., a space station). If an astronaut cannot accurately visualize the station, navigation of the station may cause delays and frustration. There may also be consequences for space habitat design if squared volumes do not look square to astronauts. Virtual reality training may be a way to train the astronauts to compensate for such altered spatial representation.

Further investigations carried out in space will perhaps reveal that other higher cortical functions are impaired in weightless conditions. The combination of virtual reality with multi-EEG recordings (for the measurement of evoked-related potentials and brain mapping), both equipments being soon available on ISS, should provide exciting results on the adaptive mechanisms of cerebral functions in absence of gravity.

Chapter 8

COUNTERMEASURES

Mitigating the harmful effects of prolonged exposure to space radiation and weightlessness is one the most significant challenges that must be addressed to realize the long-duration planetary exploration missions currently envisioned. Given the fact that the astronaut explorers who will undertake these missions will be exposed to these deleterious effects for up to several years while they travel to and from Mars, it is of extreme importance that effective countermeasures are identified, developed, tested, and proven prior to undertaking such challenging missions (Clément & Bukley 2007).

In the framework of President's Bush "Vision for Space Exploration" (NASA 2004) NASA experts have identified the risks of crew exposure to space travel during exploration missions. The *Bioastronautics Critical Path Roadmap* defined crew health and performance standards to set acceptable medical risks for long-duration ISS, Moon and Mars missions (NASA 2005) (Figure 8-01). Based on recommendations made in an earlier report (Ball 2001) and follow-up research, a total of 45 major risks to human health were reviewed and prioritized[10]. A considerable number of risks addressed CNS dysfunctions, such as impaired sensorimotor capabilities, motion sickness and human performance failure due to degradation of sleep and cognitive capabilities. Experts from other space agencies came to similar conclusions (Horneck *et al.* 2006).

The stated goal of the Bioastronautics Roadmap was "to reduce risk through effective and efficient mitigation solutions developed from a focused research and technology development strategy" (NASA 2005). Accordingly, the Roadmap provided information that helped establish tolerances for humans exposed to the effects of space travel. Although this approach was later abandoned for a lack of resources (Longnecker & Molins 2006), it had the merit to focus on countermeasures aimed at maintaining crew health and function within identified limits. This chapter reviews those countermeasures currently in use on board spacecraft for the maintenance of CNS functions, as well as those being developed for future exploration missions.

Figure 8-01. Current plans for exploration missions include short-term missions to the Moon (2020), followed by permanent lunar outposts (2030) in preparation for a human mission to Mars (2040). Photo credit: ESA/P. Carril.

---

[10] The complete list of risks and research questions can be downloaded at the following URL: http://bioastroroadmap.nasa.gov/User/discipline.jsp (accessed 10 October 2007)

G. Clément, M.F. Reschke, *Neuroscience in Space*.
DOI: 10.1007/978-0-387-78950-7_8, © Springer Science+Business Media, LLC 2008

1   WHAT IS A COUNTERMEASURE?

The life support systems on board a spacecraft prevent the space environment from degrading the crews' health status by providing or emulating Earth-normal cabin condition, including normal temperature, pressure, gas concentration, and shielding to minimize radiation from reaching the crew. In an analog manner, countermeasures include procedures, devices, or therapies to prevent or minimize adverse health and medical events resulting from short- or long-duration exposure to microgravity.

Ideally, countermeasures will prophylactically intervene to prevent illness, injury and pathophysiology that would result in behavioral and performance degradation. Countermeasures must provide acceptable mission and performance preflight, during the flight and during the postflight recovery as well. Countermeasures include the definition of preflight selection criteria, the methods that train or prepare for known adaptive responses to altered gravity and a return to Earth, the procedures that are prescribed throughout the mission for minimizing the health impact of these adaptive responses, and the treatments that restore and correct a deficit that only become evident during or after the mission. A countermeasure prescription is a direction for using a countermeasure, including the modality (e.g., hardware device, drug, procedure), duration, intensity and frequency, as well as the physiological monitoring equipment and parameters necessary to gauge that countermeasure's effectiveness (Clément 2005).

The countermeasure procedures currently utilized during space flight are:

a.   Pharmacological – anti-motion sickness, anti-orthostatic intolerance and anti-bone loss drugs.

b.   Exercise – treadmill, cycle ergometer and resistance exercise; isometric, isokinetic, isotonic, and concentric protocols (Figure 8-02).

c.   Dietary – fluid loading, mineral and metabolic supplements.

d.   Environmental – lighting, oxygen pre-breathing prior to EVA.

e.   Mechanical – lower body negative pressure, anti-gravity suit.

f.   Psychological – ground support, biofeedback.

g.   Special training – preflight adaptation trainer, neutral buoyancy laboratory, parabolic flight, virtual reality, skill maintenance.

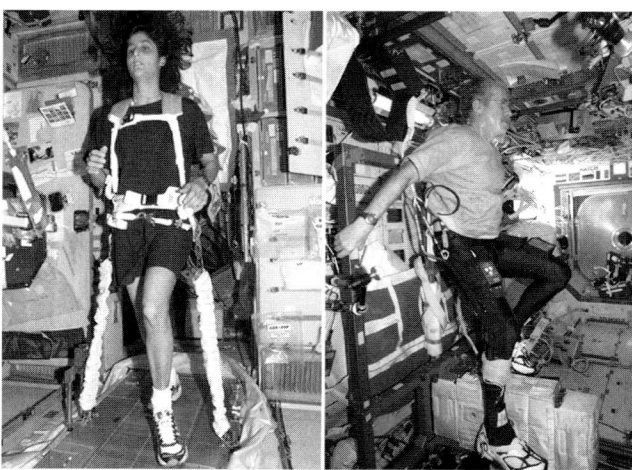

Figure 8-02. Astronauts exercising on board the ISS. Prescription for long-duration missions requires three hours of daily exercise on the treadmill (left), cycle ergometer (right) and other resistive exercise devices. Photo courtesy of NASA.

Most of these countermeasures have usually been developed to address one particular physiological or medical problem. One difficulty is that the normal sequence of adaptive physiological events to space flight conditions, the normal variability between individuals and the interaction among individual countermeasures may confound evaluation of their effectiveness.

A related problem is how to evaluate the effectiveness of a countermeasure? What are acceptable limits of loss in nervous system functions? Is it based on the average or relative loss identified from previous studies or missions? Is it relative to the required ability to perform? Or just how the crewmembers feel? Should the objective of a flight countermeasure be to maintain the physiological functions as before flight or to maintain physiological functions required for a minimum safety level without compromising the long-term health of the astronaut? Lack of agreed-upon measures of effectiveness is perhaps the most significant deficiency in enabling accurate assessment of a countermeasure and comparing information across simulation studies and flight data. Flight surgeons, astronauts and mission planners must be involved in establishing the operational requirements for successful countermeasures.

One promising approach to the countermeasure issue is the well known but never implemented artificial gravity approach. If most of the physiological problems are due to the absence of gravity, then clearly the fix is to put gravity back. However, whether replacing gravity will fully restore Earth-like health in space remains to be investigated. Few systematic gravity-dose-response studies have been performed, both on Earth and in space. These studies are a prerequisite to determine the artificial gravity prescription, i.e., the gravity dose level, duration and frequency, required for an acceptable standard of the CNS functions affected by space flight. Like many other countermeasures, artificial gravity carries with its implementation the question of dual adaptation. A countermeasure that only maintains Earth-like stasis does not aid in allowing crewmembers to develop "space normal" performance.

2   COUNTERMEASURES FOR SPACE MOTION SICKNESS

Space flights have often been described as camping trips: healthy people being in space for a short period of time with some basics amenities, but accepting many of the physiological changes that occur. "Perhaps the nutrition isn't the same as in the normal daily diet, but there are enough nutrients to get through, and you accept the physiological consequences of the colder weather and the elements" (Bungo 1989). Invariably, a few individuals will be systematically incapacitated at the beginning of the mission, as is the case for a high altitude expedition. This incapacitation is often the result of *space motion sickness* (SMS) (see Chapter 3, Section 1). Vomiting and inability to perform some of the tasks at hand are the most severe manifestations of this condition. This condition has been routinely accepted as part of the camping trip. The good news is that, so far, there have been enough other crewmembers not suffering from SMS to execute required procedures. It is also fortunate that there has been sufficient recovery time and a dearth of absolutely mission and time critical tasks so that SMS has had only limited effects on flights so far. However, it clearly is a significant physiological phenomenon, which should be tackled by countermeasures (Bungo 1989).

Space neuroscience studies have been historically associated with finding a cure for SMS. The disruptive nature of SMS, occurring as it does during the early, critical stages of a mission, as well as the occurrence of spatial disorientation occurring throughout the flight and during return to Earth, have led to a variety of approaches for the prevention or control of these effects. Although only limited success has been achieved to date, research directed toward prevention of SMS has proceeded along four broad lines of inquiry: selection, training, and pharmacological and mechanical countermeasures. Each of these is described below[11] in terms of its potential for symptom control and the problems associated with its use.

2.1 Selection

In the Russian space program, a significant aspect of screening and selection criteria are devoted to the vestibular system. Khilov (1975) recommended a number of tests designed to select the best candidates for aviation and space flight. He contended that the most suitable individuals would exhibit the smallest magnitude of the following responses:

a.   Nystagmus and dizziness elicited by electrical stimulation and rotation.
b.   Degree of torso deflection from vertical upon assuming an upright position after rotation in a 90-deg forward head and body tilt position.
c.   Susceptibility to the linear up and down motion of a two-axis swing.
d.   Autonomic response to variable acceleration in a centrifuge.
e.   Sensitivity to Coriolis acceleration.
f.   Elicitation of motion sickness symptoms by "double rotation", i.e., rotation of a Bàràny chair at the end of a centrifuge arm.

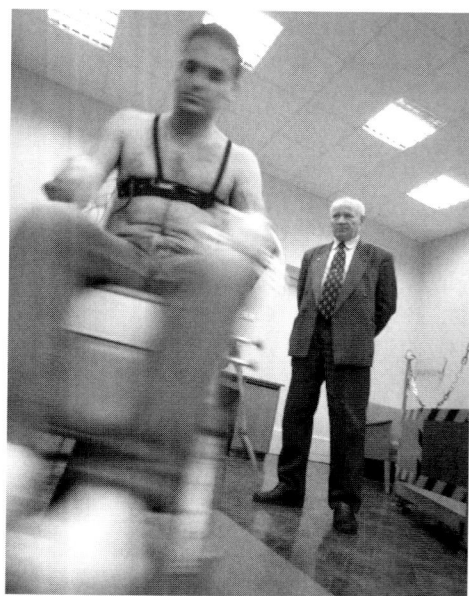

Figure 8-03. Cosmonaut vestibular training at Star City in Moscow, Russia. During passive rotation in yaw the cosmonauts tilt their head in pitch and in roll until they develop symptoms of motion sickness. These sessions are required daily during the last couple of weeks prior to a space flight on board the Soyuz vehicle. Photo courtesy of CNES.

Selection tests of the cosmonauts have traditionally included sessions of Coriolis and cross-coupled angular accelerations using rotating chairs and parallel swings (Figure 8-03). The objective of these tests is to select individuals with a higher tolerance

---

[11] The following sections derive from a chapter written by Reschke *et al.* (1996) for a joint U.S.-Russian textbook on Space Biology and Medicine.

to vestibular stimulation. Despite this vestibular screening, and the extensive preflight training described in the following section, the incidence of SMS symptoms in Russian space crews is not different from that in U.S. space crews, who are not screened for motion sickness resistance. In the past, some U.S. astronauts, however, have participated on a voluntary basis as test subjects in experiments on motion sickness susceptibility prior to their space flight (see *Preflight Adaptation Training* below). Among those crewmembers that had completed preflight Coriolis tests, no correlation has been found between test tolerance and SMS susceptibility (Reschke 1990).

Flight surgeons discuss the natural course of SMS and *postflight motion sickness* (PFMS) (see Chapter 3, Section 1.1) with the entire crew during the course of routine medical training. A session is held three to six months prior to flight with a refresher session ten days before launch. Crewmembers are reminded to avoid provocative stimuli, such as pitch or roll head movements and unusual visual orientation. In practice, avoiding such stimuli involves performing movements slowly, "bending down" to access an item at knee level, and maintaining a visual vertical reference rather than floating upside down during the first hours and days of the mission. The crew is also taught to avoid excessive heat and noxious odors. Maintaining adequate hydration in flight during the course of symptoms is also emphasized.

Flight controllers and planners are educated on the effects of SMS on crew performance. Mission flight rules currently prohibit scheduling of critical activities like Shuttle landings or extra-vehicular activities within three days of reaching orbit (Davis *et al.* 1993b). Shuttle planners also attempt to lighten crew activities during the first one to two days because of known decreased performance (Ortega & Harm 2007).

2.2   Vestibular Training

Vestibular training techniques investigated thus far have been based on one of two suppositions, either that adaptation to stressful motion can be hastened by previous exposure to conflicting sensory inputs or that symptoms can be avoided by learned control of autonomic responses.

2.2.1   Provocative Sensory Conflicts

The Russian program uses preflight vestibular training quite extensively. This training is based on the belief that adaptation to stressful motion can be hastened by previous exposure to conflicting sensory inputs. Hypothesizing that motion sickness susceptibility is proportional to the amplitude of vestibular signals, Khilov (1974) proposed that the most effective method of increasing tolerance to motion sickness is to habituate the vestibular system, thus reducing the magnitude of vestibular inputs.

During the *cosmonaut vestibular training* (CVT), cosmonauts seated in a chair rotating in yaw execute pitch head movements until either vomiting is reached or a predetermined number of head movements is performed (see Figure 8-03). This training is achieved through daily sessions of repeated exposure to body yaw rotation with chair acceleration and deceleration of 180 deg/sec^2. The number of sessions ranges from four to more than twelve, depending on the initial susceptibility of cosmonauts to this Coriolis-induced sickness. Ground-based studies have shown that only about half of the subjects exposed to CVT manifest a reduction in the severity of motion sickness symptoms (Clément *et al.* 2001a).

In Russia, the CVT generally takes place during the last couple of weeks before launch (Krioutchkov *et al.* 1993). One hypothesis is that SMS results from a sensory conflict between semicircular canals and otolith signals during head movements in pitch or roll that naturally stimulate the otolith organs on Earth (see Chapter 3, Section 1.5.2). The rationale behind the CVT is that tolerance to the conflict between semicircular canals and nonconforming otolith cues generated by Coriolis and cross-coupled angular accelerations on Earth would transfer to weightlessness conditions (Popov *et al.* 1970).

However, many experiments have demonstrated that adaptation to one sensory conflict situation does not necessarily transfer to other conflict situations, particularly when the specific conflict differs considerably from one situation to the other (see Lackner & DiZio 2006 for review). The efficacy of CVT against SMS has never been proven. Similar preflight training on rotating chairs was once used in the American space program, but was abandoned when it failed to reduce SMS incidence. In fact, although the cosmonauts continue to be exposed to CVT, the incidence of SMS symptoms in Russian and American space crews is quite similar (Davis *et al.* 1988).

There is some indication that, on Earth, the repeated exposure to passive body rotation elicits a gradual decrease in the intensity of vestibular responses, a phenomenon known as vestibular habituation (Collins 1973). Beyond its effects on motion sickness symptoms, the pattern of repetitive angular velocity steps during rotator starts and stops might also be responsible for the habituation of vestibular nystagmus and sensation of rotation in cosmonauts before they go to space. One important implication of the CVT concerns the use of "habituated" cosmonauts as subjects for adaptive studies of vestibular functions to weightlessness. Markedly lower vestibulo-ocular and perceptual responses in cosmonauts compared to control subjects and astronauts have been attributed to a potential effect of the CVT (Clarke *et al.* 1993, Wetzig *et al.* 1993, Clément *et al.* 2001).

There are also anecdotal reports that Russian cosmonauts execute head and body movements during the last few days on a long-duration mission, as a prophylactic measure for postflight motion sickness (Linenger 2001). This method is also occasionally used by astronaut-pilots prior to their space flight. For example, retired Shuttle Astronaut Mike Mullane recalls, "to be SMS-free was considered so important, many astronauts attempted inoculations. When it was first assumed the problem was related to Earth-based motion sickness (later disproved), astronauts would perform stomach-churning acrobatics in T-38 jets in the days prior to a launch. I was flying in Story Musgrave's backseat when he decided to prep his body for an upcoming mission. He asked ATC [Air Traffic Control] for a block of altitude and then went into a series of spiraling rolls and violent maneuvers that alternately had me slammed into my seat at 4 g and lifted from it in negative g" (Mullane 2006, p. 107).

Other methods did not prove successful: "Another equally ineffective attempt at SMS inoculation was to sleep on an inclined bed with your head lower than your feet. This became popular when the flight surgeons hypothesized that the fluid shift of weightlessness might be causing the inner ear to be disturbed, inducing vomiting. [...] By sleeping in a bed with bricks under the foot posts to tilt the head down, it was thought the resulting fluid shift to the upper body would somehow prepare it for weightlessness and eliminate SMS. It didn't. Some of those practicing head-down sleep still got sick in space, suggesting that those head-downers who didn't vomit had probably been immune anyway" (Mullane 2006, p. 107).

2.2.2 Preflight Adaptation Training

Another approach is to develop preflight adaptation training based on duplicating, to the extent possible, the sensory conditions encountered during space flight (Harm & Parker 1994). Preflight adaptation training is based on the following postulates:

    a.   Microgravity is a form of sensory stimulus rearrangement to which astronauts adapt (sensory conflict theory).

    b.   Adaptation may result from sensory compensation or reinterpretation (OTTR and sensory compensation hypotheses).

    c.   Because the central nervous system is "plastic", people can learn and store perceptual, sensory and sensorimotor responses appropriate to different sensory stimulus conditions and they can learn to invoke these alternative responses almost immediately when the conditions change; i.e., they develop "dual-adapted" states.

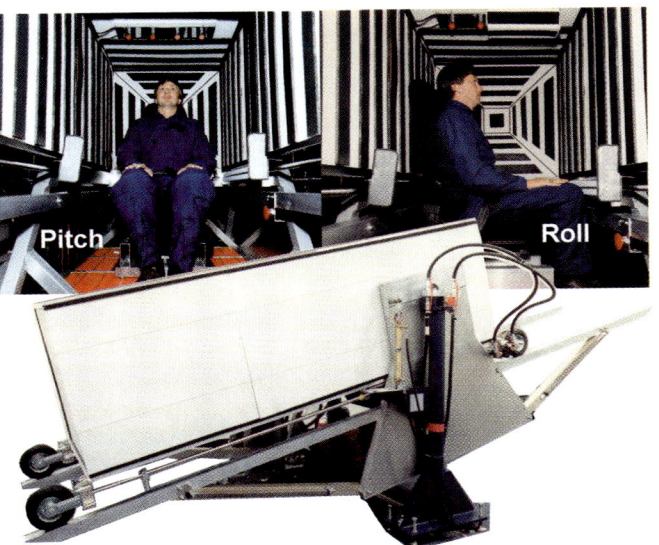

Figure 8-04. Tilt-Translation Device *used for preflight adaptation training at NASA. Before the flight, crewmembers are passively tilted in roll or in pitch while exposed to a lateral translation of the visual scene (such as on the left panel) or to a forward-backward translation, respectively, in order to induce a reinterpretation of their otolith signals by the visual system. Photo courtesy of NASA.*

The *Preflight Adaptation Trainer* (PAT) at NASA Johnson Space Center was a first attempt to provide astronauts with demonstrations of and experience with altered sensory stimulus rearrangements that produce perceptual illusions similar to those experienced during space flight. Subjects were passively tilted in pitch or roll, while simultaneously presented with a visual scene that was translating in the antero-posterior or lateral direction, respectively (Figure 8-04). Such combination of stimuli was designed to evoke reinterpretation of otolith tilt signals as linear motion and to elicit perceived self- or surround-motions that are present in the weightless environment of orbital flight (Reschke *et al.* 1988, Harm & Parker 1994).

Exposure to these altered sensory stimulus rearrangements was intended to demonstrate sensory phenomena likely to be experienced in-flight and immediately postflight, alter sensorimotor reflexes, and eliminate or reduce SMS and orientation and motion disturbances. It was hypothesized that, after training in these devices, compensatory eye movement responses, postural muscle reflexes, and self-motion and orientation experiences in relation to visual scene movements and position, would be appropriate to the weightless-adapted state. In other words, new sensory and sensorimotor programs appropriate for both microgravity and the immediate postflight period would be developed.

Several crewmembers of Shuttle missions were exposed to this training before and after space flight. A significant reduction (19%-54% depending on the symptoms) in the severity of SMS symptoms was observed in those crewmembers who were exposed to the training, by comparison with those who were not exposed to it (Harm *et al.* 1999). Comparisons of similarities and differences in perceptual experiences associated with space flight and the PAT device also indicated the following:

a.   The stimulus rearrangement conditions in the PAT device produced perceptual experiences quite similar to those associated with space flight.

b.   Postflight exposure to the PAT device could re-elicit symptoms identical or similar to SMS symptoms.

c.   Postflight exposure to the PAT device could re-elicit visual disturbances similar to those experienced in-flight.

d.   Postflight exposure to the PAT device motion profiles elicited greater perceived linear self-motion than preflight exposure.

The results of this experiment indicate that it is possible to create sensory rearrangements on Earth that mimics some of the conditions of space flight. The next step would be to repeat exposure to these conditions preflight, so that astronauts could develop sensorimotor programs appropriate for microgravity and learn to rapidly switch from one-g to zero-g programs and vice versa. If crewmembers can indeed become dual-adapted, their adaptation to microgravity and re-adaptation to Earth would be facilitated. This would reduce both SMS and PFMS (Harm *et al.* 1998).

The "z-axis aligned gravito-inertial force" or ZAG (see Chapter 4, Section 5.4.1) is another PAT device currently under investigation. Even though gravity cannot be eliminated on Earth, by keeping the gravity vector constant with respect to the trainee as the trainee changes orientations within a simulated environment, its contribution to spatial orientation in that environment can be negated (see Figure 4-13, right). This is similar to microgravity in that angular head movements can be made in pitch and roll without a changing gravity vector.

2.2.3 Biofeedback Training

The autonomic nervous system initially responds to motion stress with sympathetic activation. Following termination of the stimulus, a decrease in sympathetic activity is noted which may indicate a parasympathetic rebound. If nausea and vomiting are parasympathetic reactions to sympathetic activation, then motion sickness symptoms should be prevented by training individuals to maintain their autonomic responses at baseline levels (Cowings *et al.* 1986).

In biofeedback training, instrumental information about selected autonomic activities is provided to the subject with a visual or auditory "reward" presented for producing the desired response (e.g., decreased heart rate). The experimenter usually suggests how the desired response might be evoked in order to reduce the time required for trial and error. Autogenic training employs a collection of cognitive imagery techniques to produce the desired change in autonomic activity. Self-suggestion exercises are utilized to produce certain body sensations that correspond to the desired changes in physical parameters. *Autogenic feedback training* (AFT) combines the above techniques. Cognitive imagery is used to produce the desired changes with immediate sensory feedback on success via instrumental readouts. Verbal reinforcement from the experimenter also increases the effectiveness of the training (Cowings 1990).

AFT has been found to significantly increase tolerance to a rotating chair stimulus when compared to control groups with no training. Subjects given AFT had improved motion sickness resistance. The effectiveness of AFT was not due to decreased attention to symptoms as subjects receiving AFT performed better than subjects distracted by an alternative cognitive task during rotation tests. While highly susceptible individuals appeared to have a greater magnitude of autonomic response to motion sickness, both high and moderately susceptible subjects benefited from AFT and both groups showed the same rate of improvement. Men and women showed equal gains in motion tolerance following AFT. A training schedule of one session per day (six hours total) was as effective in raising motion sickness tolerance as a schedule involving two or more sessions per day for the same total number of training hours. AFT also increased tolerance to a vertical motion test, but did not significantly alter resistance to motion sickness induced by a test combining optokinetic stimulation with rotation (Cowings 1990).

In the U.S. Air Force program, 79% of aviators refractory to other anti-motion sickness treatments were successfully returned to flying status following an AFT treatment protocol (Jones *et al.* 1985). Dobie *et al.* (1987) did not note any increase in motion sickness resistance following biofeedback training but did observe increased tolerance with cognitive-behavioral therapy designed to allay anxiety about the development of motion sickness. The lack of improvement following biofeedback training may have been due to the endpoint used, the physiological parameters chosen for training, or insufficient linkage of autonomic control to the motion sickness environment. The success of the cognitive-behavioral therapy technique was likely due to adaptation produced by exposure to the rotating chair stimulus during desensitization sessions.

The Russians have also experimented with autonomic response training techniques. "Adaptive biocontrol with feedback", which involves control and stabilization of galvanic skin response and skin temperature, was reported to be effective in reducing symptoms and increasing resistance to experimentally-induced motion sickness (Aizikov *et al.* 1991).

Two crewmembers in the Space Shuttle program were given AFT preflight and two other crewmembers on the same flight served as untrained controls. Training focused on the physiological parameters for which each individual exhibited the largest response on a baseline rotation test. Twelve AFT sessions for a total of six hours of bidirectional training were administered. Following training, another rotating test was performed. One crewmember exhibited an increased resistance to motion sickness with a concomitant decrease in autonomic stress responses. The other trained crewmember showed a moderate increase in motion sickness tolerance with mixed success in achieving autonomic control. In-flight, this crewmember experienced one severe episode of SMS while the crewmember exhibiting greater autonomic control reported no severe symptoms. The two untrained crewmembers had multiple episodes of severe symptoms despite the administration of anti-motion sickness drugs. The poor response of the one crewmember to AFT may have been due to the less than optimal training schedule forced by a launch delay (Cowings 1990).

Since the perception of vestibular stimulation is unchanged by AFT, this type of training must interrupt the autonomic response after the sensory conflict has already occurred. Raising the threshold for autonomic activation by sensory conflict may inhibit motion sickness symptom development (Cowings & Toscano 1982). AFT and parasympatholytic drugs like scopolamine may achieve the same effect by different actions: AFT by reducing sympathetic activity thereby eliminating the parasympathetic reaction and its resultant symptoms, and scopolamine by reducing the parasympathetic response to the increased sympathetic activity that has already occurred (Cowings *et al.* 1986).

2.3   Pharmacological Countermeasures

2.3.1   Ground-Based Studies

Over the years, a number of drugs have been tested for their effectiveness in preventing motion sickness. While some drugs have been found to be generally effective, no drug or drug combination has yet been identified which provides protection from motion sickness for all individuals.

Potentially effective anti-motion sickness drugs have been tested using a wide variety of motion provocation conditions, symptom-scoring techniques, and test endpoint criteria. Anti-motion sickness drug testing has been conducted on board ships, on the land, and on board aircraft. Testing at sea typically utilizes a cruise ranging in duration from hours to weeks. Ground-based tests usually involve stimulation of the semicircular canals and otoliths (Coriolis and cross-coupling) by requiring the subject to make standardized head movements out of the axis of rotation while in a revolving room or chair. Other ground-based motion environments include vertical oscillation and tilting to simulate sea motion. Aerobatic maneuvers and parabolic flight are motion environments tested in aircraft.

While the incidence of motion sickness may be determined in a variety of ways, most drug research determines the degree of motion sickness experienced by scoring the severity of symptoms. For example, a scale assigns point values to certain symptoms recognized as indicative of motion sickness (e.g., sweating, pallor, nausea) allowing the determination of a total sickness level by summing the point scores of the symptoms observed (Graybiel *et al.* 1968). The use of double-blind techniques in anti-motion sickness drug experiments, as in others drug experiments, is necessary to avoid experiment bias. Due to the large inter-subject variability in susceptibility to motion sickness, inclusion of one or more placebo treatment(s) for each subject enhances the data interpretation by allowing a subject's response to a drug to be compared to his or her own baseline susceptibility. This method also aids in identifying adaptation effects. Intra-subject variability is reduced by performing all of a subject's tests at the same time of day and by allowing an adequate amount of time between test sessions to elapse thus reducing the possibility of adaptation to the motion environment.

Wood & Graybiel (1968) found that ranking a number of drugs with respect to their effectiveness in preventing the symptoms of motion sickness tended to group the drugs according to principal pharmacological action. Numerous studies have found scopolamine, an anticholinergic (parasympatholytic) drug, to be the most effective in treating motion sickness (Wood *et al.* 1986). Scopolamine in a dose of 0.6 to 1.0 mg was found 90% effective in preventing motion sickness, whereas other anticholinergics, such as atropine, hyoscine aminoxide, phenglutarmide, orphenadine and benztropine, ranged in effectiveness from 50% to 74%.

Although most of the antihistamines tested for anti-motion sickness properties have had some benefit, they generally provide less protection than scopolamine. Promethazine, the most effective of the antihistamines, approaches scopolamine in efficacy, with a 25 mg dose of promethazine approaching a 0.6 mg dose of scopolamine in efficacy (Wood & Graybiel 1972). The few sympatholytic drugs that are effective against motion sickness were of only marginal benefit and had less effect than the least effective antihistamine.

The ability of sympathomimetic drugs to prevent motion sickness was first discovered when amphetamine was combined with scopolamine to counteract the sedation caused by the latter. Experimental control subjects taking amphetamine alone also exhibited increased tolerance to motion. Wood & Graybiel (1972) found that combining a parasympatholytic drug (scopolamine) with a sympathomimetic produced either an additive effect (ephedrine) or a synergistic effect (amphetamine). These combinations were far more effective than any single drug.

Anti-motion sickness drug research has been reviewed by Reschke *et al.* (1994) and by Wood (1990). The vast majority of anti-motion sickness drugs have been administered orally. Because the duration of action of these drugs is typically brief, frequent dosing is needed if the motion is expected to last for extended periods. An additional complication is the reduction in gastric motility characteristic of acute motion sickness (Reason & Brand 1975).

Drugs must be given prophylactically to avoid decreases in drug absorption. For this reason, alternate routes of administration, such as transdermal application, suppositories and intramuscular (IM) injections have been investigated (Davis *et al.* 1993b). The transdermal application delivers the drug via a patch placed in the post-auricular area prior to exposure to motion. Absorbed through the skin, the dose is delivered from the surface of the patch over a 72-hour period. By contrast, IM injection allows the administration of anti-motion sickness drugs after the onset of motion sickness symptoms, thereby eliminating administration of drugs to those who do not require them.

Another factor complicating the search for motion sickness remedies is the occurrence of side effects, which can preclude using the most effective dose or dosing schedule. For example, scopolamine can cause dry mouth, drowsiness, vertigo, and blurred vision, with higher dose levels associated with increased side effects. Sympathomimetic drugs have been associated with anxiousness.

The mechanism(s) of action of the effective anti-motion sickness drugs is unclear. Money (1970) in noting that some drugs in certain drug groups prevented motion sickness, proposed that the action(s) of the drugs responsible for their anti-motion sickness efficacy may not be related to the action that resulted in those drugs being grouped together in the first place.

2.3.2 SMS Medication during Space Flight

A reported 30% of all Shuttle crewmembers[12] have received medication for relief of SMS in-flight (Santy & Bungo 1991). The most detailed reports of the incidence of SMS symptoms and use of anti-motion sickness drugs come from the Skylab missions. Most astronauts on these missions took a dose of scopolamine-dexedrine (0.4 mg / 5 mg) as a preventive measure. This "scope-dex" combination had been found to be effective in seasickness and other types of terrestrial motion sickness (Wood & Graybiel 1968). When used in the U.S. space program, scope-dex prevented development of SMS symptoms early; however, it was discovered during the early Shuttle program to produce rebound illness when withdrawn. That is, the combination of drugs or just scopolamine alone has a state dependency associated with its use for motion sickness, and astronauts developed SMS symptoms later in the flight when the drug was no longer taken (Homick *et al.* 1983). Reports are also available on the prophylactic and in-flight use of 0.4 mg scopolamine taken orally with 2.5 to 5.0 mg d amphetamine for 19 Space Shuttle crewmembers (Davis *et al.* 1993a). Only three of these 19 crewmembers experienced no SMS symptoms, seven crewmembers developed symptoms while taking the medication, and nine experienced delayed symptoms on the second flight day or later. So, scope-dex seemed to delay adaptation to microgravity and is no longer used (Davis *et al.* 1993a). These observations are consistent with ground-based research findings (Wood *et al.* 1986).

Currently there are no known ground-based tests that predict SMS susceptibility (Reschke 1990, Reschke *et al.* 1994). The best predictor is prior history of SMS. Therefore, NASA flight surgeons recommend that first time flyers forego drug prophylaxis to determine whether the astronaut will need medications on future flights. However, Shuttle commanders, pilots and flight engineers are not allowed to take anti-motion sickness medications prior to launch due to the potential risk of drug side effects impairing piloting performance. Mission specialists can take an oral phenergan-dexedrine (25 mg / 5 mg) combination, or phenergan alone, prophylactically in the final hours prior to launch. Anecdotally, the efficacy of this regimen seems acceptable, despite concerns about performance effects (Graybiel & Knepton 1977, Hordinsky *et al.* 1982).

Intramuscular promethazine has also been used successfully to treat SMS symptoms during the flight. The first in-flight IM injection of promethazine (50 mg) was reported in 1991 (Bagian 1991). The severe SMS symptoms of the recipient resolved completely, with no subsequent recurrence. Davis *et al.* (1993b) later reported that, of 20 crewmembers given 25 to 50 mg of promethazine IM (adjusted for body weight) on the first flight day (FD1), 25% were still classified as "sick" on the second flight day. In contrast, 50% of the 74 crewmembers reporting SMS on the first day of flight who did not receive IM promethazine were still "sick" on FD2, a statistically significant difference. Some of those not treated with IM promethazine had been treated either before or during flight with other anti-motion sickness medications without success. Ninety percent of those who received IM promethazine reported relief from SMS symptoms within 1 to 2 hours of dosing; only three crewmembers needed a second

---

[12] Today it is recognized that both the incidence of SMS and drug treatment is much higher than that originally reported in the Santy and Bungo (1991) report, approaching perhaps 70 to 80% of all crewmembers. However, this data is not directly available due to an organization that prevents much of this data being reported from the NASA flight surgeons to the research laboratories.

dose. Three of the IM promethazine recipients reported drowsiness after administration; however, the injection often is given immediately before the sleep period. An IM injection of 25 to 50 mg of promethazine is now the recommended treatment for moderate to severe cases of SMS in the U.S. space program (Davis *et al.* 1993b, Jennings 1998). Clearly, SMS is a self-limited illness that most will overcome quickly as they adapt to the microgravity environment. Typically a single dose of IM promethazine will resolve the acute symptoms. Very rarely will moderate to severe cases require IM medication beyond flight day two (Davis *et al.* 1993b).

Promethazine does not seem to delay adaptation (Davis *et al.* 1993b) and may actually hasten adaptation to provocative motion (Lackner & Graybiel 1994). Promethazine under a suppository form has reportedly resolved SMS symptoms effectively as well (Davis *et al.* 1993b). It is best administered in the "pre-sleep" period of the flight day in order to reduce possible side effect (e.g., drowsiness or lethargy) impacts on mission activities, although drowsiness has been reported infrequently (Bagian & Ward 1994, Davis *et al.* 1993b). Moreover, an already ill crewmember will feel better after treatment, and this may help limit negative effects on performance. In addition, the excitement of space flight and engagement in critical tasks may also help to counteract the soporific effects of the medication (Lackner & Graybiel 1994).

Other anti-motion sickness drugs have been used during flight as well. Matsnev *et al.* (1983) reported that the administration of dimenhydrinate to eight cosmonauts before launch and during the early portion of the mission decreased but did not eliminate symptoms (Matsnev & Bodo 1984). Metoclopramide and naloxone have both been used in the U.S. space program, but neither has shown evidence of beneficial effect (Thornton *et al.* 1987).

The variable success of anti-motion sickness drugs administered in-flight may be due to changes in drug absorption or metabolism by such factors as dehydration, reduced gastrointestinal motility, alterations in body chemistry related to adaptation to the weightless environment, changes in cabin pressure, and disruption of normal sleep-wake cycles. These factors may influence both the dosage of drugs and the route of administration (Pavy-LeTraon *et al.* 1997). In microgravity, blood flow increases in the upper part of the body and decreases in the lower part. Relative disuse of muscle groups can cause atrophic changes as well. As a result of these two effects, the blood available and the amount of atrophy that has taken place at a possible injection site will influence the bioavailability of medication from an IM injection. For example, intra-muscular promethazine for motion sickness usually is given in the arm rather than in the hip in space, with good results.

Other physiological changes may also affect medication absorption and metabolism. The movement of oral medications out of the stomach may be decreased by the weightlessness of the gastric contents in space, and intestinal absorption rates may be reduced by blood and other fluid shifts to other areas of the body. Fluid shifts may also affect the bioavailability of medications sensitive to the first pass effect in the liver, where metabolism occurs (Saivin *et al.* 1997). Finally, renal excretion rates may be influenced by microgravity

The concomitant use of drugs for other indications is another confounding factor (Santy & Bungo 1991). The recent success of IM promethazine is encouraging; however, whether the effectiveness of IM promethazine vs. other anti-motion sickness drugs "is due to the pharmacologic effect of promethazine itself or the IM route of administration and its effect on bioavailability" is unknown (Bagian 1991).

Currently, NASA flight surgeons make use of medications postflight as needed to treat moderate to severe symptoms of *postflight motion sickness* (PFMS). Meclizine (Antivert, 25-50 mg) appears to be an effective treatment provided that the crewmember can tolerate oral medications. However, rigorous studies have not been done to definitively establish this. Promethazine (25-50 mg in IM or suppository form) is quite effective and is indicated for uncontrollable or large volume emesis. Fluids are administered as needed, either orally or intravenously to replace lost volume and maintain hydration. Unlike for SMS, PFMS occurs in a setting of acute relative dehydration and cardiovascular compromise, and the threshold for administering intravenous fluid supplementation should be accordingly lower. It is important to judge the relative effects of cardiovascular compromise and PFMS during this period prior to initiating treatment. This is helped by simple orthostatic assessment of pulse and blood pressure between recumbent and sitting or standing positions if tolerated, as well as observations of head movement limitation.

Often the managing flight surgeon guides the crewmember in gentle challenges to re-adaptation to Earth gravity, e.g., making small but progressive head movements. It is also helpful to avoid making large, rapid head movements, particularly in pitch and roll, during the early postflight period. Extra caution is recommended immediately post-landing during de-suiting procedures (Ortega & Harm 2007).

PFMS is also self-limited, but recovery time tends to be related to time on orbit, lengthening with longer duration missions. Return to flying duties typically happens within seven days following missions of less than two weeks in length. Some reports from the Russian program indicate that complete normalizing of vestibular and sensorimotor function following multiple month flights may not occur for several months. However, adequate, independent function usually returns within days to weeks. The NASA-Mir protocol considered returning U.S. astronauts to flying duties on an individual basis at 30 days postflight. Also, it is important to know that certain motion stimuli may cause "relapse" or "toggling" to an earlier stage of re-adaptation days to weeks after return from space flight (Reschke *et al.* 1996).

2.4   Mechanical Countermeasures

The Russians have investigated the ability of various mechanical and electrical devices to alleviate the SMS symptoms. These devices, designed to prevent the complete adaptation of the body to weightlessness, are intended to counteract deconditioning of the body during long-duration missions, as well as relieve SMS during the first days of flight. However, the small number of individuals tested as well as the lack of control subjects makes it difficult to accurately determine their efficacy.

2.4.1   Pressurized Insoles

The *Cupola SAND-501* is a pair of sandals with spring-loaded insoles containing a cuff that can be inflated to 20-60 mm Hg with an attached bulb and manometer. A Cuban cosmonaut wore these sandals for four hours on the first day of his mission on board the Salyut-6 space station. He claimed that a 20 mm Hg pressure on the soles of his feet created a sensation of heaviness in the lower limbs, causing the disappearance of the inversion illusion (see Chapter 7, Section 2.1.1) that he had experienced without the sandals. The inversion illusion reportedly recurred one to two hours following the removal of the sandals. From the third to the sixth day on board Salyut-6, the lower pressure was insufficient to eliminate the inversion illusion, so the cosmonaut increased the pressure to 60 mm Hg and wore the sandals for six hours. Nevertheless, he continued to experience the inversion illusion. This observation indicates that sensory habituation, or an inhibition of the adaptation process, may have occurred with the use of the sandals (Hernandez-Korwo *et al.* 1983).

2.4.2   Load Suits

The crew of Soyuz-13 first wore load suits. These suits are flight-type suits with elastic bands of adjustable tension in the area of the chest, back, abdomen, side, and leg seams. These suits impart a load on the body to compensate for the lack of gravity. The suits were worn from the first day of flight and were removed or the tension lessened only at night. The cosmonauts reportedly had a "high opinion" of the suits. The crew of Soyuz-14/Salyut-3 also wore the suits. While the cosmonauts considered wearing the load suits "pleasant", the still experienced inversion illusions, headward fluid shifts and symptoms of SMS (Gurovskiy *et al.* 1975, Vorobyev *et al.* 1976).

2.4.3   Pneumatic Occlusion Cuff and LBNP

A pneumatic occlusion cuff was placed on the hip by Soyuz-38 cosmonauts to reduce or prevent the headward shift of body fluids that occurs in weightlessness, which was once thought to be a potential contributor to the development of SMS symptoms by Russian scientists. The cuff, worn for 20 to 30 minutes at -40 to -60 mm Hg, reportedly decreased dizziness, illusions, nausea, and the sensation of "head pulsation." Exposure of the Soyuz-38 crew to *lower body negative pressure* (LBNP) of -25 mm Hg was also reported to have "a positive influence on the health of the cosmonauts" (Matsnev *et al.* 1983, Gorgiladze & Bryanov 1989). Unfortunately, these reports were not very specific about the effects of the cuff and LBNP on SMS symptoms.

Figure 8-05. French astronaut Michel Tognini on board Mir is wearing a head cap attached to his hip with rubber cords. This "Chapka" device was supposed to help reducing the head movements responsible for space motion sickness for the first days in space. To our knowledge, it is no longer used. Photo courtesy of CNES.

2.4.4   Neck Pneumatic Shock Absorber

The *neck pneumatic shock absorber* (NPSA) device is basically a cap with rubber cords that provides a load to the cervical vertebrae and neck muscles (Figure 8-05). When worn, the cosmonauts must stretch their neck muscles to maintain an erect head position. The cap also mechanically restrains the crewmembers from tilting their head in pitch and roll (Matsnev *et al.* 1983). The individual wearing the device sets the cord tension. The NPSA was designed to be worn during working hours for the first three or four days of a mission. It was used on the Soyuz-T3, -49, -40, and -T7 spacecraft as well as on the Salyut-6 and 7 orbital stations. The cosmonauts reported the NPSA to be effective in alleviating dizziness, illusions, discomfort, and nausea with no adverse impact on performance. The effectiveness of this device was attributed to a "normalization of the vestibulo-cervical reflex system" (Matveyev 1987). However, the forced restriction of head movements with this device was most likely responsible of its effectiveness against SMS, since head movements are known to exacerbate symptoms.

2.4.5 Electrical Devices

Electrical devices that pass an electrical current through the body have also been explored. Ground-based studies indicated increased tolerance to experimentally induced motion sickness following electrical stimulation using two electrodes, one placed on the forehead and one applied in the area of the mastoid (Melnik *et al.* 1986, Polyakov 1987). Galvanic vestibular stimulation has also been used on Earth to stimulate the vestibular receptors and elicit symptoms of motion sickness and postural imbalance (Sévérac 1992, Moore *et al.* 2006). However, electrical stimulation operates via mechanisms that are as yet unclear and this technique has not been tested in the space environment.

3   COUNTERMEASURES FOR BEHAVIOR AND PERFORMANCE

There are a number of methods used for preparing the crewmembers to perform meaningful tasks essential to mission safety and mission completion. These methods rely on training or in-flight sessions where the crewmembers are evaluated on their performance. The objective is to decrease the tasks errors by identifying them, analyzing the error patterns, and establish a new training aimed to "design out" error. Although some on-board simulations to maintain task skills exist, most of this research is done on the ground in training simulators.

3.1   Intra-Vehicular Activity Training

For example, during the early U.S. space program, in order to familiarize the crew with spatial disorientation, the Mercury astronauts received training in the *Multiple Axis Space Test Inertia Facility* (MASTIF) located at NASA Lewis Research Center in Cleveland, Ohio (Figure 8-06). The purpose of this training was twofold: (a) to give the astronauts familiarization with the physiological and psychological effects of tumbling; and (b) to recover from tumbling when it occurs. A slow build-up of axes and rates was used to a maximum of 30 rpm rotating about all three axes. The astronauts in all cases were able to stop tumbling in a relatively short period of time, using the Mercury-type rate indicator and hand controller.

An *Air Lubricated Free Axis* (ALFA) trainer also utilized a periscope display or a window with a simulated Earth horizon for controlling actual capsule attitudes and rates during orbit and retrorocket firing. The astronauts controlled the capsule by a Reaction Control System consisting of pressurized air reaction control nozzles.

Figure 8-06. The Multiple Axis Space Test Inertia Facility *(MASTIF) was an enormous set of three concentric cages, called gimbals, one inside each other. The engineers could program the machine to rotate just one or all three cages, simulating a completely out-of-control capsule, pitching, rolling and yawing through space. The astronauts had to learn to use a hand control stick that released spurts of gas that acted as a brake against the rotating motion of the gimbals. The goal was to stop the cockpit from tumbling and bring it to a complete stop. Photo courtesy of NASA.*

Centrifuge training provides a full-scale simulation of the launch and ascent phase and the aborts associated with each type of trajectory. The primary purpose of this training is to give the astronauts training in capsule attitude and rate control, monitoring normal sequencing functions, and rectifying emergency problems while being exposed to environmental conditions that might be associated with the flight. The environmental factors emphasized during these simulations are acceleration, reduced cabin and suit pressure conditions, and the effects of these conditions on astronaut performance. The astronauts also receive additional training with voice communications and code. Further evaluation of the controls, couch, vehicle lighting and instrument design is also accomplished during these simulations.

The astronauts receive familiarization with weightlessness by being flown as passengers through several parabolic trajectories in KC-135, DC-9, or Illyouchin type aircraft. The duration of weightlessness is about 20 seconds, and the number of parabolas per flight varies from 30 to 40, depending of the type of aircraft being used. During these flights, the astronauts receive experience in orienting their body and limbs, donning and doffing space suits, and eating and drinking.

3.2   Extra-Vehicular Activity Training

In preparation for the lunar missions, Apollo astronauts practiced setting up of lunar surface experiments, collecting lunar samples and simulated traverses during extra-vehicular activity training (Figure 8-07) In some of the training sessions, crewmembers were attached to a 6-deg of freedom harness connected to counterweight so that the net mass of the crew and its tools was equivalent to that of the lunar gravity environment (see Figure 1-06).

The demands of extra-vehicular activities now placed on astronauts are unprecedented. The EVA hours required for the ISS assembly alone are estimated to exceed the total number of EVA hours during the first 15 years of the space program by a factor of three. Structural components are added to the ISS in a rigorously planned assembly sequence. Therefore, the ISS assembly is a choreographed event that requires intense task-based training. The astronauts' movements, the robotic manipulations, and the assembly sequences are not only thoroughly planned, but also exhaustively rehearsed. By far, the majority of ISS assembly training is accomplished in the *Neutral Buoyancy Laboratory* (NBL), which is a giant swimming pool containing ISS mockups located at the Johnson Space Center. The NBL provides an environment where the EVA rehearsal is controlled and well understood, allowing training for tasks that require team-coordinated moving of massive, but weightless objects. However, the dynamics between weightless conditions and underwater simulation are different (see Chapter 1, Section 4.2, p. 17), and could pose a problem for task training.

Maintenance or repair tasks performed during the operational phase of the ISS is more likely to be skill-based rather than preplanned, rehearsed tasks. The same holds for tasks to be performed during a Moon or Mars mission. Even for the most highly trained astronauts, task-specific training for a prescribed maintenance scenario will not remain fresh because of the long mission durations. The crew may be required to apply certain procedures long after the training period by relying on their memory. On-orbit training could provide an effective way to prepare the astronauts who must perform complicated or unpredictable maintenance and repair tasks in microgravity. This training would be based on real-time communications between the EVA crew and ground support personnel.

Figure 8-07. Two members of the Apollo-16 lunar landing mission participate in lunar surface extra-vehicular activity simulation training at the Kennedy Space Center, Florida. Photo courtesy of NASA.

Virtual reality is another mode in which this training could be accomplished. NASA is already using a virtual reality system that allows astronauts to practice the careful placing of their hands and feet in rock-climbing fashion for EVA training (Figure 8-08). The virtual reality system augments visual simulation with a unique haptic simulator called "Charlotte." The device is named after the storybook spider character in E. B. White's classic children's book, *Charlotte's Web*. Pairs of astronauts move the simulated object by its handles while the system simulates the dynamics and drives the motors that move the objects appropriately. Users report very high fidelity for masses of 200 kg up.

The three-dimensional architecture and inconsistency of the visual vertical of adjacent quarters and modules, combined with the limited visual experience of crewmembers, is a major cause of the spatial disorientation problems. Astronauts normally see the interior of a spacecraft from a variety of body orientations and viewpoints currently not simulated on the ground. It requires cognitive skills to interrelate cues perceived in a body-centered (intrinsic) frame of reference built up directly through navigation as well as through an overall (extrinsic) frame of reference defined by the spacecraft. Astronauts can either learn this interrelationship in-flight or develop the required cognitive knowledge prior to flight via virtual reality simulation. Researchers at the *National Space Biomedical Research Institute* are developing a virtual reality-based training method for astronauts as a countermeasure to this in-flight spatial disorientation and navigation. Using a virtual 3D space station model, the subjects will be able to explore a virtual ISS while wearing a head-mounted display with a head tracker (Figure 8-08). They will learn each module separately, or the whole ISS at once. Visibility will sometimes be obstructed by smoke. It is expected that after this preflight training, astronauts should show quantitatively superior spatial knowledge and navigation skills (Aoki & Oman 2007).

Some astronauts have already participated in a series of training sessions using a laboratory version of this system. During one Spacelab mission one astronaut practiced moving along the walls and ceiling, attempting to view those features as a floor. He practiced moving into the tunnel, which connects the Spacelab to the middeck, then turning around, reemerging into the Spacelab, and identifying his orientation after re-emergence. Following the mission, he used the virtual reality simulator and attempted to perform the same activities. Preflight, this astronaut reported some orientation and motion difficulties in the virtual Spacelab environment. He found it difficult to view the Spacelab walls as a floor and he reported that the hand controller was difficult to use. One day after the mission, he remarked that it was easy to perceive the ceiling or walls as a floor and that "locomotion" through the virtual Spacelab using the hand controller was "intuitive." A week after landing, he reported greater difficulty in mentally rotating the virtual Spacelab and the hand controller became less intuitive than it had been immediately postflight (Harm & Parker 1993, Parker & Harm 1993).

3.3 Psychological Training

The psychological effects of isolation and confinement on human behavior and performed are studied during ground-based studies in closed quarters. Some of the Skylab astronauts and Mir cosmonauts were subjects in such studies prior to their mission. The major objective of these simulations is to study the organization of the crew's activity and its interaction with the ground segment, as well as expose the crew to the psychological issues (e.g., task errors, crew conflict, cultural and personal isolation) that they might encounter during the actual space mission. Such an experiment simulating a human mission to Mars will start in the fourth quarter of 2007 at the *Institute of Biomedical Problems* of the Russian Academy of Sciences. Six volunteers will be selected internationally for this MARS-500 project. The experiment will last for 520 days and may be extended to 700 days. During this time investigators-volunteers will live and work in a special complex built to the anticipated dimensions of the foreseen actual missions, and will examine in detail the crews' ability to communicate with mission control and families via e-mail and satellite links.

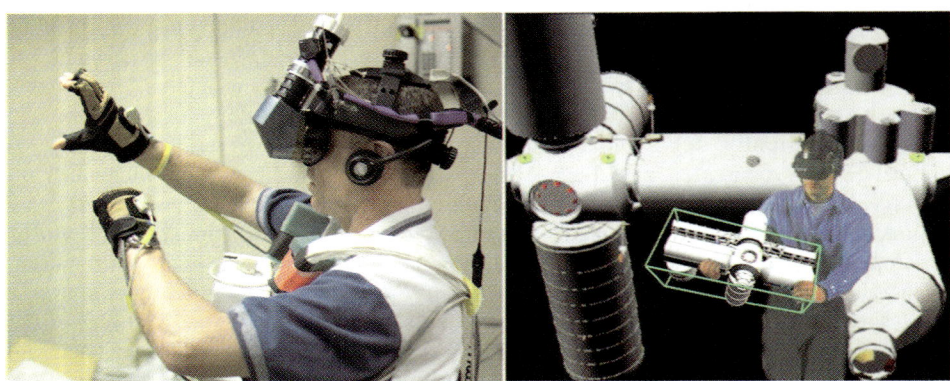

Figure 8-08. Left. An astronaut uses virtual reality hardware to rehearse some of his duties on the upcoming mission to the ISS. He is wearing special gloves, stereoscopic headphones and a head motion tracker while looking at computer displays simulating actual movements around the various locations on the ISS hardware with which he will be working. Right. Virtual reality can also be used as a navigation training tool, by helping memorize ISS map and enhancing mental representation and rotation. Photos courtesy of NASA.

4   ARTIFICIAL GRAVITY

Ongoing efforts are now focused on preparing for human interplanetary missions to Mars in the not-too-distant future. Mission durations will be measured in years. The explorers will face severe physiological deconditioning due to weightlessness. Detrimental effects on human health and performance as a result of exposure to microgravity have been identified in space experiments, ground-based research, and operational flight experience. This is the case even when currently available countermeasures are used. As we advance from the ISS to the Moon and then on to Mars missions of increasingly extended duration with multiple gravity levels (Figure 8-09), the requirements for effective countermeasures become more complex. The transit phases of a Mars mission will involve very long-duration exposure to microgravity, much longer than the duration of the current expeditions on board the ISS. Ground-support equipment and personnel will not be available at the landing site to assist the Mars travelers after their arrival, as is the case for returning ISS crewmembers (Figure 8-10). Generally, undesirable effects become more severe with extended exposure duration. Duration of exposure is directly related to a specific mission scenario and will affect crew physiology, mission performance, and postflight recovery (Aerospace Medicine Advisory Committee 1992).

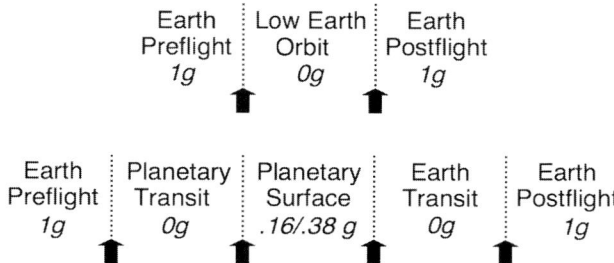

Figure 8-09. There will more transition periods between gravity levels during the missions to the Moon or Mars than during the current missions in low Earth orbit. The neurovestibular responses are principally affected by these transition periods.

Space biomedical researchers have been working for many years to develop countermeasures to reduce or eliminate the deconditioning associated with prolonged weightlessness. Despite these countermeasures, most astronauts experience problems with balance and spatial orientation during the first few days after landing. They also risk muscle tears and bone fractures and therefore must exercise an added degree of caution during their recovery period. Given that the purpose of a human mission to Mars is not to go there and simply survive, more effective countermeasures or combinations of countermeasures must be developed to address the effects of long-term exposure to microgravity. Astronauts arriving at Mars in a weakened physical condition who can't manage to ambulate would hardly be able to successfully execute an exploration mission. They would be at risk in the event of a sensorimotor performance failure during piloting, extra-vehicular activity, or remote guidance tasks. Until the problems associated with microgravity exposure are overcome, such missions cannot be seriously considered.

A number of different countermeasures have been employed in an attempt to mitigate the effects of human exposure to microgravity, generally aiming to stimulate a particular physiological system (Sawin *et al.* 1998). These countermeasures, although inefficient and onerous for some astronauts, are reasonably effective against some of the muscle and cardiovascular losses, but have only limited effectiveness in countering the full range of sensory, sensorimotor and cognitive changes during space flight.

Artificial gravity is the simulation of the pull of gravity on board a manned spacecraft by the steady rotation or linear acceleration of all or part of the vehicle (Stone 1973). Artificial gravity represents an alternative approach to addressing the problems of microgravity-induced effects on the human body. Rather than addressing each individual system in a piecemeal fashion, which is only valid if the principle of superposition holds for the combined effect of these interacting subsystems, artificial gravity stimulates all of the physiological systems simultaneously by reproducing the normal Earth gravitational environment. All physical and physiological systems are challenged. Bones are stressed, antigravity muscles are called into action, the otoliths of the vestibular system are stimulated in a manner similar to that on Earth, and the cardiovascular system is similarly stressed. Obviously, artificial gravity cannot address all of the problems associated with long duration space flight, in particular that of radiation exposure, altered day-night cycles, and the psychological issues that will no doubt arise from extended confinement and isolation. It does, however, offer a countermeasure with the possibility to address the debilitating and potentially fatal problems of bone loss, cardiovascular deconditioning, muscle weakening, sensorimotor and neurovestibular disturbances, and regulatory disorders. Because artificial gravity addresses all such systems across the board, it can be considered as an integrated countermeasure (Clément & Pavy-Le Traon 2004).

Figure 8-10. An ISS crewmember is being transported to a helicopter by ground personnel (right) after the Soyuz capsule that returned him from the ISS had landed in the Kazakhstan desert (left). Photo courtesy of NASA/Bill Ingalls.

Our space flight experience with the Mir and the ISS space stations indicates that changes in the neurovestibular system are reversible upon return to normal Earth gravity after missions lasting up to several months. However, re-adaptation can take several weeks. It is not certain that 0.38 g will be sufficient for re-adapting these physiological functions while on Mars. It is also not certain that the Moon gravity (0.16 g) will be sufficient to prevent the deconditioning of body functions due to reduced gravity. The Apollo program demonstrated the ability for humans to successfully perform short-duration mission tasks and extra-vehicular activities on the lunar surface. However, gravity thresholds for biological processes have not been determined. Therefore, the impact of extended duration exposure to 0.16 g or 0.38 g on the Moon and Mars surfaces, respectively, on the physiological responses is unknown. Consequently, the requirements for countermeasures in Moon and Mars bases cannot be fully defined.

Human centrifuges could be technologically and programmatically feasible as countermeasures for the reduced gravity within planetary bases or on board a Mars transit vehicle. The two options for artificial gravity in the transit vehicle are either long-radius continuous rotation of the habitat or short-radius intermittent exposure to centrifugal forces with an embarked centrifuge. Both approaches significantly impact transit vehicle design and mission operations. In a planetary base, however, the only option is intermittent artificial gravity, or gravity augmentation.

Scientific evidence on the merit of artificial gravity loading under these conditions is needed (Aerospace Medicine Advisory Committee 1992). To determine the best technique for implementing artificial gravity in a space mission, a complex set of trade studies must be executed. The parameters that must be considered include, but are not limited to, vehicle design, engineering costs, mission constraints, countermeasure efficacy, reliability requirements, and vehicle environmental impacts.

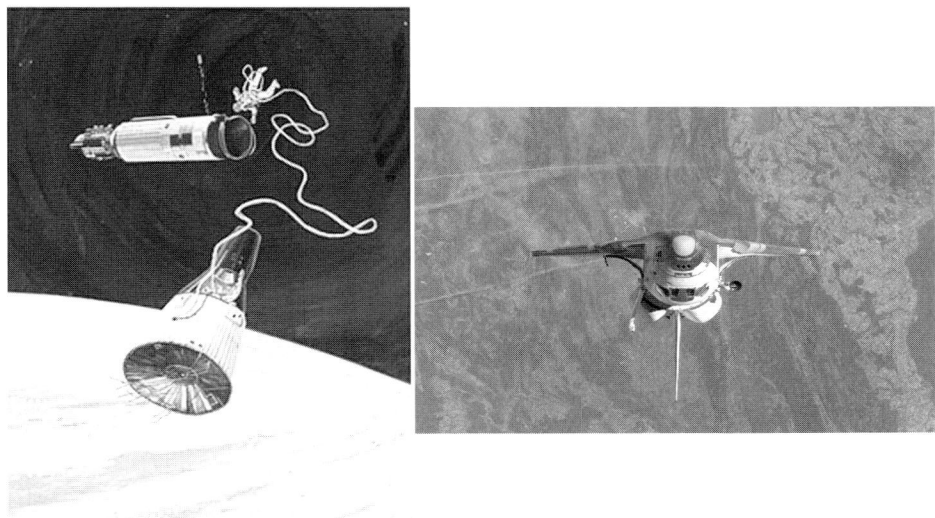

Figure 8-11. Left. When the Gemini-11 capsule and the Agena were docked together by a tether, a slow cartwheel motion of 55 deg/min rate was generated, causing the first demonstration of artificial gravity in space. Right. The Space Shuttle already does a single back flip as it approaches the ISS to allow the station crew to photograph the heat shield on the Orbiter's belly. In that maneuver, the Space Shuttle rotates at only 0.11 rpm. A pitch rate 21 times as fast, i.e., 2.31 rpm, would be required to generate an artificial gravity of just 0.1 g in the middeck of the Space Shuttle.

From a physiological countermeasure perspective, a good solution is to provide artificial gravity continuously throughout the mission (Figure 8-11). This approach would most likely reduce or eliminate physiological deconditioning, improve human factors (e.g., spatial orientation, hygiene, food preparation, work efficiency), facilitate more efficient medical operations and equipment usage (e.g., countermeasure applications, surgery, cardio-pulmonary resuscitation), and provide a more habitable environment (e.g., management of liquids and contaminant). However, these benefits would need to be weighed against technical risks and uncertainties. These include engineering challenges such as system functional, performance and operational requirements; engineering and architectural designs; fluid management mechanisms; and propulsion system options. Furthermore, human factors and physiological issues will certainly result from the deactivation of artificial gravity once the space vehicle arrives in the vicinity of Mars. Considering that half of all astronauts require one to three days to adapt to microgravity, a similar period of adaptation is expected when artificial gravity is removed. Therefore, a complete set of trade studies cannot be fully executed and analyzed until after further physiological research is completed and vehicle design options are evaluated.

However, there are a number of engineering challenges associated with generating artificial gravity through the use of very large rotating vehicles or the application of very high linear acceleration, given the engines required to accomplish this. Such designs are not likely to be realized in the near future. In the case of a crewed Mars spacecraft, the structure required would be prohibitively large, massive and certainly not energy efficient. An alternative approach being explored is to provide astronauts with a small spinning bed. They would lay on their back with their head near the center of rotation and their feet pointing radially outward. Thus, their lower body could be loaded for a specified period of time each day in approximately the same way as under normal Earth gravity. While not expected to be as efficient a solution from a physiological standpoint, given the gravity gradient effects and intermittent exposure, this procedure may prove effective. The engineering costs and design risks would certainly be lower as compared to designing a rotating spacecraft.

There is little information available on the physiological effects of the effects of different levels or duration of gravity loading. It would seem prudent to identify how variables levels of gravity can be used to normalize physiological processes. Martian EVAs in 0.38 g in an encumbering space suit will provide a significant amount of exercise. Lunar 0.16 g data from long-duration (3-6 months) expeditions will also provide potentially important data points for establishing the minimum level of gravity and exercise required. In the interim, it is important that artificial gravity studies take place and bone, muscle, cardiovascular, and neurovestibular function data from long-duration ISS astronauts in zero-g and lunar EVA astronauts in 0.16 g be carefully assessed. A research program should identify the gravity levels that are necessary to maintain affected tissue and physiological systems, determine how these loads should be applied (e.g., continuous vs. intermittent centrifugation), and provide protocols to minimize or eliminate undesirable side effects.

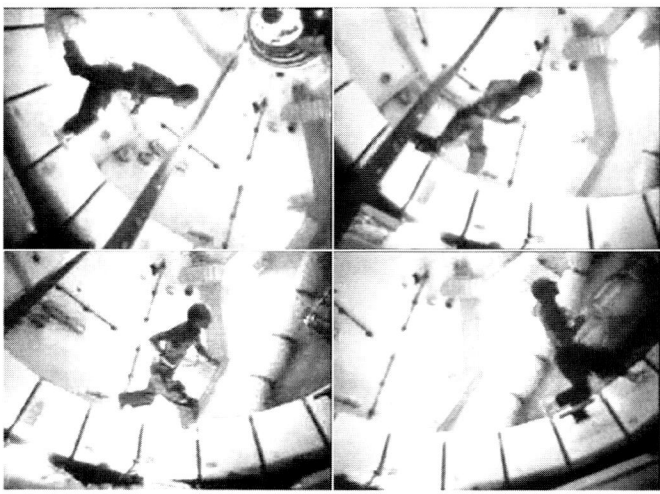

Figure 8-12. The diameter of the Skylab workshop was wide enough for the astronauts to generate the right amount of centrifugal force that allowed them to run along the rim of the cylindrical module. This activity provided them with both artificial gravity and exercise. Photo courtesy of NASA.

Artificial gravity should also be integrated with other countermeasures such as exercise (Figure 8-12), sensorimotor training and pharmacological prescriptions. Additionally, these studies will contribute to our fundamental understanding of the effects of gravitational and Coriolis forces on physiological systems.

A combination of short- and long-duration studies using ground centrifuges (Figure 8-13) and slow rotating rooms could be useful in answering many of the key questions concerning the application period, frequency, and intensity of centrifugal force. Chronic exposure to Coriolis forces necessary to operate in artificial gravity will have to be studied (Lackner & DiZio 2003). The physiological responses to transitions between artificial gravity, microgravity, and Moon or Mars gravity have to be investigated (Young et al. 2001). These studies would be useful in assessing whether dual adaptation to a rotating and a non-rotating environment is possible. The outcomes of these studies are essential to developing the artificial gravity prescription to be used during long-duration space missions (Clément & Bukley 2007).

Figure 8-13. Artificial gravity is being tested during bed rest by spinning volunteers on a short-radius centrifuge. Because of the deconditioning that takes place during bed rest, subjects after bed rest typically experience orthostatism intolerance when moving from lying down to standing upright. Prescriptions including intermittent spinning on the centrifuge (1 hour/day with 1 g at the heart level) were shown to be effective against this deconditioning. This photograph shows the ESA short-radius centrifuge utilized during bed rest studies at MEDES in Toulouse. Photo courtesy of CNRS/Sébastien Godefroy.

5   SUMMARY

Space motion sickness is operationally significant during the transition periods between gravito-inertial force levels, i.e., insertion into weightlessness and during the return to Earth, because of its time course and nature. Treatment for the disorder is symptomatic. SMS is typically self-limiting, abating in two to three days in most, but not all, crewmembers. The risk of SMS during EVA and for piloting tasks is of particular concern.

The current drug of choice for treatment of "moderate" to "severe" in-flight SMS is intra-muscular injection of promethazine. However, in spite of the use of promethazine, SMS remains the most overt medical phenomenon during the first few days of space flight. Consequently, more research is needed in this area. In particular, it is important to understand the reasons why some astronauts recover from SMS within a few days and others do not.

Operational constraints imposed by SMS will be more of a concern for the early lunar missions than they have been for long-duration ISS increments because of their relatively short duration. Frank sickness manifested by vomiting is easily diagnosed. However, the effects of low-level symptoms, similar to the sopite syndrome, and the drugs used to treat SMS on astronaut performance are more subtle and insidious. These effects overlap with the effects of sleep loss and stress that are also of concern for performance and behavior. It is often noted that the twelve Apollo crewmembers that reached the lunar surface made no reports of SMS once they landed on the Moon. However, at that time there were also no reports of any postflight motion sickness, which is now widely recognized, and may be even more prevalent than on-orbit SMS.

Research is ongoing to develop new, faster acting, less-painful-to-administer formulations of anti-motion sickness drugs. Work on new methods for quantifying performance deficits, on understanding how visual and motion cues contribute to SMS etiology and on the physiology and pharmacology of the vestibular-emetic linkage is also in progress. There are also interesting new results on the genetic basis of susceptibility. We could also do a better job of teaching crewmembers what is already known about the causes of space sickness, and about effective pharmacologic and non-pharmacologic means of controlling it.

Unlike its investment in vestibular and oculomotor research, NASA has made only a modest investment in motion sickness studies over the past 25 years. Even a small investment in new pharmacologic and non-pharmacologic approaches to mitigate SMS may pay off significantly with improved overall performance and increased flexibility of scheduling, particularly during the first three days of flight. It is also unfortunate that medical representation to the astronaut office appears to be satisfied with the current drug treatment regime. The effect of such complacency is to hamper or block fresh research.

Strategies for mitigating the effects of sleepiness and circadian disruption on an astronaut's performance and alertness levels need to be developed and evaluated. Neuroscientists and spacecraft designers could collaborate on the configuration of interior architectural features, work areas, and the relative orientations of adjacent or docked spacecraft modules to minimize some of the SMS effects, as well as spatial disorientation problems. Might virtual reality training techniques, which astronauts currently use to plan their spacewalks, be used to reduce the incidence of visual reorientation and inversion illusions while working inside a spacecraft? Can individual performance on operationally relevant three-dimensional orientation, navigation and tele-operation tasks be predicted based on simple tests of individual mental rotation and perspective-taking skills? Answers to these questions can be directly applied to the design of the future *Crew Exploration Vehicle*, the *Lunar Surface Access Module* (Figure 8-14), and eventually the *Mars Transfer Vehicle* habitats. These answers also apply to the physical arrangement of ground simulators, and to the development of virtual reality-based techniques for preflight orientation and navigation training for astronauts (Oman 2007).

With all of the human space flight experience gained over the past forty-five years, no completely effective single countermeasure, or combination of countermeasures, exists. If a crew of astronauts were to embark on a journey to Mars today, the suite of piece-meal countermeasures currently employed would leave them in a state of complete inoperability after their six-month exposure to weightlessness en route to their destination. Long-duration planetary exploration missions will require on-orbit countermeasures to maintain dual-adaptation to both zero-g and one-g. Many believe that adaptation to Mars gravity and re-adaptation to Earth gravity would be enhanced by frequent exposure to simulated gravitational states on board the spacecraft en route to and from Mars. This would require some type of onboard human-rated centrifuge or complete spacecraft rotation to produce an inertial force similar to gravity, which would be also coupled with physical countermeasures to maintain bone and muscle mass. This artificial gravity solution, while potentially effective, raises a number of operational, engineering and physiological issues that will need to be addressed. The physiological responses to continuous exposure to anything other than 1 g are unknown. Research is needed to identify the minimum level, duration, and frequency of gravity level required to maintain normal physiological function, as well as the importance of a gravity gradient across the body (Clément & Bukley 2007).

Figure 8-14. The Orion vehicle is part of the NASA Constellation Program to send human explorers back to the Moon and then onward to Mars and other destinations in the solar system. This artist view depicts the Orion docked with a lunar landing module in lunar orbit. Orion will be the Earth re-entry vehicle for lunar and Mars returns. Photo courtesy of NASA.

Chapter 9

A VISION FOR SPACE NEUROSCIENCE

The most overt physiological problems of space flight are disorientation, perceptual illusions, space motion sickness in-flight and immediately after landing, and locomotion problems postflight, as reported by nearly every astronaut returning from a space mission. These problems are generally most acute during transitions between gravitational force levels that are, unfortunately, the times when physical and cognitive performance is critical for safety and mission success. Postflight symptoms are more severe after 3-6 month Mir and ISS flights than on 1-2 week Shuttle missions, demonstrating that some components of neurovestibular adaptation to microgravity take place over time scales of months, rather than weeks. The construction of the ISS means that EVAs and tele-operation of robot arms and vehicles have become increasingly more common. These activities will also be critical for the construction of lunar outposts and planetary exploration. They also represent significant sensory and motor integration challenges. Sensorimotor disturbances are anticipated whenever astronauts must make a sudden transition from zero-g to the partial gravity of planetary environments, or from zero-g to an artificial gravity environment.

The previous chapters of this book have provided a survey of what is known of the deleterious effects of space flight on perception, sensorimotor coordination and cognitive performance. In this last chapter, we have tried to identify relevant research questions in space neuroscience as they relate to the current ISS missions and to the future human exploration missions. It is also taken as a given that understanding the effects of reduced gravity on the human organism is predicated on an understanding of normal physiological functioning in a one-g environment. Therefore, we have also examined ground-based studies of the relevant central and peripheral systems that are compromised by exposure to reduced gravity.

Figure 9-01. Artist view of the ISS in its configuration following the STS-115 Shuttle mission in September 2006. Photo courtesy of NASA.

G. Clément, M.F. Reschke, *Neuroscience in Space*.
DOI: 10.1007/978-0-387-78950-7_9, © Springer Science+Business Media, LLC 2008

1   BASIC SPACE NEUROSCIENCE RESEARCH

Neuroscience questions focus on understanding basic mechanisms associated with perception as it is influenced by gravity, effects of altered gravity on changes in biological rhythms, definition of appropriate neuronal models for understanding central processing in altered gravity, and understanding how signals from different receptors are involved in spatial orientation when motion and gravity change. Another unique opportunity for neuroscience is the influence of gravity (or its apparent absence) on development. There are potentially many important questions in this area, of which we will mention only a few. For example, is there a critical period for development of anti-gravitational reflexes, similar to the critical period for development of vision? Secondly, are there structural changes in the organization of the nervous system of an organism that develops in space? Are there synaptic or structural changes in an organism after being in space for long periods of time? These problems have serious practical implications, but could also be important theoretically for understanding how we develop antigravity postural mechanisms.

The scientifically important research into the influence of partial gravity on animals and cells is also fundamental to understanding the problem of human deconditioning and survival in weightlessness, on the Moon or on Mars.

1.1   Neuroscience Research on Board the ISS

Beyond the immediate use of the ISS to answer some of the more pressing issues in human physiology associated with the future exploration missions, is the larger question of the continued need for a microgravity laboratory that will support fundamental research.

After the success of the Apollo program, when NASA managers decided to go ahead with a space station in preparation for future planetary exploration and long-duration missions, their approach was very pragmatic. This space station was Skylab (see Figure 2-03), a relatively large station even by today's standards, which re-used some of the Saturn-5 elements and was visited by crews three times between 1973 and 1974. Flight duration was incremental: the first manned Skylab mission lasted 28 days, the second 59 days, and the third and last one 84 days. Skylab logged about 2,000 hours of scientific and medical experiments. The plan was to collect enough data during one month, then two months, and then three months in space, and to extrapolate the changes for longer durations (White 2007). The later Mir missions, some of which have kept cosmonauts in low Earth orbit for over a year, and now the ISS missions lasting up to six months indicate, however, that extrapolation may not be reliable for all physiological functions given the limited data available. For example, bone loss does not appear to show a plateau as predicted by the Skylab data and the immune system response seems compromised only after a few months in orbit (Clément 2005).

The bulk of data collected during space life experiments in general and neuroscience in particular comes from space flight lasting 1-2 weeks. Except for a few experiments carried out on the first Space Shuttle flights, virtually nothing is known about the very short-term adaptive processes, i.e., during the first minutes or hours of entry into weightlessness. Yet the relative changes in sensorimotor and cognitive functions and their operational significance during this period will be of great importance if commercial space tourism is to be developed for sub-orbital flight of very short duration. On the other end, there is little information on very long-duration space flights, although the ISS will greatly help to close that gap when assembly is complete (Figure 9-01).

The results described in the previous chapters have shown that there is a transient state of adaptation of sensorimotor and cognitive functions after 1-2 weeks in space. The understanding of the CNS adaptation to space flight will not be complete until there is an assessment of the mechanisms which underlie the stabilized adapted state to this condition. There are many examples of ground-based adaptive studies showing that mechanisms of long-lasting adaptation cannot be understood with data from short-lasting adaptation. Actual data during long-duration missions is required.

Also, in studying the problems of long-term adaptation or return re-adaptation, it is necessary to have a reference point. After a one or two-week space flight re-adaptation is rather rapid: it takes literally minutes or hours for some responses to be modified. After longer flight, it may take days or weeks or months before responses return to what they were before flight. However, to understand the re-adaptation mechanisms, it is necessary to collect data just after landing to have the earliest point. This valuable data point is typically denied for a number of reasons: (a) the experiment is not performed because it is assumed that it may make the returning crewmember motion sick; (b) it has been deemed more important that blood draws occur pror to any other experimental activities is allowed; and (c) operationally, retrieval of crewmembers from the landing vehicle is delayed due to public relation requirements.

Compared to the previous flight opportunities for space experiments, ISS presents unique advantages: extended duration, technically advanced facilities and sufficient sample size on a given crewmember. Our ability to clearly interpret past life sciences flight experiments and provide countermeasures has often been limited because there are too few replications of experiments. ISS capabilities promised to provide the time and number of individuals or specimens for clear interpretation of statistically valid results. Costs have prevented this advantage from becoming a reality. As long as we are limited to one or two flights per year and two crewmembers per flight the number of subjects will continue to plague all of life sciences research interests.

Although the ISS is potentially the ideal laboratory for research into all of the microgravity related issues challenging long duration exploration, it has not yet been used effectively for several reasons. While under construction, and with a limited crew of three, no time is available for intense human research. Only a few of the planned scientific racks for human physiology research are on board and sample return is currently very limited. In our view, the keys to fulfilling the potential of ISS for space research includes the support of peer-reviewed, mission-oriented flight experiments directed at finding a solution to the key challenges for exploratory missions, the provision of a full resident crew of six or seven, including astronauts trained and capable of doing biomedical studies as well as serving as test subjects, and the facilities to perform key experiments at various gravity levels The utilization of the ISS scientific laboratories will also require the regular upload and download of software, data and research specimens, even after discontinuation of Shuttle flights (Figure 9-02).

1.2 Time Course of Adaptation

The main results from neuroscience experiments in space to date have shown that sensorimotor adaptation to microgravity follows several time courses, from a few hours or days for motion sickness to several months for spatial orientation. These time courses presumably correspond to different underlying mechanisms. These mechanisms can range from simple parametric changes of the various sensorimotor subsystems within their physiological range (e.g., gain of the VOR), to substitution mechanisms that allow a sensorimotor module belonging to the normal repertoire of an individual to take over the function of a defective subsystem (e.g., the saccadic system substituting the smooth pursuit system or VOR), and to sensorimotor learning based on the available repertoire (e.g., mental representation, mental rotation). Reflexes, sensorimotor functions and perceptions are organized on the basis of egocentric or allocentric reference frames. In terrestrial gravity, the brain uses an allocentric reference frame, presumably gravity-based, to code visuospatial memories during whole-body tilts (Pelt *et al.* 2005). In space, the apparent absence of gravity induces a preponderance of the egocentric reference. However, the brain can define information in multiple frames of reference depending on sensory inputs and task demands, so it is probable that different reference frames are used for different tasks.

These remarks imply several important consequences in the design of future research programs in space neuroscience. The first consequence is that it will be necessary to study both low-level mechanisms, such as postural reflexes in order to assess, for example, the functional integrity of gravity-receptors, and high-level perceptual processes, such as cognitive functions, in order to assess adaptation due to central processing. Each individual may adopt different strategies for adaptation that may involve changes at a very elementary level, i.e., the reflex, or may require reinterpretation of the information provided by some receptors, i.e., the otoliths or the Golgi tendon organs. It may also involve high-level cognitive mechanisms, i.e., the internal model of gravity or body scheme, which by definition will involve several sensors and multimodal interactions. The so-called *reflexes* are not rigidly defined. There are goal- and context-dependent reflexes. The same holds true for percepts that can vary according to the sensorimotor context. Therefore, space neuroscience experiments must cover the traditional experiments on reflex responses (posture, eye movements) and spatial orientation investigations, as well as cognitive functions such as the mental representation of body and environment that, even if not directly affected by the apparent absence of gravity, could be involved in novel learned mechanisms.

The problem of understanding long-term adaptation is not so much the definition of reflex changes of reflexes within each so-called *reflex module,* as much as it is understanding the need to switch sensorimotor strategies to cope with deficiencies within each module. Cognitive processes, for example, participate in the regulation of the low-level neuronal operations during adaptation (Berthoz 1989). When the CNS is faced with sensory conflict, the first thing that is done is to try to use parametric adjustments within the subsystem. For instance, if there is a problem of VOR or optokinetic function, then changes in gain are made within the normal physiological range. If that does not work, eventually a decision is made to suppress one of the subsystems and to substitute it with another. The CNS will "look" in the available repertoire of subsystems for one which has the same dynamic capacities, geometrical features, and so on, or a combination of them. This implies, at least in humans, cognitive factors. To orchestrate this new combination of subsystems, eventually one has to take into consideration the mental representation of space. The brain must constantly monitor the position and movement of the body in relation to nearby objects. The effective "piloting" of the body requires an integrated representation of body (the *body scheme*) and of the space around the body (the *peripersonal space*). Recent results from neurophysiology, neuropsychology, psychophysics, and neuroimagery in both human and non-human primates support the existence of an integrated representation of visual, somatosensory and auditory peripersonal space. Such a representation involves primarily visual, somatosensory and proprioceptive modalities, which operates in body part-centered reference frames, and demonstrates significant plasticity (Holmes & Spence 2004).

Long-duration changes could then be called upon for the reinforcement of adaptive processes. If you want to solve a problem for one minute, you may not actually induce major synaptic changes. If you have to do it over and over, several days a month, two or three months, a year, eventually deep reorganization, such as provided by unique neural plasticity, may be required. The general image, which is probably very naïve, is that the first steps of adaptation are like a learning process – very active with control of new solutions. Eventually, learning will lead to complex new strategies. This implies synaptic and neurochemical plastic changes (Berthoz 1989, p. 153). Changes may occur in the sensory end organs themselves, in neural components such as hair cells, synapses or afferent terminals, or in the afferent signal. Sensory end organs and effector changes may also be mediated through secondary mechanisms like calcium loss or through CNS pathways. Unfortunately, the adaptivity or maladaptivity of these changes cannot be determined without combining observations of anatomical changes with perceptual and behavioral responses.

Figure 9-02. The International Space Station flying over the Pacific Ocean as seen from the Space Shuttle. The ISS is designed to be a permanent orbiting research facility. Its major purpose is to perform world-class science and research that only the microgravity environment can provide. Photo courtesy of NASA.

1.3   Tools for Space Neuroscience Research

The finest improvement that we can make for experimental evaluation of neuroscience function (and related physiology) in microgravity is a switch that allows gravity to be turned on and off. A centrifuge provides this switch (Reschke *et al.* 1998). Small centrifuges for lower organisms, cell cultures and small plants and animals will be available on the ISS and will provide the ability to manipulate gravitational levels from near 0 g to 2 g for a broad diversity of species. Unfortunately, such an important neuroscience facility is not available for humans. The ability to centrifuge subjects is of particular interest because one can give steady-state linear acceleration to the otolith organs. This allows generating chosen levels of artificial gravity. A human-rated centrifuge on board the ISS, with the capability of providing otolith, visual, auditory, proprioceptive and tactile stimulation, would represent a much needed tool for research on multisensory interaction in spatial orientation. As a major facility, this centrifuge should be supported with appropriate hardware for the reliable and accurate recording of eye and head movements in three dimensions, techniques for recording subjective self-orientation and self-motion and recording sensorimotor responses, and equipment capable of monitoring central and autonomic nervous system responses (Clément 1998). This facility would allow control of gravity as a variable in a manner analogous to the way light intensity, temperature, nutrient levels, drug dosages, etc, have been manipulated to elucidate the fundamental mechanisms and processes involved in the structure and function of living systems.

Furthermore, the obtained data with an on-board centrifuge will allow making predictions of the effects of Moon and Mars gravity levels. Artificial gravity using short-radius intermittent centrifugation has been proposed as a countermeasure to reduce the cardio-vascular, cardio-pulmonary and muscle deconditioning that occurs during long-term space flight (Clément & Bukley 2007). A human-rated centrifuge could be used for the study of intermittent artificial gravity inside the ISS. Ground studies will determine the potential of artificial gravity for preventing many of the microgravity related deconditioning issues. Although early positive results will guide missions planners regarding artificial gravity, only flight tests with numerous (tens) of astronauts for extended periods (several months) will allow this zero-g antidote prescription to be proven and applied to a Mars mission in conjunction with other countermeasures.

1.4   Integrative Physiology

Previous studies have been made with the implicit hypothesis that adaptive changes to microgravity would occur within given sensorimotor modules, such as visuo-motor, vestibulo-motor, etc. This is quite simplistic[13] because sensorimotor adaptation to the removal of such a fundamental force as gravity must imply multimodal reorganizations, reinterpretations, rearrangements, or re-calibrations. These multimodal interactions should perhaps be studied in experiments designed to investigate a wider type of adaptive responses, commonly referred as *integrative physiology*.

---

[13] When he returned from the first Spacelab mission in 1983, European Astronaut Wubbo Ockels made the following remark, "All the experiments we did were interesting, but I have the feeling that in space very deep changes occur and that we have just not done the right experiments to understand these changes."

Integrative physiology examines functions in the context of whole animals or humans and corresponding models of normal function or disease. Research that integrates the information from genetic, biochemical, physiological, and pathological influences is essential for our overall understanding of a system. It is understood that it is necessary to develop research programs in the neuroscience to study changes in CNS functions that occur in unusual acceleration environments. However, when possible, experiments should also integrate measurements of the cardiovascular, musculo-skeletal and autonomic nervous system functions in order to study the changes at a multisystem level. These programs should also be integrated with studies of behavior and performance that in turn could be used to develop potential countermeasures, e.g., exposure to artificial gravity.

1.5   Critical Questions in Space Neuroscience

Over the past twenty years, numerous reports have been published related to requirements for neuroscience research in space. For example, reports from the Committee on Space Biology and Medicine (Goldberg 1987), the NASA Life Sciences Strategic Planning Study Committee (Robbins 1988), the Aerospace Medicine Advisory Committee of the NASA Advisory Council (1992), the U.S. National Research Council (1991, 1998), and the ESA Microgravity Advisory Committee (Clément & Reschke 1996) have proposed recommendations for basic research topics in space neuroscience.

Based on these previous reports and the personal view from the authors, the sections that follow identify the major objectives of a space neuroscience research program and the critical questions that should be investigated both in normal and altered gravity for each of the four principal areas of research: gravity-sensing receptors, motor systems, spatial orientation, and cognition.

1.5.1   Gravity-Sensing Receptors

While there are organs developed specifically for detecting changes in the gravito-inertial force environment, there are perhaps no specific gravity receptors *per se*. Gravity is detected via many types of sensory receptors. The presence of different combinations and types of input from this ensemble of receptors generates a sense of motion or a gravitational loading that enables motor control systems to accommodate and to readily respond to gravito-inertial forces (Edgerton & Roy 2000).

Objective

a.   Understand the effect of different gravitational environments on the structure and function of gravity-sensing receptors.

Critical Questions

a.   What are the structure-function relationships of the otolith organs and semicircular canals, including development, plasticity and degeneration?

b.   What are the relevant sensors for posture, body movement and spatial orientation, including the transduction process?

c.   What are the biophysical and physiological mechanisms of vestibular hair cell transduction and the physiology and pharmacology of transmission?

d.   Do long-duration missions and the subsequent decrease in static afferent information to the vestibular, proprioceptive and somatosensory systems result in morphological changes in mechanoreceptors; are these changes prevented or reversed by Mars or artificial gravity?

e. Does gravito-inertial environments lead to changes in neuronal networks, and how are the networks identified?

1.5.2 Motor Systems

The morphology and physiology of all motor systems have evolved such that the repertoire of movement paradigms enabled survival by accommodating the physical properties of the environment, including gravity. There are numerous examples of highly integrated motor control mechanisms that are clearly designed to function in a one-g environment. One means of understanding how these control mechanisms function is to investigate their routine accommodations of daily function when the gravitational environment has changed.

Objectives

a. Determine the characteristics of motor control of gaze, posture and locomotion in altered gravity.

b. Determine how sensory inputs and coordination of muscular actions are reorganized during and after space flight.

c. Determine changes in oculomotor, somatomotor and autonomic systems in microgravity.

d. Understand the neural maps and physiological signals controlling motion in three-dimensional space under normal conditions and in the context of adaptation to altered gravity.

e. How can we differentiate between changes that may be initiated from the bottom-up as opposed to a head-down change? How do these different adaptation strategies function in different gravitational environments?

Critical Questions

a. How does gaze stabilization change in altered gravitational states? What is the most appropriate three-dimensional model of the angular and linear VOR and of central vestibular processing that will account for alterations in eye movements in microgravity? What are the characteristics of gaze and eye-head coordination with varying visual, vestibular, proprioceptive, and somatosensory inputs?

b. What are the optimal procedures for adaptation to 0.38-g and re-adaptation to 1-g environments after adaptation to microgravity?

c. What are the neural pathways that control the autonomic and endocrine responses characteristic of motion sickness and what are the pharmacological and physiological properties of these pathways?

d. What adaptive processes modify the control of various motor systems? What is the dynamic range of adaptation of motor responses in altered gravitational states?

e. What models of sensorimotor transformation can be used to most accurately predict motor behavior in altered gravitational states?

f. How do neural mechanisms regulate homeostatic processes? For example, what is the role of otolith input in regulating changes in cardiovascular function, such as orthostatic changes, heart rate and baroreceptor responses? How are these possible changes integrated?

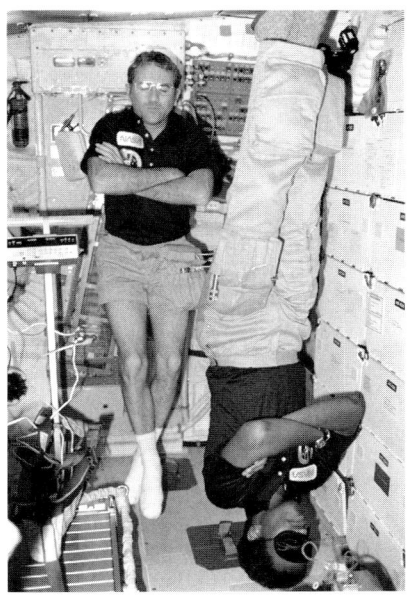

Figure 9-03. On the Space Shuttle middeck, Commander Richard Truly and Mission Specialist Guy Bluford sleep in front of forward lockers and galley. Truly sleeps with his head at the "ceiling" and his feet to the "floor." Bluford, wearing sleep mask (blindfold), is oriented with the top of his head at the "floor" and his feet on the "ceiling." Photo courtesy of NASA.

1.5.3 Spatial Orientation

The human nervous system has evolved to respond to gravitational cues, which must be integrated with other reference frames within a complex sensorimotor system. The gravitational context clearly influences visual processing and vestibular control (Figure 9-03). Space flight studies offer the unique possibility to probe the spatial orientation function by allowing an uncoupling of sensory inputs, for example those responsible for erroneous illusions of self-position and motion and space motion sickness.

Objectives

a. Understand the central neural mechanisms that contribute to spatial orientation.

b. Understand how signals from multiple senses related to gaze, body orientation, and motion are integrated at various sites in the CNS.

c. Understand the central processing that leads to SMS and the potential differences between the various gravitational and visual environments responsible for SMS.

d. Understand the neural basis for the adaptive response to altered sensory environments.

e. Develop models of central processing that can be used as heuristic and productive tools for future experiments.

f. Implement pharmacological studies in order to provide a rational basis for developing drug therapies for space motion sickness.

g. Develop, test and validate countermeasures for neurosensory aberrations caused by exposure to microgravity.

Critical Questions

a. Are there changes in the processing of signals from the semicircular canals or otolith organs that occur with adaptation? Do these changes take place within the vestibular nuclei, cerebellar structures, or other related brainstem and cortical structures? What is the time course of such changes and do they correlate with space motion sickness?

b. What are the circuitry and signals in the vestibular nuclei and brainstem that generate a gravito-inertial frame of reference? What are the roles of the different regions of the cerebellum? Do thalamo-cortical systems play a role in generating this reference?

c. Are there different molecular signals that modulate or evoke motion sickness? Are these signals or neural (synaptic transmitters) or neuroendocrine (hormonal) origin? What changes in the release of these messengers can be correlated with space motion sickness?

d. How are receptors for anti-motion sickness drugs distributed within central vestibular and other pathways?

e. What processes explain the altered perceptions of joint and body position in microgravity?

e. At what sites do signals from the different receptors involved in gaze, body orientation, posture, and motion converge? What are the characteristics of this signal integration?

f. Does altered gravity lead to changes in neural control of biological rhythms, such as sleep and temperature?

g. What neuronal models can be used to understand central processing and adaptation in altered gravitational states?

Figure 9-04. The ISS crew spends their day working on science experiments that require their input, as well as monitoring those that are controlled from the ground. They also take part in medical experiments to determine how well their bodies are adjusting to living with no gravity for long periods of time. Photo courtesy of NASA.

1.5.4 Cognition

A growing literature on the physiology of the vestibular system has demonstrated the existence of projections from the vestibular nuclei to the cerebral cortex. Cognitive deficits such as poor concentration, short-term memory loss and alteration in mental representations are known to occur frequently among patients with vestibular abnormalities and, to some extent, astronauts during space flight. To date, direct scientific study of such deficits has been limited during space missions.

Objectives

a.   Understand how adaptive changes in the vestibular, proprioceptive, somatosensory, and visual systems lead to changes in cognitive functions.

b.   Determine the perceptual processes, neurophysiological mechanisms and cortical structures underlying the mental representation of tri-dimensional space and self- and surround motion.

c.   Determine the changes that occur in CNS activity during the process of adaptation to altered gravitational conditions.

Critical Questions

a.   What are the psychophysical correlates and neural basis for perception of position and motion?

b.   What psychophysical correlates can best be used to describe spatial orientation? What are the cortical and subcortical neural correlates of intrinsic and extrinsic orientation?

c.   Does a change in vestibular input lead to changes in visual and auditory localization and multisensory spatial orientation?

d.   What ground-based paradigms and models are most effective in evaluating interactions of angular and linear acceleration, proprioception, somatosensory, and visual inputs in determining orientation in a three-dimensional environment? How do these interactions change in altered gravity?

f.   What perceptual and performance changes are produced by drugs used in the treatment of motion sickness?

g.   Can humans adapt to continuous or intermittent centrifugation during long-duration missions?

2   APPLIED SPACE NEUROSCIENCE RESEARCH

All the space faring nations are currently engaged in space exploration research, which brings with it specific mission definitions, corresponding timelines, and focused research objectives (Figure 9-04). The International Space Station is a key element in supporting this focused research. What is not evident is the amount of ISS scientific research and development that is targeted towards exploration objectives.

Nevertheless, one of the major applications of human physiology research on board the ISS is the development and validation of countermeasures to the detrimental effects of reduced gravity on crewmembers. Because most of the evaluations of potential countermeasures require long-duration exposure to weightlessness, the process of accumulating sufficient data and exploring the relevant variables is very time consuming. Initial results for countermeasure evaluation, for example, might only begin to be accumulated after 3-4 sessions of 4-6 months each. Early positive results would obviously influence both Mars mission designs and even continuing ISS crew health protection. To reach a valid scientific consensus about particular protocols, however, complete exploration might take 8-10 test missions, or up to five years to finish. Finally, it seems prudent to complete a full-length on-orbit simulation of at least the mission to Mars, if not the entire round trip, before embarking on that voyage of exploration. Obviously a lunar base could form a key portion of this simulation, along with the ISS (Figure 9-05).

A NASA recent report has identified the critical gaps in knowledge and technology as they relate to the "Vision for Space Exploration" program (Hudy & Woolford 2005). Dr. Robert Welch, from NASA Ames Research Center, contributed to this report by proposing research and technology priorities for addressing the risks to astronaut performance readiness during each of the key phases of exploration missions. The section below summarizes these research questions regarding human factors issues relevant to neuroscience research.

2.1   Training

The validity of *preflight adaptation training* (PAT) poses the most serious gap in knowledge and technology relevant to the period prior to launch. Generally speaking, PAT is the process of adapting astronauts to neurovestibular or other effects of microgravity or partial gravity before they depart for space. This procedure is not yet fully validated, even though several studies have been carried out at NASA on a few volunteer astronauts (Harm & Parker 1993).

With any form of PAT, a serious concern is whether the astronauts are being adapted to the microgravity environment to which they will actually be exposed. The procedure might do more harm than good if this is not the case. Generalizing any form of PAT to a range of sensory environments, rather than to a very specific one, is critical. The notion that adaptation to one altered environment (e.g., the one provided by a given version of PAT) makes astronauts more adaptable to a somewhat different environment (e.g., microgravity) is called *adaptive generalization*. The possibility of adaptive generalization has been examined and supported by the research of Welch *et al.* (1993). Other authors demonstrated that adaptation to a variety of optically produced visual rearrangements increased the ability to adapt to a novel rearrangement, further confirming the process of adaptive generalization (Bock *et al.* 2001, Roller *et al.* 2001).

Previous preliminary PAT studies indicate that further investigation is warranted, as some form of PAT may prove to be a useful countermeasure for the initial effects of microgravity (or other sub-terrestrial gravities). A research priority then is to reevaluate the PAT procedure to determine if it's application provide some level of inoculation of astronauts against some of the perceptual and behavioral effects, to include SMS that they experience when first entering microgravity.

Behavior and performance are other areas in which preflight training would be value added for planetary missions. Although data are limited, behavioral problems, including fatigue, irritability, depression, anxiety, mood fluctuations, boredom, tension, social withdrawal, and motivational changes have been documented during long-duration missions (Kanas & Manzey 2003). There have also been reports of instances of hostility between flight crewmembers and space and ground crews. The psychosocial dynamics of small groups of humans living in confined and isolated environments for prolonged periods are not well understood.

Critical issues associated with crew composition, structure, and training protocols include interactive crew behavior, performance, and enhanced communications. Psychological tests designed to "select out" and "select in" candidates on the basis of personal and group interviews and group compatibility analysis should be included in the crew selection criteria. Training should include a confinement period of candidate crews in a high fidelity Mars transfer vehicle mock-up, and possibly testing for a long-duration exposure on board the ISS to test crew compatibility and performance, and interactions with ground support (Aerospace Medicine Advisory Committee 1992).

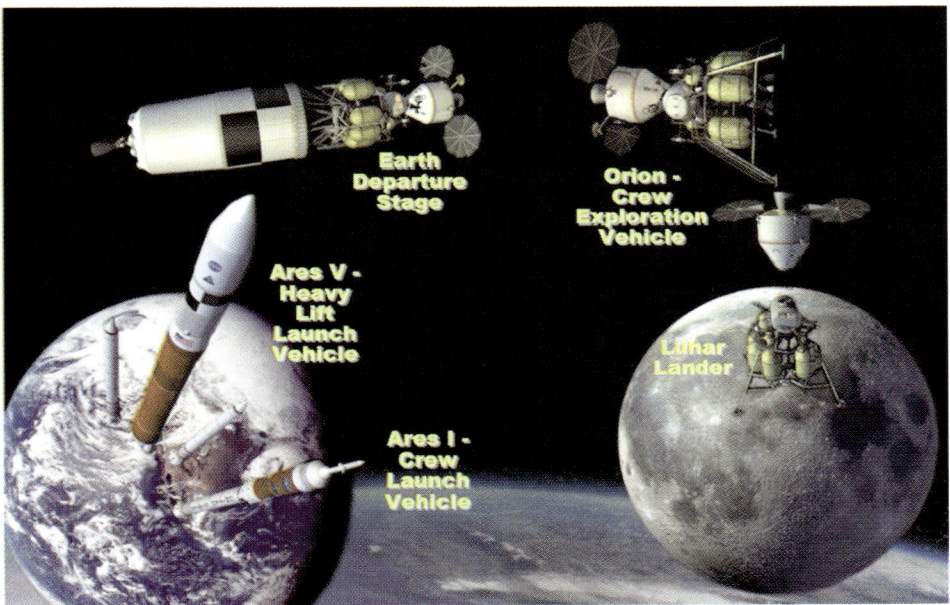

Figure 9-05. Project Constellation *is a NASA program to create a new generation of spacecraft for human space flight, consisting primarily of the* Ares I *and* Ares V *launch vehicles, the* Orion *crew capsule, the* Earth Departure Stage *and the* Lunar Surface Access Module*. The* Earth Departure Stage *(EDS) will be launched on an* Ares V *rocket. The* Orion *spacecraft will launch separately on an* Ares I *rocket, rendezvous and dock with the EDS, which will then be configured for the journey to the Moon. Photo courtesy of NASA.*

2.2   Transit Earth-Mars

Determining how the CNS functions adapt and re-adapt to various gravitational environments is a major goal as we commit to interplanetary missions of long duration. Of primary importance is to understand the acute and long-term central and peripheral nervous system adaptation to the space environment and to develop adequate physiological and performance countermeasures to the crew.

Over the past forty-five years, efforts in space neuroscience have been directed at understanding the acute changes that occur in the neurovestibular and sensorimotor systems, mostly during short-duration space missions (see Table in Chapter 2, Section

4). However, very few experiments have been performed during the first minutes or hours of adaptation to microgravity and re-adaptation to Earth gravity. This is a shortcoming of all the research that has been performed during the Space Shuttle program. Major research emphasis should be placed on obtaining an understanding of the acute changes that occur during the first few minutes and hours of space flight, and immediately after landing. These periods are characterized by sudden changes in gravitational levels, which have an impact on sensorimotor functions. The results of this research will be useful since the exploration missions will include several transitions between gravitational levels (1 g to 0 g en route to Mars; 0 g to 0.38 g when landing on Mars; 0.38 g to 0 g during the departure from Mars; and 0 g to 1 g when returning to Earth).

On the other end of the spectrum, only a limited number of controlled scientific experiments have been performed during and after very long-duration (> 6 months) missions on board the Russian Mir space station and the ISS. The proposed long stays in a lunar outpost and Mars missions of durations ranging up to three years make it imperative that research in the neuroscience begins to concentrate on the long-term chronic effects of exposure to microgravity on the CNS. The ISS would also be used as a testbed for validating new, more efficient countermeasures against the detrimental effects of long-term exposure to microgravity.

2.2.1 Adaptation Procedures

Naturally occurring adaptation processes generally overcome most of the perceptual and perceptual-motor effects of microgravity. However, the crewmembers are in jeopardy while this process occurs because of a decreased ability to effectively respond in an emergency situation. Furthermore, if it really is true that the ability to follow procedures and understand instructions is reduced by the distraction of adapting to microgravity (i.e., the "space stupids"), then their cognitive abilities may also be jeopardized. In any event, the sooner adaptation is complete, the better. With respect to training procedures for accelerating and maximizing adaptation and dual adaptation to microgravity as well as to the sub-terrestrial gravities of the Moon and Mars, a serious gap in our current training technology exists.

Existing knowledge about the optimal procedures for adapting to altered sensory environments (Welch *et al.* 2003) can be used to build such testing procedures. These should be implemented with astronauts at the earliest possible date. Factors that should be incorporated include: (a) active interaction with the altered g environment; (b) immediate sensory feedback from these activities; (c) error-corrective feedback; (d) distributed practice; and, when possible, (e) incremental exposure.

It is important to note that such training exercises will simultaneously serve as tests of the degree to which astronauts have adapted to the microgravity environment. They would be authorized to engage in EVAs and other activities that a non-adapted astronaut probably should not attempt only when they have "passed" these tests at an acceptable level, unless an emergency situation dictates otherwise.

Studies should be conducted to investigate whether or not humans will be able to maintain a dual state of adaptation as well as if they will suffer from motion sickness when transitioning from microgravity to artificial or reduced gravity if artificial gravity is adopted as a physiological countermeasure in transit to Mars or during return to Earth. This holds regardless of whether an on-board short-radius centrifuge or a spinning spacecraft providing intermittent or continuous artificial gravity is employed.

There are a number of areas related to performance and group functioning that should be investigated. These include microgravity and other space environmental effects on fundamental behavioral processes such as perception, sensation, learning, and motors skills; physiological changes and reliable correlates of performance; circadian rhythms, sleep patterns and work-rest schedules; and individual and team motivation and coping strategies for environmental stressors. Supporting research should also include workloads, schedules, interactions with ground support teams, nonintrusive performance data collection, and modeling of complex performance (Aerospace Medicine Advisory Committee 1992).

2.2.2   Reduced Dynamic Visual Acuity

The deficits in dynamic visual acuity when first entering microgravity, which are presumed to occur based on the substantial postflight reductions observed, should be carefully measured (Bloomberg *et al.* 1997). Using larger-than-normal print for signs and other reading material and advising astronauts to stabilize themselves whenever they are attempting to make fine visual discriminations are possible countermeasures.

2.2.3   Size-Distance Illusions

Astronauts tend to underestimate the size and distance of objects in space, be they looking out of a porthole or engaging in an EVA. Microgravity or to the absence of atmospheric perspective distance cues in the vacuum of space could be the cause of these misperceptions. Although these illusions are commonly reported on Earth in the clear air of the desert and thus should be even more dramatic in the complete absence of an atmosphere, no tests have ever been performed. This knowledge gap must be filled. Assuming that such illusions do occur, a potentially useful countermeasure is found in the work of Ferris (1973) showing that errors in absolute distance estimates of objects viewed underwater can be largely corrected by means of informational feedback.

Figure 9-06. During an extra-vehicular activity, an astronaut hanging "upside-down" at the end of the remotely-operated robotic arm is inspecting the insulating tiles of the Space Shuttle. Episodes of height vertigo have been reported during unfamiliar visual orientations. Photo courtesy of NASA.

2.2.4 Height Vertigo

Height vertigo is a debilitating experience anecdotally reported during some EVAs (Figure 9-06). Therefore, a systematic assessment of the frequency, characteristics and necessary conditions for this condition is needed. Oman (2001) has suggested several potential countermeasures to deal with height vertigo. These include recommending that while on an EVA, the astronauts face the Shuttle or ISS as much as possible and work "right side up" relative to these spacecraft while the Earth nadir is in the upper visual field. He has also proposed that body, hand and foot restraints might further serve to mitigate this problem.

2.2.5 Mental Rotation

When an observer views an object that is upside down or skewed relative to its familiar up-down orientation, he may have problems even recognizing it. Because observers have difficulty recognizing inverted faces (Thompson 1980) and have difficulty determining whether the person is smiling or frowning, it seems likely that the same phenomena will occur in microgravity, along with difficulty in reading non-verbal facial cues, all of which could cause serious problems in face-to-face communication (Cohen 2000). The problem is further compounded by the well-known facial puffiness caused by microgravity-induced fluid shifts. When crewmates are tilted or upside down relative to each other, this recognition difficulty may cause potentially serious problems to occur. Therefore, a systematic measurement of the ability of astronauts to recognize objects with a familiar up and down when viewed from unfamiliar orientations is required.

2.2.6 Individual Differences

Microgravity affects astronauts in significantly different ways, depending on the individual. A method to predict these observed individual differences among astronauts and determine the rate and level of adaptation that is achieved is clearly needed. The capability to successfully make such predictions would make it possible to specifically tailor countermeasures (e.g., training procedures) to the individual astronaut (Reschke *et al.* 1998).

2.3   Living on Mars

2.3.1 Landing on Mars

Part of the Martian landing sequence requires aero-braking, which will expose the astronauts to high g forces along their body x-axis (chest-to-back) for a short period of time. Under nominal conditions, this maneuver will be solely under computer control. However, any departure from nominal conditions may require astronauts to engage themselves in the control of the landing sequence. During this phase of the mission, there is a potential for impaired sensorimotor function to interfere with any of the operational tasks during landing due to spatial disorientation and perceptual illusions, which may occur during and after g-transitions. In particular, the hand-eye coordination is likely to be prone to errors of accuracy and reaction time (Cohen 1970). The location and layout of critical spacecraft flight control elements, like switches, joysticks, and flat panel displays, will be critical in determining or mitigating the extent of these errors.

Figure 9-07. This illustration gives an artist's impression of what a human expedition to the surface of Mars might look like. Hardware seen here includes the Mars explorer, a traverse vehicle and a habitation module. The astronauts-geologists sample rocks for a possible evidence of preserved fossils. Photo courtesy of NASA/Pat Rawlings.

2.3.2   Effects of Long-Term Exposure to 0.38 g

The ability to perform post landing tasks may be compromised by impaired movement and coordination caused by long-term exposure to microgravity. The limited long-duration space flight experience on Mir and the ISS indicates that astronauts and cosmonauts experience postural imbalance and locomotor uncoordination upon return to Earth. These symptoms are due not only to changes in muscle strength, but also to the lasting effects of adaptation to both sensory inputs and central motor programs. It is uncertain as to what the potential effects of long-term exposure to microgravity on perception and sensorimotor performance will be. In contrast to the reversible effects of adaptation, not much is known about the possibility that chronic exposure to microgravity can cause permanent (i.e., structural) changes in those sensory organs that mediate perceptual and sensorimotor coordination. Long-term studies of these capacities should be undertaken using the ISS crews, if at all possible, in a timeframe that would allow useful data to be derived in support of the Mars missions. We still do not know the time required to fully re-adapt to preflight (1 g) performance following very long periods of exposure to microgravity. It could be possible that re-adaptation will take longer in a 0.38-g environment. Furthermore, it is unknown as to whether up to 18 months of exposure to 0.38 g will allow a normal re-adaptation of perceptual-motor coordination and neuro-vestibular functioning (Figure 9-07).

One striking aspect of impaired sensorimotor performance following transitions of gravitational environments is the large individual variability observed. A few crewmembers are temporarily incapacitated while others are relatively unaffected. Restriction of operational tasks for a period of time following landing, a common countermeasure used to mitigate risks following return to Earth, may compromise mission objectives for exploration missions. If an emergency during the landing on Mars demanded quick egress from the space vehicle, then a prolonged period of postural inactivation and impairment of locomotion would be totally unacceptable. Moreover, this condition could also seriously impair the ability of astronauts to accomplish mission tasks or even care for themselves in the Mars gravitational environment.

What is probably more important, given the much greater distance from Earth support, is that astronauts be properly prepared for EVAs on Mars, more so than on the Moon. During a recent summit to review medical concerns with Apollo crewmembers, there was an interest expressed over planetary surface rover operations (Scott Wood, personal communication). Misperception of terrain slope and vehicle tilt angle led to the sensation that one was at risk of falling out of the rover (Figure 9-08). Adaptation to novel patterns of sensory cues experienced during motion on Mars surface can impair driving performance during rover operations. The ability to maintain orientation and control of rovers can be further compromised by visual background cues (lighting conditions, shadows, textures) that are different from terrestrial driving conditions. There is also a significant likelihood that the astronauts could underestimate the size or distance of objects during clear atmospheric conditions and the opposite effects during dust storms.

Another concern is the loss of dynamic visual acuity during head movements, as discussed in previous sections. Head-eye coordination is critical during locomotion, piloting task and vehicle operation. The vertical vestibulo-ocular reflex makes a significant contribution to the maintenance of dynamic visual acuity. Changes in gaze stabilizing reflexes have been observed following short-duration space flights, and these changes become more pronounced with increasing mission duration. Recent studies have also documented cognitive changes in the mental representation of the three-dimensional space during exposure to altered gravitational environments.

Figure 9-08. Geologist-Astronaut Harrison H. Schmitt is photographed standing next to a huge, split boulder during the third Apollo-17 extra-vehicular activity at the lunar Taurus-Littrow landing site. The Lunar Roving Vehicle *is in the left foreground. Note the considerable slope of the terrain. Because of the reduced gravitational field on the Moon, the astronauts reported difficulties in estimating the true vertical. Photo courtesy of NASA.*

Before a mission to Mars can safely be undertaken, the adaptive processes of the sensory, motor and cognitive systems to microgravity need to be better understood, and countermeasures must be devised for a faster re-adaptation of the CNS functions that are expected to occur following the transitions between various gravitational environments. In particular, investigations should address the following issues:

a. Motion sickness upon return to a gravitational environment (postflight motion sickness) needs to be better understood and mitigation strategies developed.

b. The dynamic range of the adaptation of sensorimotor responses in various gravitational environments needs to be identified. This may be accomplished by using a centrifuge on board the ISS or in a Moon habitat. Accurate predictions of the effects Mars gravity may be accomplished via modeling.

c. It is not known if permanent functional deficits result from the decrease in afferent input to the vestibular, proprioceptive and somatosensory systems as a function of the adaptation associated with long exposure to 0 g or 0.38 g.

d. Morphological or structural changes in CNS and neuromuscular functions that may account for these deficits need to be identified.

e. The procedures that produce rapid and complete adaptation to Martian gravity and Earth gravity after exposure to microgravity must be validated. This may be accomplished using Martian gravity simulation by executing parabolic flight maneuvers on Earth, or using a centrifuge on board the ISS or in a Moon habitat.

Figure 9-09. Astronauts looking through the porthole of Space Shuttle. Space is a hazardous environment. Resident space station crews regularly monitor very small features and changes around the globe and click nearly 5000 pictures of our planet during a six-month mission. The spatial resolution of the images approaches the highest spatial resolution of color images now available from commercial remote sensing satellites. Photo courtesy of NASA.

2.4   Returning to Earth

The potential behavioral effects of the six months of microgravity to which astronauts will be exposed are the main issues relevant to the Earth-Mars transit period. That is, assuming that they will not be exposed to an artificial gravity environment via a rotating spacecraft.

Identifying the most effective and efficient procedures for re-adapting astronauts to Earth's gravity after landing is an obvious research and technology gap. These procedures would likely be advocated for all of the other gravitational transitions described above. The re-adaptation procedures could include exercises aimed at the re-acquisition of terrestrial posture, balance, and locomotion. Some astronauts already do such exercises on their own before leaving the spacecraft. One could imagine, for example, a brief, standardized series of specific hand-eye and locomotion exercises that the astronauts could do immediately upon landing. Astronauts might be in jeopardy in the absence of such procedures, especially in an emergency situation coincident with the landing.

3   CONCLUSION

Earth orbit has become a busy arena of human activity in the forty-five years since the cosmonauts and astronauts made their first brief travels into space. More than 450 people (about ten per year on average) have traveled into orbit on U.S., Russian and, most recently, Chinese spacecraft during this time. The first astronauts traveled stuffed into capsules barely large enough to accommodate their bodies, eating squeeze-tube food and peering out at the Earth through tiny portholes (Figure 9-09). Their flights lasted only a matter of hours. Astronauts are routinely launched seven at a time these days. They spend two weeks working on board the Space Shuttle, or six months living on board the International Space Station. In the years to come, future explorers will spend several months or years exploring the Moon or Mars.

Throughout the brief history of space flight there has been significant improvements in spacecraft and a commensurate increase in the numbers of people traveling into orbit. Each successive spacecraft, from Voskhod through the International Space Station, has been larger, more comfortable and more capable. The ISS provides comforts similar to those on Earth, none of which were available when human space flight first began. We are now learning how to live and work in space, not just to travel there for short excursions.

The ISS is considered as a stepping-stone for human forays further out into the solar system, be that establishing a lunar base, planetary travel, or travel beyond the solar system. The neurosensory system will be in a "no gravity" environment for a considerable period of time. It will be exposed not only to the altered sensations of that environment, but also the unique gravitational field found on a new planetary surface.

No operational countermeasures currently exist for postflight motion sickness (other than drugs which further complicate the adaptation process), orthostatic intolerance, or simple locomotor function (e.g., walking and standing up-right) on return to gravity. The degree to which there will be postflight sensorimotor disturbances following transit to and landing on Mars is also unknown. However, most mission scenarios envision six-month transit times and postflight disturbances would likely be similar to those observed following ISS missions. Thus, ISS is currently a testbed for improving training methods, countermeasures and treatments associated with adaptation to changes in gravity.

We still have so much to learn about the complex functioning of the central nervous system, even in the one-g environment of Earth (Figure 9-10). Space medicine, particularly as it relates to space exploration, is still in its infancy. We have the unique opportunity to take the initial steps of attempting to clarify which nervous system changes could be most critical in long-term exposure to space and to identify the best research strategies (Igarashi *et al.* 1987).

The most basic neuroscience questions must be answered so that we can minimize risks and optimize crew performance during transit and planetary operations. The study results will certainly find other applications in medicine and biotechnology. We are in a unique position to understand how Earth's gravitational environment has shaped the evolution of sensory and motor systems.

Figure 9-10. On January 6th, 2005, the NASA's JPL exploration rover called Opportunity *found an iron meteorite on Mars, the first meteorite of any type ever identified on another planet. Because of its shape, the pitted, basketball-size object was nicknamed "The Brain" by the science team. Photo courtesy of NASA.*

REFERENCES

Aerospace Medicine Advisory Committee (1992) *Strategic Considerations for Support of Humans in Space and Moon/Mars Exploration Missions*. NASA Advisory Council, Washington DC

Agrell B, Dehlin O (1998) The clock-drawing test. *Age & Ageing* 27: 399-403

Aizikov GS, Kreidich YuV, Grigoryan RA (1991) Sensory interaction and methods of non-medicinal prophylaxis of space motion sickness. *Physiologist* 34: S220-223

Albery WB, Martin ET (1996) Development of space motion sickness in a ground-based human centrifuge. *Acta Astronautica* 38: 721-731

Allum H, Graf W, Dichgans J, *et al.* (1976) Visual-vestibular interactions in the vestibular nuclei of the goldfish. *Exp Brain Res* 26: 463-485

Aoki H, Oman CM (2007) Virtual reality-based preflight astronaut 3D navigation training. MIT, Cambridge. Retrieved 16 October 2007 from URL: http://www.nsbri.org/Research/Projects/viewsummary.epl?pid=222

Anderson DJ, Reschke MF, Homick JL, *et al.* (1986) Dynamic posture analysis of Spacelab crewmembers. *Exp Brain Res* 64: 380-391

André-Deshays C, Israël I, Charade O, *et al.* (1993) Gaze control in microgravity. 1. Saccades, pursuit, eye-head coordination. *J Vestib Res* 3: 331-344

Angelaki DE, McHenry MQ, Dickman JD, *et al.* (1999) Computation of inertial motion: Neural strategies to resolve ambiguous otolith information. *J Neuroscience* 19: 316-327

Anken RH, Rahmann H (1999) Effect of altered gravity on the neurobiology of fish. *Naturwissenschaften* 86: 155-167

Arrott AP, Young LR (1986) M.I.T./Canadian vestibular experiments on the Spacelab-1 mission: 6. Vestibular reactions to lateral acceleration following ten days of weightlessness. *Exp Brain Res* 64: 347-357

Arrott AP, Young LR, Merfeld DM (1990) Perception of linear acceleration in weightlessness. *Aviat Space Environ Med* 61: 319-326

Assaiante C, Amblard B (1993) Ontogenesis of head stabilization in space during locomotion in children: influence of visual cues. *Exp Brain Res* 93: 499-515

Augurelle AS, Thonnard JL, White O, *et al.* (2003) The effects of a change in gravity on the dynamics of prehension. *Exp Brain Res* 148: 533-540

Azzam EI, de Toledo SM, Little JB (2001) Direct evidence for the participation of gap junction-mediated intercellular communication in the transmission of damage signals from particle irradiated to nonirradiated cells. *Natl Acad Sci USA* 98: 473-478

Bacal K, Billica R, Bishop S (2003) Neurovestibular symptoms following space flight. *J Vestib Res* 13: 93-102

Bagian JP (1991) First intramuscular administration in the U.S. space program. *J Clin Pharmacol* 31: 920

Bagian JP, Ward DF (1994) A retrospective study of promethazine and its failure to produce the expected incidence of sedation during space flight. *J Clin Pharmacol* 34: 649-651

Baker JT, Nicogossian AE, Hoffler GW, *et al.* (1977) Changes in the Achilles tendon reflexes following Skylab missions. In: *Biomedical Results from Skylab*.

Johnston RS, Dietlein LF (eds) NASA, Washington DC, NASA SP-377, pp 131-135

Balaban CD, Porter JD (1998) Neuroanatomic substrates for vestibulo-autonomic interactions. *J Vestib Res* 8: 7-16

Ball JR (2001) *Safe Passage: Astronaut Care for Exploration Missions*. Institute of Medicine. National Academy of Sciences, Washington, DC

Baloh RW, Richman L, Yee RD, *et al.* (1983) The dynamics of vertical eye movements in normal human subjects. *Aviat Space Environ Med* 54: 32-38

Bàràny R (1906) Untersuchungen über den vom Vestibularapparat des Ohres Reflektorisch Ausgelôsten Rhythmischen Nystagmus und seine Begleiterscheinungen. *Monatsschrift fur Ohrenkeilkunde und Laryngo-Rhinologie* 40: 193-207

Barfield W, Cohen M, Rosenberg C (1997) Visual and auditory localization as a function of azimuth and elevation. *Int J Aviat Psychol* 7: 123-138

Baumgarten von RJ (1986) European experiments in the Spacelab mission 1. Overview. *Exp Brain Res* 64: 239-246

Baumgarten von RJ, Baldrighi G, Shillinger G (1972) Vestibular behavior of fish during diminished g-force and weightlessness. *Aerospace Med* 43: 626-632

Baumgarten von RJ, Wetzig J, Vogel H, *et al.* (1982) Static and dynamic mechanisms of space vestibular malaise. *Physiologist* 25: S33-S36

Baumgarten von RJ, Benson A, Berthoz A, *et al.* (1986) European experiments on the vestibular system during the Spacelab D-1 mission. In: *Proceedings of the Norderney Symposium on Scientific Results of the German Spacelab Mission D-1*. Norderney, Germany, 27-29 August 1986. Sahm PR, Jansen R, Keller MH (eds) pp 477-490

Baumgarten von RJ, Boehmer G, Brenske A, *et al.* (1987) A nonthermoconvective mechanism generating caloric nystagmus: Similarities between pressure-induced and caloric nystagmus in the pigeon. In: *The Vestibular System: Neurophysiologic and Clinical Research*. Graham MD, Kemink JL (eds) Raven Press, New York, pp 51-59

Benson AJ (1974) Modification of the response to angular accelerations by linear accelerations. In: *Handbook of Sensory Physiology, Volume VI. Vestibular System, Part 2: Psychophysics, Applied Aspects and General Interpretations*. Kornhuber HH (ed) Springer-Verlag, Berlin, pp 281-320

Benson AJ (1988) Spatial disorientation: Common illusions. In: *Aviation Medicine*. Ernsting J, King P (eds) Butterworths, London, Chapter 21, pp 297-317

Benson AJ (1990) Sensory functions and limitations of the vestibular system. In: *Perception and Control of Self Motion*. Warren R, Wertheim AH (eds) Lawrence Erlbaum Assoc, Hillsdale, pp 145-169

Benson AJ, Viéville T (1986) European vestibular experiments on the Spacelab-1 Mission: 6. Yaw axis vestibulo-ocular reflex. *Exp Brain Res* 64: 279-283

Benson AJ, Kass JR, Vogel H (1986) European vestibular experiments on the Spacelab-1 Mission: 4. Thresholds of perception of whole-body linear oscillation. *Exp Brain Res* 64: 264-271

Benson AJ, Guedry FE, Parker DE, *et al.* (1997) Microgravity vestibular investigations: perception of self-orientation and self-motion. *J Vestib Res* 7: 453-457

Berger M, Mescheriakov S, Molokanova E, *et al.* (1997) Pointing arm movements in short- and long-term spaceflights. *Aviat Space Environ Med* 68: 781-787

Berthoz A (1989) Possible active contribution of neuronal mechanisms controlling gaze in the compensation of vestibular dysfunction. In: *Proceeding, Workshop on Nervous System Plasticity in Relation to Long-Term Exposure to Microgravity Environment.* Houston, 9-10 October 1987. Igarashi M, Nute NG, MacDonald S (eds) NASA Space Biomedical Research Institute, USRA Division of Space Biomedicine, Houston, pp 43-50

Berthoz A, Brandt T, Dichgans J, *et al.* (1986) European vestibular experiments on the Spacelab-1 mission: 5. Contribution of the otoliths to the vertical vestibulo-ocular reflex. *Exp Brain Res* 64: 272-278

Berthoz A, Pozzo T (1988) Intermittent head stabilization during postural and locomotory tasks in humans. In: *Posture and Gait: Development, Adaptation and Modulation.* Amblard B, Berthoz A, Clarac F (eds) Elsevier, Amsterdam, pp 189-198

Berry CA, Homick JL (1973) Findings on American astronauts bearing on the issue of artificial gravity for future manned space vehicles. *Aerospace Med* 44: 163-168

Biaggioni I, Costa F, Kaufmann H (1998) Vestibular influences on autonomic cardiovascular control in humans. *J Vestib Res* 8: 35-41

Bles W, Van Raaij JL (1988) Pre- and postflight postural control of the D1 Spacelab mission astronauts examined with a tilting room. *Report TNO-IZF* 1988-25

Bles W, De Graaf B (1993) Postural consequences of long duration centrifugation. *J Vestib Res* 3: 87-95

Bles W, De Graaf B, Bos JE, *et al.* (1997) A sustained hyper-G load as a tool to simulate space sickness. *J Gravit Phys* 4: 1-4

Bles W, Bos JE, De Graaf B, *et al.* (1998) Motion sickness: Only one provocative conflict? *Brain Res Bull* 47: 481-487

Bloomberg JJ, Mulavara AP (2003) Changes in walking strategies after spaceflight. *IEEE Eng Med Biol Mag* 22: 58-62

Bloomberg JJ, Peters BT, Smith SL, *et al.* (1997) Locomotor head-trunk coordination strategies following space flight. *J Vestib Res* 7: 161-177

Bloomberg JJ, Layne CS, McDonald PV, *et al.* (1999) Effects of space flight on locomotor control. In: *Extended Duration Orbiter Medical Project Final Report 1989-1995.* Sawin CF, Taylor GR, Smith WL (eds) NASA, Washington DC, NASA SP-1999-534, Chapter 5.5, pp 1-7

Bock O (1998) Problems of sensorimotor coordination in weightlessness. *Brain Res Rev* 28: 155-160

Bock O, Howard IP, Money KE, *et al.* (1992) Accuracy of aimed arm movements in changed gravity. *Aviat Space Environ Med* 63: 994-998

Bock O, Fowler B, Comfort D (2001) Human sensorimotor coordination during space flight: An analysis of pointing and tracking responses during the Neurolab Space Shuttle mission. *Aviat Space Environ Med* 72: 877-883

Bock O, Schneider S, Bloomberg JJ (2001) Conditions for interference versus facilitation during sequential sensorimotor adaptation. *Exp Brain Res* 138: 359-365

Bos JE, Bles W (2002) Theoretical considerations on canal-otolith interaction and an observer model. *Biol Cyber* 86: 191-207

Bos JE, Bles W, De Graaf B (2002) Eye movements to yaw, pitch and roll about vertical and horizontal axes: Adaptation and motion sickness. *Aviat Space Environ Med* 73: 436-444

Bottini G, Sterzi R, Paulesu E, *et al.* (1994) Identification of the central vestibular projections in man: a positron emission tomography activation study. *Exp Brain Res* 99: 164-169

Bower B (1992) Infants signal the birth of knowledge. *Science News* 142: 325

Boyd CA, Noble D (1993) *The Logic of Life: The Challenge of Integrative Physiology*. Oxford University Press, Oxford

Bracchi F, Gualtierotti T, Morabito A, *et al.* (1975) Mutliday recordings from the primary neurons of the stratoreceptors of the labyrinth of the bullfrog. *Acta Oto-Laryngol* (Suppl) 334: 3-27

Brandt T, Dieterich M (1999) The vestibular cortex. Its locations, functions and disorders. *Ann NY Acad Sci* 871: 293-312

Briegleb WJ, Neubert J, Schatz A, *et al.* (1988) Light microscopic analysis of the gravireceptor in Xenopus larvae developed in microgravity. *Adv Space Res* 9: 241-244

Brooks VB (1986) Locomotion. In: *The Neural Basis of Motor Control*. Oxford University Press, New York, pp 181-194

Brown EL, Hecht H, Young LR (2003) Sensorimotor aspects of high-speed artificial gravity: I. Sensory conflict in vestibular adaptation. *J Vestib Res* 12: 271-282

Bryanov II, Yemel'yanov MD, Matveyev AD, *et al.* (1976) *Space Flights in the Soyuz Spacecraft: Biomedical Research*. Leo Kanner Associates, Redwood City

Bryanov II, Gorgiladze GI, Kornilova LN, *et al.* (1986) Vestibular function. In: *Results of Medical Research Performed on the Salyut 6-Soyuz Orbital Scientific-Research Complex*. Gazenko OG (ed) Meditsina, Moscow, pp. 169-185 and 248-256

Buckey JC, Lasley EN (2005) Astronauts study the brain in space. *Cerebrum* 7: 7-27

Buckey JC, Lane LD, Levine BD, *et al.* (1996) Orthostatic intolerance after spaceflight. *J Appl Physiol* 81: 7-1

Bungo MW (1989) Welcoming remarks. In: *Proceeding, Workshop on Nervous System Plasticity in Relation to Long-Term Exposure to Microgravity Environment, October 1987*. Igarashi M, Nute NG, MacDonald S (eds) NASA Space Biomedical Research Institute, USRA Division of Space Biomedicine, Houston, pp 5-6

Burrough B (1998) *Dragonfly: NASA and the Crisis Aboard Mir*. Harper-Collins, New York

Burton AW, Pick Jr HL, Holmes C, *et al.* (1990) The independence of horizontal and vertical dimensions in handwriting with and without vision. *Acta Psychologica* 75: 201-212

Calvin WH (1995) Cortical columns, modules and Hebbian cell assembles. In: *The Handbook of Brain Theory and Neural Networks*. Arbib MA (ed) MIT Press, Cambridge, pp 269-272

Cavagna GA, Heglund NC, Taylor CR (1977) Mechanical work in terrestrial locomotion: two basic mechanisms for minimizing energy expenditure. *Am J Physiol* 233: R243-R261

Cavagna GA, Willems PA, Heglund NC (1998) Walking on Mars. *Nature* 393: 636

Cernan E, Davis R (1999) *The Last Man on the Moon*. Easton Press, Norwalk, CT

Chéron G, Leroya A, De Saedeleera C, *et al.* (2006) Effect of gravity on human spontaneous 10-Hz electroencephalographic oscillations during the arrest reaction. *Brain Res* 1121: 104-116

Cheung BS, Hofer K (2005) Desensitization to strong vestibular stimuli improves tolerance to simulated aircraft motion. *Aviat Space Environ Med* 76: 1099-1104

Clark B, Graybiel A (1966) Factors contributing to the delay in the perception of the oculogravic illusion. *Am J Psychol* 79: 377-388

Clark JB (2007) Health threats in human spaceflight. Paper presented at the *16th IAA Humans in Space Symposium*. Beijing, China, 20-24 May 2007

Clarke AH (1998) Vestibulo-oculomotor research and measurement technology for the space station era. *Brain Res Rev* 28: 173-184

Clarke AH (2007) Ocular torsion response to active head-roll movement under one-g and zero-g conditions. *J Vestib Res*, in press

Clarke AH, Engelhorn A (1998) Unilateral testing of utricular function. Exp Brain Res (1998) 121: 457-464

Clarke AH, Halswanter T (2007) The influence of microgravity on the orientation of Listing's Plane. *Vision Res*, in press

Clarke AH, Teiwes W, Scherer H (1992) Variation of gravitoinertial force and its influence on ocular torsion and caloric nystagmus. *Ann NY Acad Sci* 656: 820-822

Clarke AH, Teiwes W, Scherer H (1993a) Evaluation of the three-dimensional VOR in weightlessness. *J Vest Res* 3: 207-218

Clarke AH, Teiwes W, Scherer H (1993b) Vestibulo-oculomotor testing during the course of a spaceflight mission. *Clin Investigat* 71: 740-748

Clarke AH, Grigull J, Müller R, *et al.* (2000) The three-dimensional vestibulo-ocular reflex during prolonged microgravity. *Exp Brain Res* 134: 322-334

Clarke AH, Ditterich J, Druen K, *et al.* (2002) Using high frame rate CMOS sensors for three-dimensional eye tracking. *Behav Res Methods Instrum Comput* 34: 549-560

Clément G (1998) Alteration of eye movements and motion perception in microgravity. *Brain Res Rev* 28: 161-172

Clément G (2003) A review of the effects of space flight on the asymmetry of vertical optokinetic and vestibulo-ocular reflexes. *J Vestib Res* 13: 255-263

Clément G (2005) *Fundamentals of Space Medicine*. Microcosm Press, El Segundo and Springer, New York

Clément G (2007) *Mental Representation of Spatial Cues During Spaceflight*. International Space Station Medical Project (ISSMP). Available on line at: http://hrf.jsc.nasa.gov/science/default.asp?e_id=19 (Accessed 16 October 2007).

Clément G, Berthoz A (1990) Cross-coupling between horizontal and vertical eye movements during optokinetic nystagmus and optokinetic afternystagmus elicited in microgravity. *Acta Otolaryngol (Stockh)* 109: 179-187

Clément G, Lathan CE (1991) Effects of static tilt about the roll axis on horizontal and vertical optokinetic nystagmus and optokinetic after-nystagmus in humans. *Exp Brain Res* 84: 335-341

Clément G, Lestienne F (1988) Adaptive modifications of postural attitude in conditions of weightlessness. *Exp Brain Res* 72: 381-389

Clément G, Reschke MF (1996) Neurosensory and sensory-motor functions. In: *Biological and Medical Research in Space: An Overview of Life Sciences Research in Microgravity*. Moore D, Bie P, Oser H (eds) Springer-Verlag, Heidelberg, Chapter 4, pp 178-258

Clément G, Pavy-Le Traon A (2004) Centrifugation as a countermeasure during actual and simulated spaceflight: A review. *Eur J Appl Physiol* 92: 235-248

Clément G, Slenzka K (2006) *Fundamentals of Space Biology. Research on Cells, Animals, and Plants in Space.* Microcosm Press, El Segundo and Springer, New York

Clément G, Bukley A (2007) *Artificial Gravity.* Microcosm Inc, Hawthorne and Springer, New York

Clément G, Berthoz A, Lestienne F (1987) Adaptive changes in perception of body orientation and mental image rotation in microgravity. *Aviat Space Environ Med* 58 (Suppl 9): A159-A163

Clément G, Gurfinkel VS, Lestienne F, *et al.* (1984) Adaptation of posture control to weightlessness. *Exp Brain Res* 57: 61-72

Clément G, Gurfinkel VS, Lestienne F, *et al.* (1985) Changes of posture during transient perturbations in microgravity. *Aviat Space Environ Med* 56: 666-671

Clément G, Viéville T, Lestienne F, *et al.* (1986) Modifications of gain asymmetry and beating field of vertical optokinetic nystagmus in microgravity. *Neurosci Lett* 63: 271-274

Clément G, André-Deshays C, Lathan CE (1989) Effects of gravitoinertial force variations on vertical gaze direction during oculomotor reflexes and visual fixation. *Aviat Space Environ Med* 60: 1194-1198

Clément G, Reschke MF, Verrett CM, *et al.* (1992a) Effects of gravitoinertial force variations on optokinetic nystagmus and on perception of visual stimulus orientation. *Aviat Space Environ Med* 63: 771-777

Clément G, Wood SJ, Reschke MF (1992b) Effects of microgravity on the interaction of vestibular and optokinetic nystagmus in the vertical plane. *Aviat Space Environ Med* 63: 778-784

Clément G, Popov KE, Berthoz A (1993) Effects of prolonged weightlessness on human horizontal and vertical optokinetic nystagmus and optokinetic after-nystagmus. *Exp Brain Res* 94: 456-462

Clément G, Petropoulos A, Darlot C, *et al.* (1995) Eye movements and motion perception during off-vertical axis rotation (OVAR) at small angles of tilt after spaceflight. *Acta OtoLaryngol (Stockh)* 115: 603-609

Clément G, Wood S, Reschke MF, Berthoz A, Igarashi M (1999) Yaw and pitch visual-vestibular interaction in weightlessness. *J Vestib Res* 9: 207-220

Clément G, Deguine O, Parant M, *et al.* (2001a) Effects of cosmonaut vestibular training on vestibular function prior to space flight. *Eur J App Physiol* 85: 539-545

Clément G, Moore ST, Raphan T, *et al.* (2001b) Perception of tilt (somatogravic illusion) in response to sustained linear acceleration during space flight. *Exp Brain Res* 138: 410-418

Clément G, Maciel F, Deguine O (2002) Perception of tilt and ocular torsion of normal human subjects during eccentric rotation. *Otol Neurotol* 23: 958-966

Clément G, Berthoz A, Cohen B, *et al.* (2003) Perception of the spatial vertical during centrifugation and static tilt. In: *The Neurolab Spacelab Mission: Neuroscience Research in Space.* Buckey J, Homick J (eds) US Government Printing Office, NASA Johnson Space Center, Houston, NASA SP-2003-535, pp 5-10

Clément G, Deguine O, Bourg M, *et al.* (2007) Effects of vestibular training on motion sickness, nystagmus and subjective vertical. *J Vestib Res*, in press

Clément G, Denise P, Reschke MF, *et al.* (2007) Human ocular counter-rotation and roll tilt perception during off-vertical axis rotation after space flight. *J Vestib Res*, in press

Clement J (1982) Students' preconceptions in introductory mechanics. *Am J Phys* 50: 66-71

Cohen B, Kozlovskaya I, Raphan T, *et al.* (1992) Vestibulo-ocular reflex of Rhesus monkey after space flight. *J Appl Physiol* 73 (Suppl 2): 112S-1120S

Cohen MM (1970) Hand-eye coordination in altered gravitational fields. *Aerospace Med* 41: 647-649

Cohen MM (2000) Perception of facial features and face-to-face communications in space. *Aviat Space Environ Med* 71: A51-A57

Colby CL, Goldberg ME (1999) Space and attention in parietal cortex. *Ann Rev Neurosci* 22: 319-349

Collewijn H, van der Steen J, Ferman L, *et al.* (1985) Human ocular counterroll: Assessment of static and dynamic properties from electromagnetic scleral coil recordings, *Exp Brain Res* 59: 185-196

Collins M (1989) *Carrying the Fire: An Astronaut's Journey.* Bantam Books, New York

Collins M (1990) *Mission to Mars: An Astronaut's Vision of our Future in Space.* Grove Weidenfeld, New York

Collins WE (1973) Habituation of vestibular responses: An overview. In: *Fifth Symposium on the Role of the Vestibular Organs in Space Exploration.* NASA SP-314, Washington DC, pp 157-193

Connors MM, Harrison AA, Akins FR (1985) *Living Aloft: Human Requirements for Extended Spaceflight.* NASA, Washington, DC, NASA SP-483

Conrad N, Klausner HA (2005) *Rocket Man. Astronaut Pete Conrad's Incredible Ride to the Moon and Beyond.* New American Library, New York

Cooper HS (1976) *A House in Space.* Rhinehart & Winston, New York

Cornilleau-Peres V, Droulez J (1990) Three-dimensional motion perception: Sensorimotor interactions and computational models. In: *Perception and Control of Self-Motion.* Warren R, Wertheim AH (eds) Lawrence Erlbaum, New Jersey, pp 81-100

Correia MJ, Perachio AA, Dickman JD, *et al.* (1992) Changes in monkey horizontal semicircular afferent responses following space flight. *J Applied Phys* 73: 121S-131S

Courtine G, Pozzo T (2004) Recovery of the locomotor function after prolonged microgravity exposure. I. Head-trunk movement and locomotor equilibrium during various tasks. *Exp Brain Res* 158: 86-99

Courtine G, Papaxanthis C, Pozzo T (2002) Prolonged exposure to microgravity modifies limb endpoint kinematics during the swing phase of human walking. *Neurosci Lett* 332: 70-74

Cowings PS (1990) Autogenic-feedback training: A treatment for motion and space sickness. In: *Motion and Space Sickness.* Crampton GH (ed) CRC Press, Boca Raton, pp 353-372

Cowings PS, Toscano WB (1982) The relationship of motion sickness susceptibility to learned autonomic control for symptom suppression. *Aviat Space Environ Med* 53: 570-575

Cowings PS, Suter S, Toscano WB, *et al.* (1986) General autonomic components of motion sickness. *Psychophysiology* 23: 542-551

Crampton GH (ed) (1990) *Motion and Space Sickness.* CRC Press, Boca Raton

Dai M, Raphan T, Cohen B (1991) Spatial orientation of the vestibular system: dependence of optokinetic after nystagmus (OKAN) on gravity. *J Neurophysiol* 66: 1422-1439

Dai M, McGarvie L, Kozlovskaya IB, *et al.* (1994) Effects of space flight on ocular counter-rolling and the spatial orientation of the vestibular system. *Exp Brain Res* 102: 45-56

Dai M, Raphan T, Kozlovskaya IB, *et al.* (1996) Modulation of vergence by off-vertical yaw axis rotation in the monkey: Normal characteristics and effects of space flight. *Exp Brain Res* 111: 21-29

Dai M, Kunin M, Raphan T, *et al.* (2003) The relation of motion sickness to the spatial-temporal properties of velocity storage. *Exp Brain Res* 151: 173-189

Dai M, Raphan T, Cohen B (2006) Effects of baclofen on the angular vestibulo-ocular reflex. *Exp Brain Res* 171: 262-271

Davis JR, Vanderploeg JM, Santy PA, *et al.* (1988) Space motion sickness during 24 flights of the Space Shuttle. *Aviat Space Environ Med* 59: 1185-1190

Davis JR, Jennings RT, Beck BG (1993a). Comparison of treatment strategies for space motion sickness. *Acta Astronautica* 29: 587-591

Davis JR, Jennings RT, Beck BG, *et al.* (1993b) Treatment efficacy of intramuscular promethazine for space motion sickness. *Aviat Space Environ Med* 64: 230-233

DeJong H, Sondag E, Kuipers A, *et al.* (1996) Swimming behavior of fish during short periods of weightlessness. *Aviat Space Environ Med* 67: 463-466

Denise P, Darlot C, Droulez J, *et al.* (1988) Motion perceptions induced by off-vertical axis rotation (OVAR) at small angles of tilt. *Exp Brain Res* 73: 106-114

Diamond SG, Markham CH (1991) Prediction of space motion sickness susceptibility by disconjugate eye torsion in parabolic flight. *Aviat Space Environ Med* 59: 1158-1162

Diamond SG, Markham CH (1998) The effect of space missions on gravity-responsive torsional eye movements. *J Vestib Res* 3: 217-231

Dichgans J, Brandt T (1978) Visual-vestibular interaction: Effects on self-motion perception and postural control. In: *Handbook of Sensory Physiology.* Held R, Leibowitz HW, Teuber H-L (eds) Springer, New York, Volume 8, pp 755-804

Dijk DJ, Neri DE, Wyatt JK, *et al.* (2003) Sleep, circadian rhythms and perfomance during Space Shuttle missions. In: *The Neurolab Spacelab Mission: Neuroscience Research in Space.* Buckey JC, Homick JL (eds) NASA, Washington DC, NASA SP-2003-535, pp 211-221

DiZio P, Lackner JR (1988) The effects of gravitoinertial force level and head movements on post-rotational nystagmus and illusory after-rotation. *Exp Brain Res* 70: 485-495

DiZio P, Lackner JR (1992) Influence of gravito-inertial force level on vestibular and visual velocity storage in yaw and pitch. *Vision Res* 32: 123-145

DiZio P, Lackner JR (1995) Motor adaptation to Coriolis force perturbations of reaching movements: Endpoint but not trajectory adaptation transfers to the non-exposed arm. *J Neurophysiol* 74: 1787-1792

DiZio P, Lackner JR, Evanoff JN (1987) The influence of gravitoinertial force level on oculomotor and perceptual responses to sudden stop stimulation. *Aviat Space Environ Med* 58: A224-A230

Doba N, Reis DJ (1974) Role of the cerebellum and the vestibular apparatus in. regulation of orthostatic reflexes in the cat. *Circulation Res* 40: 9-18

Dobie TG, May JG, Fischer WD, *et al.* (1987) A comparison of two methods of training resistance to visually-induced motion sickness. *Aviat Space Environ Med* 58: A34-A41

Donelan JM, Kram R (2003) Exploring dynamic similarity in human running using simulated reduced gravity. *J Exp Biol* 203: 2405-2415

Draeger J, Wirt H, Schwartz R, *et al.* (1986) Messung Des Augeninnendrucks Unter MG-Bedingungen. *Naturwissenschaften* 73: 450-452

Duntley SQ, Austin RW, Taylor JH (1966) Experiment S-8/D-13, visual acuity and astronaut visibility. In: *Gemini Mid-Program Conference.* NASA, Washington DC, NASA SP-121, pp 435-454

Eckart P (1996) *Space Flight Life Support and Biospherics.* Microcosm Press, El Segundo and Kluwer Academic Publishers, Dordrecht

Eddy DR, Schiflett SG, Schlegel RE, *et al.* (1998) Cognitive performance aboard the Life and Microgravity Spacelab. *Acta Astronautica* 43:193-210

Edgerton RV, Roy RR (2000) Gravitational biology of the neuromotor systems: A perspective to the new era. *J Appl Physiol* 89: 1224-1231

Eichenbaum T, Howard IP, Dudchenko P, *et al.* (1999) The hippocampus, memory, and place cells: Is it spatial memory or a memory space? *Neuron* 23: 209-226

Eliott AR, Shea SA, Dijk DJ, *et al.* (2001) Microgravity reduces sleep disordered breating in humans. *Am J Respir Crit Med* 164: 478-485

Erp van JB, van Veen HA (2006) Touch down: The effect of artificial touch cues on orientation in microgravity. *Neurosci Lett* 404: 78-82

Evans JM, Stenger MB, Moore FB, *et al.* (2004) Centrifuge training increases presyncopal orthostatic tolerance in ambulatory men. *Aviat Space Environ Med* 75: 850-858

Fernández C, Goldberg J (1976a) Physiology of peripheral neurons innervating otolith organs of the squirrel monkey. I. Response to static tilts and to long-duration centrifugal force. *J Neurophysiol* 39: 970-984

Fernández C, Goldberg J (1976b) Physiology of peripheral neurons innervating otolith organs of the squirrel monkey. II. Directional selectivity and force-response relations. *J Neurophysiol* 39: 985-995

Fernández C, Goldberg J (1976c) Physiology of peripheral neurons innervating otolith organs of the squirrel monkey. III. Response dynamics. *J Neurophysiol* 39: 996-1008

Ferris SH (1973) Improvement of absolute distance estimation underwater. Long-term effectiveness of training. *Percept Mot Skills* 36: 1089-1090

Findlay JM, Walker R (1999) A model of saccade generation based on parallel processing and competitive inhibition. *Behav Brain Sci* 22: 661-674

Fisk J, Lackner JR, DiZio P (1993) Gravitoinertial force level influences arm movement control. *J Neurophysiol* 69: 504-511

Fitton B, Battrick B (2001) *A World Without Gravity.* ESA Publications Division, Noordwijk, The Netherlands, ESA SP-1251

Friederici AD, Levelt WJM (1987) Resolving perceptual conflicts: the cognitive mechanisms of spatial orientation. *Aviat Space Environ Med* 58: A164-169

Frost JD, Shumate WH, Salamy JG, *et al.* (1977) Experiment M133. Sleep monitoring on Skylab. In: *Biomedical Results from Skylab*. Johnston RS, LF Dietlien LF (eds) NASA Washington DC, NASA SP-377, pp 113-126

Fuglesang C, Narici L, Picozza O, *et al.* (2006) Phosphenes in low Earth orbit: Survey responses from 59 astronauts. *Aviat Space Environ Med* 77: 449-452

Fukuda T (1982) *Statokinetic Reflexes in Equilibrium and Movement*. University of Tokyo Press, Tokyo

Galambos R, Hecox KE (1978) Clinical applications of the auditory brain stem response. *Otolaryngol Clin North Am* 11: 709-722

Gauer O, Haber H (1950) Man under gravity-free conditions. *German Aviation Medicine*, World War II: 641-643

Gauthier GM, Martin BJ, Stark L (1986) Adapted head and eye movement responses to added-head inertia. *Aviat Space Environ Med* 57: 336-342

Gerathewohl S (1959) Weightlessness. In: *Man In Space - The United States Air Force Program For Developing The Spacecraft Crew*. Gantz K (ed) Duell, Sloan & Pearce, New York

Gibson JJ (1966) *The Senses Considered as Perceptual Systems*. Houghton Mifflin, Boston

Gillingham KK, Wolfe JW (1985). Spatial orientation in flight. In: *Fundamentals of Aerospace Medicine*. DeHart RL (ed) Lea & Febiger, Philadelphia, pp 299-381

Ginsburg A, Vanderploeg J (1987) Near vision acuity and contrast sensitivity. In: *Results of the Life Sciences DSO's Conducted Aboard the Space Shuttle 1981-1986*. Bungo M, Bagian T, Bowman M, Levitan B (eds) Space Biomedical Research Institute, NASA, Houston, pp 179-182

Gizzi M, Raphan T, Rudolph S, *et al.* (1994) Orientation of human optokinetic nystagmus to gravity: A model-based approach. *Exp Brain Res* 99: 347-360

Glasauer S, Mittelstaedt H (1998) Perception of spatial orientation in microgravity. *Brain Res Rev* 28: 185-193

Glasauer S, Amorim MA, Bloomberg JJ, *et al.* (1995) Spatial orientation during locomotion following space flight. *Acta Astronautica* 36: 423-431

Godwin R (ed) (1999) *Apollo 12. The NASA Mission Reports*. Apogee Books, Burlington, Ontario

Goldberg JM (1977) *A Strategy for Space Biology and Medical Science for the 1980s and 1990s*. Committee on Space Biology and Medicine. Space Science Board. National Research Council. National Academy Press, Washington DC

Goldberg J, Fernández C (1982) Eye movements and vestibular nerve responses produced in squirrel monkeys by rotations around an Earth-horizontal axis. *Exp Brain Res* 46: 393-402

Golding JF, Bles W, Bos JE, *et al.* (2003) Motion sickness and tilts of the inertial force environment: active suspension systems vs. active passengers. *Aviat Space Environ Med* 74: 220-227

Gombrich EH (1969) *Art and Illusion*. Princeton University Press, Princeton

Gontcharov IB, Kovachevich IV, Pools SL, *et al.* (2005) In-flight medical incidents in the NASA-Mir program. *Aviat Space Environ Med* 76: 692-696

Gorgiladze GI, Bryanov II (1989) Space motion sickness. *Kosm Biol Aviakosm Med* 23: 4-14

Graaf de B, Bekkering H, Erasmus C, *et al.* (1992) Influence of visual, vestibular cervical and somatosensory tilt information on ocular rotation and perception of the horizontal. *J Vestib Res* 2: 15-30

Graybiel A (ed) (1965) *The Role of the Vestibular Organs in the Exploration of Space.* NASA, Washington DC, NASA SP-77

Graybiel A (ed) (1966) *Second Symposium on The Role of the Vestibular Organs in the Exploration of Space.* NASA, Washington DC, NASA SP-115

Graybiel A (ed) (1968) *Third Symposium on The Role of the Vestibular Organs in the Exploration of Space.* NASA, Washington DC, NASA SP-152

Graybiel A (ed) (1970) *Fourth Symposium on The Role of the Vestibular Organs in the Exploration of Space.* NASA, Washington DC, NASA SP-187

Graybiel A (ed) (1973) *Fifth Symposium on The Role of the Vestibular Organs in the Exploration of Space.* NASA, Washington DC, NASA SP-314

Graybiel A (1975) Angular velocities, angular accelerations, and Coriolis accelerations. In: *Foundations of Space Biology and Medicine.* Calvin M, Gazenko OG (eds) NASA, Washington DC, Volume II, Book 1, Chapter 7, pp 247-304

Graybiel A (1980) Space motion sickness: Skylab revisited. *Aviat Space Environ Med* 51: 814-822

Graybiel A, Fregly AR (1966) A new quantitative ataxia test battery. *Acta Otolaryngol (Stockh)* 61: 292-312

Graybiel A, Knepton JC (1972) Direction-specific adaptation effects acquired in a slow rotating room. *Aerospace Med* 43: 1179-1189

Graybiel A, Knepton JC (1977) Evaluation of a new antinauseant drug for the prevention of motion sickness. *Aviat Space Environ Med* 48: 867-871

Graybiel A, Lackner JR (1979) Rotation at 30 rpm about the z-axis after 6 hours in the 10° head-down position: Effect on susceptibility to motion sickness. *Aviat Space Environ Med* 50: 390-392

Graybiel A, Kennedy RS, Guedry FE, *et al.* (1965) The effects of exposure to a rotating environment (10 rpm) on four aviators for a period of 12 days. In: *The Role of the Vestibular Organs in the Exploration of Space.* NASA, Washington DC, NASA SP-77, pp 295-338

Graybiel A, Miller EF, Billingham J, *et al.* (1967) Vestibular experiments in Gemini flights V and VII. *Aerospace Med* 38: 360-370

Graybiel A, Wood CD, Miller EF (1968) Diagnostic criteria for grading the severity of acute motion sickness. *Aerospace Med* 39: 453-455

Graybiel A, Miller EF, Homick JL (1975) Individual differences in susceptibility to motion sickness among six Skylab astronauts. *Acta Astronautica* 2: 155-174

Graybiel A, Miller EF, Homick JL (1977) Experiment M-131. Human vestibular function. In: *Biomedical Results from Skylab.* Johnston RS, Dietlein LF (eds) NASA, Washington DC, NASA SP-377, pp 74-103

Gregory RL (1965) Constancy and the geometric illusions. *Nature* 206: 745-746

Griffin, MJ, Newman MM (2004) Visual field effects on motion sickness in cars. *Aviat Space Environ Med* 75: 739-748

Grigoriev AI, Yegorov AD (1990) *Preliminary Medical Results of the 180-day Flight of Prime Crew 6 on Space Station MIR.* Fourth Meeting of the US/USSR Joint Working Group on Space Biology and Medicine, San Francisco, CA, 16-22 September 1990

Grigoryan RA, Gazenko OG, Kozlovskaya IB, *et al.* (1986) Vestibulo-cerebellar regulation of oculomotor reactions in microgravity conditions. In: *Adaptive Processes in Visual and Oculomotor Systems*. Keller EL, Zee DS (eds) Pergamon Press, New York, pp 121-128

Groen EL (1997) *Orientation to Gravity: Oculomotor and Perceptual Responses in Man*. Ph.D. Thesis, University of Utrecht

Groen EL, De Graaf B, Bles W, *et al.* (1996) Ocular torsion before and after 1 hour centrifugation. *Brain Res Bull* 40: 5-6

Groen EL, Jenkin HJ, Howard IP (2002) Perception of self-tilt in a true and illusory vertical plane. *Perception* 31: 1477-1490

Grossman GE, Leigh RJ (1990) Instability of gaze during locomotion in patients with deficient vestibular function. *Ann Neurol* 27: 528-532

Grossman GE, Leigh RJ, Abel LA (1988) Frequency and velocity of rotational head perturbations during locomotion. *Exp Brain Res* 70: 470-476

Grossman GE, Leigh RJ, Abel LA, *et al.* (1989) Performance of the human vestibuloocular reflex during locomotion. *J Neurophysiol* 62: 264-272

Gualtierotti T (1987) Vestibular integrated function and microgravity. In: *Proceedings of the Third ESA European Symposium on Life Sciences Research in Space, Graz*, Austria. ESA, Noordwijk, ESA SP-271, pp 227-232

Guedry FE (1965) Orientation of the rotation-axis relative to gravity: Its influence on nystagmus and the sensation of rotation. *Acta Otolaryngol (Stockh)* 60: 30-48

Guedry FE (1974) Psychophysics of vestibular sensation. In: *Handbook of Sensory Physiology. Vestibular System. Part 2: Psychophysics and Applied Aspects and General Interpretations*. Held R, Liebowitz HW, Teuber HL (eds) Springer Verlag, Berlin, pp 3-154

Guedry FE, Benson A (1970) Tracking performance during sinusoidal stimulation of the vertical and horizontal semicircular canals. In: *Recent Advances in Aerospace Medicine*. Busdy DE (ed) Reidel Publishing Co, Dordecht, pp 276-288

Guedry FE, Kennedy RS, Harris DS, *et al.* (1964) Human performance during two weeks in a room rotating at three rpm. *Aerospace Med* 35: 1071-1082

Gurfinkel VS, Lestienne F, Levik YS, *et al.* (1993a) Egocentric references and human spatial orientation in microgravity. II. Body-centred coordinates in the task of drawing ellipses with prescribed orientation. *Exp Brain Res* 95: 343-348

Gurfinkel VS, Lestienne F, Levik YS, *et al.* (1993b) Egocentric references and human spatial orientation in microgravity: I. Perception of complex tactile stimuli. *Exp Brain Res* 95: 339-342

Gurovskiy NN, Yeremin AV, Gazenko OG, *et al.* (1975) Medical investigations during flights of the spaceships 'Soyuz-12', 'Soyuz-13', 'Soyuz-14' and the 'Salyut-3' orbital station. *Kosm Biol Aviakosm Med* 9: 48-54

Hansen JR (1995) *Spaceflight Revolution: NASA Langley Research Center From Sputnik to Apollo*. NASA, Washington DC, NASA SP-4308

Hansen JR (2005) *First Man – The Life of Neil A. Armstrong*. Simon & Schuster, New York

Harm DL, Parker DE (1993) Perceived self-orientation and self-motion in microgravity, after landing and during preflight adaptation training. *J Vestib Res* 3: 297-305

Harm DL, Parker DE (1994) Preflight adaptation training for spatial orientation and space motion sickness. *J Clin Pharmacol* 34: 618-627

Harm DL, Zografos LM, Skinner NC, *et al.* (1993) Changes in compensatory eye movements associated with simulated stimulus conditions of space flight. *Aviat Space Environ Med* 64: 820-826

Harm DL, Bloomberg JJ, Reschke MF, *et al.* (1991) Adaptive modification of gaze control following tilt reinterpretation. *Aviat Space Environ Med* 62: 477-481

Harm DL, Parker DE, Reschke MF, *et al.* (1998) Relationship between selected orientation rest frame, circular vection and space motion sickness. *Brain Res Bull* 47: 497-501

Harm DL, Reschke MF, Parker DE (1999) Visual-vestibular integration: Motion perception reporting. In: *Extended Duration Orbiter Medical Project.* Sawin CF, Taylor GR, Smith WL (eds) NASA Johnson Space Center, Houston, NASA SP-1999-534, Chapter 5.2, pp 1-12

Haslwanter T, Curthoys IS, Black R, *et al.* (1994) Orientation of Listing's plane in normals and in patients with unilateral vestibular deafferentation. *Exp Brain Res* 101: 525-528

Hawkins R, Zieglschmid J (1975) Clinical aspects of crew health. In: *Biomedical Results of Apollo.* Johnston RS, Dietlein LF, Berry CA (eds) NASA, Washington DC, NASA SP-368, pp 43-81. Retrieved on 11 March 2007 from URL: http://history.nasa.gov/SP-368/s2ch1.htm

Heidelbaugh ND, Wescott DE, Kare MR (1975) Taste and aroma testing. In: *Skylab-4 Preliminary Biomedical Report*, NASA, Houston, JSC08818, pp 2-123-2-134

Held R, Dichgans J, Bauer J (1975) Characteristics of moving visual scenes influencing spatial orientation. *Vision Res* 15: 357-365

Helmoltz von H (1925) Physiological optics. In: *The Perceptions of Vision.* Volume 3. Southall JP (ed) The Optical Society of America, Menasha

Hernandez-Korwo R, Kozlovskaya IB, Kreydich YV, *et al.* (1983) Effect of 7-day spaceflight on structure and function of human locomotor system. *Kosm Biol Aviakosm Med* 17: 37-44

Hertwig I, Hentschel J (1989) Vestibular morphology of Xenopus laevis (Amphibian, Anura) following larval development in zero gravity (Space Shuttle, D-1 Mission). *Zool J Anat* 118: 463-472

Hess BJ, Angelaki DE (2003) Gravity modulates Listing's plane orientation during both pursuit and saccades. *J Neurophysiol* 90 1340-1345

Highstein SM, Fay RR, Popper AN (2004) *The Vestibular System.* Springer, New York

Hockey GH (1997) Compensatory control in the regulation of human performance under stress and high workload: A cognitive-energetical framework. *Biol Psychol* 45: 73-93

Hoffman RA, Pinsky LS, Osborne WZ, *et al.* (1977) Visual light flash observations on Skylab 4. In: *Biomedical Results from Skylab.* Johnston RS, Dietlein LF (eds) NASA, Washington DC, NASA SP-377, pp 127-130

Hoffman SJ, Kaplan DI (1997) *Human Exploration of Mars: The Reference Mission of the NASA Mars Exploration Study Team.* NASA Johnson Space Center, Houston, NASA SP-6107

Hofstetter-Degen K, Wetzig J, von Baumgarten RJ (1993) Oculovestibular interactions under microgravity. *Clin Investig* 71: 749-756

Holmes NP, Spence C (2004) The body schema and the multisensory representation(s) of peripersonal space. *Cogn Process* 5: 94-105

Holst von E (1935) Über den Lichtrückenreflex bei Fischen. *Pubbl Staz Zool Napoli* 15: 143-158

Holst von E, Mittelstaedt H (1950) Das Reafferenzprinzip (Wechselwirkung zwischen Zentralnervensystem und Peripherie). *Naturwissenschaften* 37: 464-476

Holland AW, Vander Ark ST (1993) *Task Analysis of Shuttle Entry and Landing Activities.* NASA, Houston, NASA-TM-104761

Homick JL, Miller EF (1975) Apollo flight crew vestibular assessment. In: *Biomedical Results of Apollo.* Johnston RS, Dietlein LF, Berry CA (eds) NASA, Washington DC, NASA SP-368, pp 323-340

Homick JL, Reschke MF (1977) Postural equilibrium following exposure to weightless space flight. *Acta Otolaryngol* 83: 455-464

Homick JL, Vanderploeg JM (1989) The neurovestibular system. In: *Space Physiology and Medicine.* 2nd Edition. Nicogossian AE, Huntoon CL, Pool SL (eds) Lea and Febiger, Philadelphia, pp 154-166

Homick JL, Reschke MF, Miller EF (1977) Effects of prolonged exposure to weightlessness on postural equilibrium. In: *Biomedical Results from Skylab* Johnston RS, Dietlein LF (eds) NASA, Washington DC, NASA SP-377, pp 104-112

Homick JL, Kohl RL, Reschke MF, *et al.* (1983) Transdermal scopolamine in the prevention of motion sickness: Evaluation of the time course of efficacy. *Aviat Space Environ Med* 54: 994-1000

Hordinsky JR, Schwertz E, Beier J, *et al.* (1982) Relative efficacy of the proposed Space Shuttle antimotion sickness medications. *Acta Astronautica* 6: 375-383

Horn ER (2003) The development of gravity sensory systems during periods of altered gravity dependent sensory input. *Adv Space Biol Med* 9: 133-171

Horn ER (2006) Microgravity-induced modifications of the vestibuloocular reflex in Xenopus laevis tadpoles are related to development and the occurrence of tail lordosis. *J Exp Biol* 209: 2847-2858

Horneck G, Facius R, Reichert M, *et al.* (2006) HUMEX - A study on the survivability and adaptation of humans to long-duration exploratory missions. Part II: Missions to Mars. *Adv Space Res* 38: 752-759

Howard IP (1982) *Human Visual Orientation.* John Wiley & Sons, Chichester

Howard IP (1986) The perception of posture, self-motion, and the visual vertical. In: *Handbook of Perception and Human Performance.* Boff KR, Kaufman L, Thomas JP (eds) John Wiley & Sons, New York, Volume 1, pp 18-1 to 18-62

Howard IP, Templeton WB (1966) *Human Spatial Orientation.* John Wiley & Sons, London

Howard IP, Childerson L (1994) The contribution of motion, the visual frame, and visual polarity to sensations of body tilt. *Perception* 23: 753-762

Howard IP, Hu G (2001) Visually induced reorientation illusions. *Perception* 30: 583-600

Howard IP, Bergstrom SS, Ohmi M (1990) Shape from shading in a different frame of reference. *Perception* 19: 523-530

Howard IP, Groen EL, Jenkin H (1997) Visually induced self-inversion and levitation. *Invest Ophtal Vis Sci* 40: S801

Hu S, Willoughby LM, Lagomarsino JJ, *et al.* (1996) Optokinetic induced taste aversions correlate with over-all symptoms of motion sickness in humans. *Percept Mot Skills* 82: 859-864

Hubbard TL (1995) Environmental invariants in the representation of motion: Implied dynamics and representational momentum, gravity, friction and centripetal force. *Psychonomic Bull Rev* 2: 322-338

Hudspeth A, Gillespie P (1994) Pulling springs to tune transduction: adaptation by hair cells. *Neuron* 12: 1-9

Hudy C, Woolford B (2005) *Space Human Factors Engineering Gap Analysis Project Final Report.* NASA Langley Research Center, Hanover, MD, NASA/TP-2007-213739

Igarashi M, Takahashi T, Kubo T, *et al.* (1978) Effect of macular ablation on vertical optokinetic nystagmus in the squirrel monkey. *J Otorhinolaryngol Relat Spec* 40: 312-318

Igarashi M, Himi T, Kulecz WB, *et al.* (1987) The role of saccular afferents in vertical optokinetic nystagmus in primates. A study in relation to optokinetic nystagmus in microgravity. *Arch Otolaryngol* 244: 143-146

Ijiri K (1995) *The First Vertebrate Mating in Space.* RICUT, Tokyo

Isableu B, Ohlmann T, Crémieux J, *et al.* (1997) Selection of spatial frame of reference and postural control variability. *Exp Brain Res* 114: 584-589

Israel I, André-Deshays C, Charade O, *et al.* (1993) Gaze control in microgravity. 2. Sequences of saccades toward memorized visual targets. *J Vestib Res* 3: 345-360

Ivanov Y, Popov V, Khachtur'yants L (1972) Motor efficiency of a cosmonaut in flight. *Procesing of Medical Information and Assessment of Work Capacity in Space Flight.* Joint Publications Research Service, Arlington, JPRS-57417

Izumi-Kurotani A, Wassersug RJ, Yamashita M, *et al.* (1992) Frog behavior under microgravity. *Proceedings of the 9th ISAS Space Utilization Symposium.* NASA, Washington DC, pp 112-114

Jeannerod M (1988) *Neural and Behavioral Organization of Goal Directed Movement.* Oxford University Press, Oxford

Jennings RT (1998) Managing space motion sickness. *J Vestib Res* 8: 67-70

Jennings RT, Davis JR, Santy PA (1988) Comparison of aerobic fitness and space motion sickness in the space shuttle program. *Aviat Space Environ Med* 58: 448-451

Johannes B, Salnitski VP, Polyakov VV, *et al.* (2003) Changes in the autonomic reactivity pattern to psychological load under long-term microgravity – twelve men during six-month spaceflights. *Aviakosm Ecolog Med* 37: 6-16

Johnston RS, Dietlein LF (1977) *Biomedical Results from Skylab.* NASA, Washington, DC, NASA SP-377

Jones TD (2006) *Sky Walking: An Astronaut's Memoir.* HarperCollins Publishers, New York

Jones W (1968) Opening remarks. In: *The Role of the Vestibular Organs in the Exploration of Space.* Graybiel A (eds) NASA, Washington DC, NASA SP-77, p. 8

Jones DR, Levy RA, Gardner L, *et al.* (1985) Self-control of psychophysiologic response to motion stress: Using biofeedback to treat airsickness. *Aviat Space Environ Med* 56: 1152-1157

Kahane P, Hoffmann D, Minotti L, *et al.* (2003) Reappraisal of the human vestibular cortex by cortical electrical stimulation study. *Ann Neurol* 54: 615-24

Kalb R, Hillman D, DeFelipe J, *et al.* (2003) Motor system development depends on experience: A microgravity study of rats. In: *The Neurolab Spacelab Mission:*

Neuroscience Research in Space. Buckey JC, Homick JL (eds) NASA, Washington DC. NASA SP-2003-535, pp 95-103

Kanas N, Manzey D (2003) *Space Psychology and Psychiatry*. Kluwer Academic Press, Dordrecht

Kanas N, Sandal G, Ritsher JB, *et al.* (2006) *Psychology and Culture during Long-Duration Space Missions*. International Academy of Astronautics Study Group on Psychology and Culture During Long-Duration Space Missions. Final Report. 15 August 2006

Kandel S, Orliaguet JP, Viviani P (2000) Perceptual anticipation in handwriting: the role of implicit motor competence. *Percept Psychophys* 62: 706-716

Kane RL, Short P, Sipes W, *et al.* (2005) Development and validation of the spaceflight cognitive assessment tool for Windows (WinScat). *Aviat Space Environ Med* 76: B183-B191

Kass JR, Bruzek W, Probst Th, *et al.* (1986) European vestibular experiments on the Spacelab-1 mission: 2. Experimental equipment and methods. *Exp Brain Res* 64: 247-254

Kaufman GD, Wood SJ, Gianna CC, *et al.* (2001) Spatial orientation and balance control changes induced by altered gravitoinertial force vectors. *Exp Brain Res* 137: 397-410

Kenyon RV, Young LR (1986) M.I.T./Canadian vestibular experiments on the Spacelab-1 mission: 5. Postural responses following exposure to weightlessness. *Exp Brain Res* 64: 335-346

Keshner EA (2004) Head-trunk coordination in elderly subjects during linear anterior-posterior translations. *Exp Brain Res* 158: 213-222

Khilov KL (1974) Some problems in evaluating the vestibular function of aviators and cosmonauts. *Kosm Biol Aviakosm Med* 8: 47-52

Khilov KL (1975) *The Cerebral Cortex in the Functions of the Vestibular Analyzer*. Meditsina, Moscow

Kingma I, Savelsbergh GJ, Toussaint HM (1999) Object size effects on initial lifting forces under microgravity conditions. *Exp Brain Res* 124: 422-428

Knierim JJ, McNaughton BL, Poe GR (2000) Three-dimensional spatial selectivity of hippocampal neurons during space flight. *Nature Neurosci* 3: 209-210

Kornilova LN, Yakovleva IY, Tarasov IK, *et al.* (1983) Vestibular dysfunction in cosmonauts during adaptation to zero-g and readaptation to 1g. *Physiologist* 26: S35-S40

Kornilova LN, Grigorova V, Bodó G (1993) Vestibular function and sensory interaction in space flight. *J Vestib Res* 3: 219-230

Kornilova LN (1997) Vestibular function and sensory interaction in altered gravity. *Adv Space Biol Med* 6: 275-313

Kozlovskaya IB, Aslanova IF, Grigorieva LS, *et al.* (1982) Experimental analysis of motor effects of weightlessness. *Physiologist* 25: 49-52

Kozlovskaya IB, Aslanova IF, Barmin VA, *et al.* (1983) The nature and characteristics of a gravitational ataxia. *Physiologist* 26: S108-S109

Kozlovskaya IB, Babaev BM, Barmin VA, *et al.* (1984) The effect of weightlessness on motor and vestibulo-motor reactions. *Physiologist* 27: S111-S114

Krioutchkov B, Morgoun V, Voronine L, *et al.* (1993) Physiological adaptation during space flights. CNES TM-90/0677/BC44, Toulouse, pp 11-47

Lackner JR (1988) Some proprioceptive influences on the perceptual representation of body shape and orientation. *Brain* 111: 281-297

Lackner JR (1989) Sensory-motor adaptation to non-terrestrial force levels. In: *Proceedings of the Symposium on Vestibular Organs and Altered Force Environments*. M Igarashi, K Nute (eds) NASA/USRA, Houston, Texas, 15 October 1987, pp 69-77

Lackner JR (1992a) Multimodal and motor influences on orientation: Implications for adapting to weightless and virtual environments. *Perception* 21: 803-812

Lackner JR (1992b) Sense of body position in parabolic flight. *Ann NY Acad Sci* 656: 329-339

Lackner JR, DiZio P (1993) Multisensory, cognitive, and motor influences on human spatial orientation in weightlessness. *J Vestib Res* 3: 361-372

Lackner JR, DiZio P (1994) Rapid adaptation to Coriolis force perturbations of arm trajectory. *J Neurophysiol* 72: 299-313

Lackner JR, DiZio P (1996) Motor function in microgravity: Movement in weightlessness. *Cur Opin Neurobiol* 6: 744-750

Lackner JR, DiZio P (1997) Sensory motor coordination in an artificial gravity environment. *J Gravit Physiol* 4: 9-12

Lackner JR, DiZio P (1998a) Spatial orientation as a component of presence: insights gained from nonterrestrial environments. *Presence* 7: 108-115

Lackner JR, DiZio P (1998b) Gravitational force background level affects adaptation to Coriolis force perturbations of reaching movements. *J Neurophysiol* 80: 546-553

Lackner JR, DiZio P (2000) Human orientation and movement control in weightless and artificial gravity environments. *Exp Brain Res* 130: 2-26

Lackner JR, DiZio P (2003) Adaptation to rotating artificial gravity environments. *J Vestib Res* 13: 321-330

Lackner JR, DiZio P (2006) Space motion sickness. *Exp Brain Res* 175: 377-399

Lackner JR, Graybiel A (1979) Parabolic flight: loss of sense of orientation. *Science* 206: 1105-1108

Lackner JR, Graybiel A (1981a) Illusions of postural, visual, and aircraft motion elicited by deep knee bends in the increased gravitoinertial force phase of parabolic flight. *Exp Brain Res* 44: 312-316

Lackner JR, Graybiel A (1981b) Variations in gravitoinertial force level affect the gain of the vestibulo-ocular reflex: Implications for the etiology of space motion sickness. *Aviat Space Environ Med* 52: 154-158

Lackner JR, Graybiel A (1985) Head movements elicit motion sickness during exposure to microgravity and macrogravity acceleration levels. In: *Vestibular and Visual Control on Posture and Locomotor Equilibrium*. Igarashi M, Black FO (eds) Karger, Basel, pp 170-176

Lackner JR, Graybiel A (1986) Head movements in non-terrestrial force environments elicit motion sickness: Implications for the etiology of space motion sickness. *Aviat Space Environ Med* 57: 443-448

Lackner JR, Graybiel A (1987) Head movements in low and high force environments elicit motion sickness: Implications for space motion sickness. *Aviat Space Environ Med* 58 (Suppl 9): A212-A217

Lackner JR, Graybiel A (1993) Perceived orientation in free fall depends on visual, postural, and architectural factors. *Aviat Space Environ Med* 54: 47-51

Lackner JR, Graybiel A (1994) Use of promethazine to hasten adaptation to provocative motion. *J Clin Pharmacol* 34: 644-648

Lacquaniti F, Maioli C (1989) The role of preparation in tuning anticipatory and reflex responses during catching. *J Neurosci* 9: 134-148

Lathan C, Wang Z, Clément G (2000) Changes in the vertical size of a three-dimensional object drawn in weightlessness by astronauts. *Neurosci Lett* 295: 37-40

Layne CS, Bloomberg JJ, McDonald PV, *et al.* (1994) Lower-limb electromyographic-activity patterns during treadmill locomotion following space flight. *Aviat Space Environ Med* 65: 449

Layne CS, McDonald VP, Bloomberg JJ (1997) Neuromuscular activation patterns during treadmill walking after space flight. *Exp Brain Res* 113: 104-116

Layne CS, Mulavara AP, McDonald PV, *et al.* (2001) Effect of long-duration spaceflight on postural control during self-generated perturbations. *J Appl Physiol* 90: 997-1006

Leach CS (1987) Fluid control mechanisms in weightlessness. *Aviat Space Environ Med* 58: A74-79

Lebedev V (1988) *Diary of a Cosmonaut: 211 Days in Space*. GLOSS Company, College Station

Leigh RJ, Zee DS (1999) *The Neurology of Eye Movements*. 3rd Edition. FA Davis Company, Philadelphia

Léone G (1998) The effect of gravity on human recognition of disoriented objects. *Brain Res Rev* 28: 203-214

Léone G, Lipshits M, McIntyre J, *et al.* (1995) Independence of bilateral symmetry detection from a gravitational reference frame. *Spatial Vision* 9: 127-137

Leigh RJ, Zee DS (1991) *The Neurology of Eye Movements*. 2nd Edition. FA Davis Company, Philadelphia

Lichtenberg BK (1988) Vestibular factors influencing the biomedical support of humans in space. *Acta Astronautica* 17: 203-206

Linenger JM (2001) *Off the Planet: Surviving Five Perilous Months Aboard the Space Station Mir*. McGraw-Hill, New York

Link M (1965) *Space Medicine in Project Mercury*. NASA, Washington DC, NASA SP-4003

Lipshits MI, Mclntyre J (1999) Gravity affects the preferred vertical and horizontal in visual perception of orientation. *NeuroReport* 10: 1085-1089

Lipshits MI, Mauritz K, Popov KE (1981) Quantitative analysis of anticipation postural components of a complex voluntary movement. *Fiziol Cheloveka* 8: 254-263

Lipshits MI, McIntyre J, Zaoui M, *et al.* (2001) Does gravity play an essential role in the asymmetrical visual perception of vertical and horizontal line length? *Acta Astronautica* 49: 123-130

Lipshits MI, Bengoetxea A, Cheron G, McIntyre J (2005) Two reference frames for visual perception in two gravity conditions. *Perception* 34: 545-555

Locke J (2003) Space motion sickness symptomatology: 20 years' experience of NASA's Space Shuttle program. Paper presented at the 74th ASMA Annual Scientific Meeting, San Antonio, TX. *Aviat Space Environ Med* 74: 101 (Abstract 225)

Longnecker DE, Molins RA (2006) *A Risk Reduction Strategy for Human Exploration of Space: A Review of NASA's Bioastronautics Roadmap*. Institute of Medicine,

National Research Council of the National Academies, The National Academies Press, Washington DC

Mader TH (1991) Intraocular pressure in microgravity. *J Clin Pharmacol* 31: 947-950

Mallis MM, DeRoshia CW (2005) Circadian rhythms, sleep, and performance in space. *Aviat Space Environ Med* 76: B94-B107

Manzey D (2000) Monitoring of mental performance during spaceflight. *Aviat Space Environ Med* 7: A69-A75

Manzey D, Lorenz B (1998) Mental performance during short-term and long-term spaceflight. *Brain Res Rev* 28: 215-221

Manzey D, Lorenz B, Schiewe A, *et al.* (1993) Behavioral aspects of human adaptation to space: Analyses of cognitive and psychomotor performance in space during an eight-day space mission. *Clin Investig* 71: 725-731

Manzey D, Lorenz B, Schiewe A, *et al.* (1995) Dual-task performance in space: results from a single-case study during a short-term space mission. *Human Factors* 37: 667-681

Manzey D, Lorenz B, Polyakov V (1998) Mental performance in extreme environments: Results from a performance monitoring study during a 438-day spaceflight. *Ergonomics* 41: 537-559

Manzey D, Lorenz B, Heuer H, *et al.* (2000) Impairments of manual tracking performance during spaceflight: More converging evidence from a 20-day space mission. *Ergonomics* 43: 589-609

Marquez JJ, Oman CM, Liu AM (2004) You-are-here maps for International Space Station: Approach and Guidelines. *SAE International* 2004-01-2584. Retrieved on 26 July 2006 from URL:

http://stuff.mit.edu/people/amliu/Papers/Marquez-YAH-2004-01-2584.pdf

Massion J, Gurfinkel VS, Lipshits MI, *et al.* (1993) Axial synergies under microgravity conditions. *J Vestib Res* 3: 275-288

Massion J, Popov K, Fabre1 JC, *et al.* (1997) Is the erect posture in microgravity based on the control of trunk orientation or center of mass position? *Exp Brain Res* 114: 384-389

Mast FW, Newby NJ, Young LR (2003) Sensorimotor aspects of high-speed artificial gravity: II. The effect of head position on illusory self-motion. *J Vestib Res* 12: 282-289

Matsakis Y, Lipshits M, Gurfinkel VS, *et al.* (1993) Effects of prolonged weightlessness on mental rotation of three-dimensional objects. *Exp Brain Res* 94: 152-162

Matsnev EI, Yakovleva IYa, Tarasov IK, *et al.* (1983) Space motion sickness: phenomenology, countermeasures, and mechanisms. *Aviat Space Environ Med* 54: 312-317

Matsnev EI, Bodo D (1984) Experimental assessment of selected antimotion drugs. *Aviat Space Environ Med* 55: 281-286

Matsuo V, Cohen B (1984) Vertical optokinetic nystagmus and vestibular nystagmus in the monkey: up-down asymmetry and effects of gravity. *Exp Brain Res* 53: 197-216

Matveyev AD (1987) Development of methods for the study of space motion sickness. *Kosm Biol Aviakosm Med* 21: 83-88

McCall GE, Goulet C, Boorman GI (2003) Flexor bias of joint position in humans during spaceflight. *Exp Brain Res* 152: 87-94

McCluskey R, Clark J, Stepaniak P (2001) Correlation of Space Shuttle landing performance with cardiovascular and neurological dysfunction resulting from space flight. *NASA Bioastronautics Roadmap*. Retrieved 26 July 2006 from URL: http://bioastroroadmap.nasa.gov/User/risk.jsp?showData=13

McDonald PV, Bloomberg JJ, Layne CS (1994) Intersegmental coordination during treadmill locomotion following space flight. *Aviat Space Environ Med* 65: 449

McDonald PV, Basdogan C, Bloomberg JJ, *et al.* (1996) Lower limb kinematics during treadmill walking after space flight. Implications for gaze stimulation. *Exp Brain Res* 112: 325-331

McDonald PV, Bloomberg JJ, Layne CS (1997) A review of adaptive change in musculoskeletal impedance during space flight and associated implications for postflight movement control. *J Vestib Res* 7: 239-250

McIntyre J, Zago M, Berthoz A, *et al.* (2001) Does the brain model Newton's laws? *Nature Neurosci* 4: 693-695

McIntyre J, Gurfinkel EV, Lipshits MI, *et al.* (2005) Measurements of human force during a constrained arm motion using a force-actuated joystick. *J Neurophysiol* 73: 1201-1222

Meck JV, Waters WW, Ziegler MG, *et al.* (2004) Mechanisms of post-spaceflight orthostatic hypotension: Low alpha1-adrenergic receptor responses before flight and central autonomic dysregulation postflight. *Am J Physiol Heart Circ Physiol* 286: 1486-1495

Melnik CG, Shakula AV, Ivanov VV (1986) The use of the electrotranquilization method for increasing vestibular tolerance in humans. *Voyen Med Zhurn* 8: 42-45

Melvill-Jones G, Watt DGD (1971) Muscular control of landing from unexpected falls in man. *J Physiol* 219: 729-741

Merfeld DM (1996) Effect of space flight on ability to sense and control roll tilt: human neurovestibular studies on SLS-2. *J Appl Physiol* 81: 50-57

Merfeld DM (2003) Rotation otolith tilt-translation reinterpretation (ROTTR) hypothesis: A new hypothesis to explain neurovestibular spaceflight adaptation. *J Vestib Res* 13: 309-320

Merfeld DM, Young LR, Oman CM, *et al.* (1993) A multidimensional model of the effect of gravity on the spatial orientation of the monkey. *J Vestib Res* 3: 141-161

Merfeld DM, Jock RI, Christie SM, *et al.* (1994) Perceptual and eye movement responses elicited by linear acceleration following space flight. *Aviat Space Environ Med* 65: 1015-1024

Merfeld DM, Teiwes W, Clarke AH, *et al.* (1996a) The dynamic contribution of the otolith organs to human ocular torsion. *Exp Brain Res* 110: 315-321

Merfeld DM, Polutchko KA, Schultz K (1996b) Perceptual responses to linear acceleration after spaceflight: Human neurovestibular studies on SLS-2. *J Appl Physiol* 81: 58-68

Merfeld DM, Zupan L, Peterka RJ (1999) Humans use internal models to estimate gravity and linear acceleration. *Nature* 398: 615-618

Merfeld DM, Park S, Gianna-Poulin C, *et al.* (2005) Vestibular perception and action employ qualitatively different mechanisms. I: Frequency response of VOR and perceptual responses during translation and tilt. *J Neurophysiol* 94: 186-198

Michel F, Schotti B (1975) *The Disconnection Syndrome in Man*. INSERM, Lyon

Miller EF, Graybiel A (1973) Perception of the upright and susceptibility to motion sickness as functions of angle of tilt and angular velocity in off-vertical rotation. In: *Fifth Symposium on the Role of the Vestibular Organs in Space Exploration.* Graybiel A (ed) NASA, Washington DC, NASA SP-314, pp 99-103

Mittelstaedt H (1983) A new solution to the problem of the subjective vertical. *Naturwissenschaften* 70: 272-281

Mittelstaedt H (1992) Somatic versus vestibular gravity reception in man. *Ann NY Acad Sci* 656: 124-139

Mittelstaedt H, Fricke E (1988) The relative effect of saccular and somatosensory information on spatial perception and control. *Adv Otorhinolaryngol* 42: 24-30

Mittelstaedt H, Glasauer S (1993) Crucial effects of weightlessness on human orientation. *J Vestib Res* 3: 307-314

Miura M, Sekitani T (1993) Follow-up of square drawing test in vestibular neuronitis. *Acta Otolaryngol (Stockh)* 503: 35-38

Money KE (1970) Motion sickness. *Physiol Rev* 50: 1-39

Money KE, Cheung BS (1983) Another function of the inner ear: Facilitation of the emetic response to poisons. *Aviat Space Environ Med* 54: 208-213

Money KE, Cheung BS (1991) Alterations of proprioceptive function in the weightless environment. *J Clin Pharm* 31: 1007-1009

Monk TH, Buysse DJ, Billy BD, et al. (1998) Sleep and circadian rhythms in four orbiting astronauts. *J Biol Rhythms* 13: 188-201

Moore S, Clément G, Raphan T, et al. (2001) Ocular counterrolling induced by centrifugation during orbital space flight. *Exp Brain Res* 137: 323-335

Moore S, Cohen B, Raphan T, et al. (2005) Spatial orientation of optokinetic nystagmus and ocular pursuit during orbital space flight. *Exp Brain Res* 160: 38-59

Moore S, MacDougall H, Peters B, et al. (2006) Modelling locomotor dysfunction vestibular stimulation following spaceflight with galvanic vestibular stimulation. *Exp Brain Res* 172: 208-220

Mori S (1995) Disorientation of animals in microgravity. *Nagoya J Med Sci* 58: 71-81

Mount Fr, Foley T (1999) Assessment of human factors. In: *Extended Duration Orbiter Medical Project Final Report 1989-1995.* Sawin CF, Taylor GR, Smith WL (eds) NASA, Washington DC, NASA SP-1999-534, Chapter 6, pp 1-15

Mueller C, Kornilova L, Wiest G, et al. (1994) Visually induced vertical self-motion sensation is altered in microgravity adaptation. *J Vestib Res* 4: 161-167

Mullane M (2006) *Riding Rockets. The Outrageous Tales of a Space Shuttle Astronaut.* Scribner, New York

Murashugi CM, Howard IP (1989) Up-down asymmetry in vertical optokinetic nystagmus and afternystagmus: Contributions of the central and peripheral retinae. *Exp Brain Res* 77: 183-192

Muybridge E (1955) *The Human Figure in Motion.* Dover Publications Inc, New York

Narici L, Belli F, Bidoli V, et al. (2004) The ALTEA/Alteino projects: Studying functional effects of microgravity and cosmic radiation. *Adv Space Res* 33: 1352-1357

National Academy of Sciences (1988) *Space Science in the Twenty-First Century: Imperatives for the Decades 1995 to 2115.* National Academy Press, New York, pp 83-128

National Aeronautics and Space Administration (1995) *Man-Systems Integration Standards.* NASA, Houston, NASA-STD-3000, revision B, Volume 1

National Aeronautics and Space Administration (2004) *The Vision for Space Exploration.* NASA, Washington DC, NASA/NP-2004-01-334-HQ

National Aeronautics and Space Administration (2005) *Bioastronautics Roadmap. A Risk Reduction Strategy for Human Space Exploration.* NASA, Houston, NASA/SP-2005-6113

Nechaev AP (2001) Work and rest planning as a way of crewmember error management. *Acta Astronautica* 49: 271-278

Necker LA (1832) Observations on some remarkable phenomena seen in Switzerland, and an optical phenomenon which occurs on viewing of a crystal or geometrical solid. *Philosophical Magazine* 3: 329-343

Newell H (1980) *Beyond the Atmosphere: Early Years of Space Science.* NASA, Washington DC, NASA SP-4211, pp 274-275

Newman DJ, Lathan CE (1999) Memory processes and motor control in extreme environments. *IEEE Trans Sys Man Cybern* 29: 387-394

Newman DJ, Jackson DK, Bloomberg JJ (1997) Altered astronaut lower-limb and mass center kinematics in downward jumping following space flight. *Exp Brain Res* 117: 30-42

Nicogossian AE (1977) *The Apollo-Soyuz Test Project. Medical Report.* NASA, Washington DC, NASA SP-411

Nicogossian AE, Leach-Huntoon C, Pool SL (1989) *Space Physiology and Medicine.* 2nd Edition. Lea & Febiger, Philadelphia

Nicogossian AE, Huntoon CL, Pool SL (1994) *Space Physiology and Medicine,* 3rd Edition. Lea & Febiger, Philadelphia

Nicogossian AE, Parker JF (1982) *Space Physiology and Medicine.* NASA, Washington DC, NASA SP-447

Noland D (2007) Mission to the Moon: How we'll go back – and stay this time. *Popular Mechanics,* March 2007. Retrieved 5 June 2007 from URL: http://www.popularmechanics.com/science/air_space/4212906.html

Nooij SAE, Bos JE (2006) Sustained hypergravity to simulate SAS: Effect of G-load and duration. In: *Proceedings of the 7th Symposium on the Role of the Vestibular Organs in Space Exploration.* ESTEC, Noordwijk, The Netherlands, 6-9 June 2006

Nooij SAE, Bos JE, Ockels WJ (2004) Investigation of vestibular adaptation to changing gravity levels on earth. *J Vestib Res* 14: 133

Oberg JE (1982) *Mission to Mars: Plans and Concepts for the First Manned Landing.* Stackpole Books, Harrisburg

Ockels WJ, Furrer R, Messerschmid E (1990) Space sickness on Earth. *Exp Brain Res* 79: 661-663

Olabi AA, Lawless HT, Hunter JB, *et al.* (2002) The effects of microgravity and space flight on the chemical senses. *J Food Sci* 67: 468-478

Oman CM (1982) A heuristic mathematical model for the dynamics of sensory conflict and motion sickness. *Acta Otolaryngol* (Suppl) 392: 1-44

Oman CM (1988) The role of static visual-orientation cues in the etiology of space motion sickness. In: *Proceeding, Symposium on Vestibular Organs and Altered Environments, October 1987.* Igarashi M, Nute NG (eds) NASA Space Biomedical Research Institute, USRA Division of Space Biomedicine, Houston, pp 25-38

Oman CM (2003) Human visual orientation in weightlessness. In: *Levels of Perception* Harris L, Jenkin M (eds) Springer Verlag, New-York, pp 375-398

Oman CM (2007) *Visual Orientation, navigation, and Spatial Memory Countermeasures*. Retrieved 16 October 2007 from URL: http://www.nsbri.org/Research/Projects/viewsummary.epl?pid=145

Oman CM (2007) Spatial orientation and navigation in microgravity. In: *Spatial Processing in Navigation, Imagery and Perception*. Mast J, Jancke L (eds) Springer, New York, pp 209-246

Oman CM, Weigl H (1989) Postflight vestibulo-ocular reflex changes in Space Shuttle/Spacelab D-1 crew. *Aviat Space Environ Med* 60: 480-489

Oman CM, Balkwill MD (1993) Horizontal angular VOR, nystagmus dumping, and sensation duration in Spacelab SLS-1 crewmembers. *J Vestib Res* 3: 315-330

Oman CM, Kulbaski M (1988) Space flight affects the 1-g postrotatory vestibulo-ocular reflex. *Adv Otorhinolaryngol* 42: 5-8

Oman CM, Bock OL, Huang JK (1980) Visually induced self-motion sensation adapts rapidly to left-right visual reversal. *Science* 209: 706-708

Oman CM, Lichtenberg BK, Money KE, *et al.* (1986) M.I.T./Canadian vestibular experiments on the Spacelab-1 Mission: 4. Space motion sickness: symptoms, stimuli, and predictability. *Exp Brain Res* 64: 316-334

Oman CM, Young LR, Watt DGD, *et al.* (1988) MIT/Canadian Spacelab experiments on vestibular adaptation and space motion sickness. In: *Basic and Applied Aspects of Vestibular Function*. Hwang JC, Daunton NG, Wilson VJ (eds) University Press, Hong Kong, pp 204-231

Oman CM, Lichtenberg BK, Money KE (1990) Space motion sickness monitoring experiment: Spacelab 1. In: *Motion and Space Sickness*. Crampton GH (ed) CRC Press, Boca Raton, pp 217-246

Oman CM, Pouliot CF, Natapoff A (1996) Horizontal angular VOR changes in orbital and parabolic flight: human neurovestibular studies on SLS-2. *J Appl Physiol* 81: 69-81

Oosterveld W, Greven A (1975) Flight behavior of pigeons in the weightless phase of parabolic flight. *Aviat Space Environ Med* 46: 713-716

Ortega HJ, Harm DL (2007) Space and entry motion sickness. In: *Space Medicine*. Barratt M (ed) *in press*

Paige GD, Seidman SH (1999) Characteristics of the VOR in response to linear acceleration. *Ann NY Acad Sci* 871: 123-135

Palmer SE (1999) *Vision Science: Photons to Phenomenology*. MIT Press, Cambridge

Paloski WH, Black FO, Reschke MF, *et al.* (1993) Vestibular ataxia following Shuttle flights: Effect of transient microgravity on otolith-mediated sensorimotor control of posture. *Am J Otolaryngol* 14: 9-17

Paloski WH, Reschke MF, Black FO (1999) Recovery of postural equilibrium control following space flight. In: *Extended Duration Orbiter Medical Project Final Report 1989-1995*. Sawin CF, Taylor GR, Smith WL (eds) NASA, Washington DC, NASA SP-1999-534, Chapter 4.1, pp 1-16

Paloski WH, Newby NJ, Hwang EY (2004) Head movement augmented postural control testing on short-duration astronauts. *Proceedings of the Aerospace Med Assoc 75th Annual Scientific Meeting*, Anchorage, 18-21 May 2004

Parker DE (1991) Human vestibular function and weightlessness. *J Clin Pharmacol* 31: 904-910

Parker DE (2003) Spatial perception changes associated with space flight: Implications for adaptation to altered inertial environments. *J Vestib Res* 13: 331-344

Parker DE, Harm DL (1993) Mental rotation: A key to mitigation of motion sickness in the virtual environment? *Presence* 1: 329-333

Parker DE, Parker KL (1990) Adaptation to the simulated stimulus rearrangement of weightlessness. In: *Motion and Space Sickness*. Crampton GH (ed) CRC Press, Boca Raton, pp 247-262

Parker DE, Reschke MF, Arrott AP, *et al.* (1985) Otolith tilt translation reinterpretation following prolonged weightlessness: Implications for preflight training. *Aviat Environ Space Med* 56: 601-609

Parker DE, Reschke MF, Ouyang L, *et al.* (1986) Vestibulo-ocular reflex changes following weightlessness and preflight adaptation training, In: *Adaptive Processes in Visual and Oculomotor Systems*. Keller E, Zee D (eds) Pergamon Press, Oxford, pp 103-108

Parker DE, Reschke MF, Arrott AP, *et al.* (1987) Otolith tilt-translation reinterpretation following prolonged weightlessness: Implications for preflight training. In: *Results of Life Sciences DSO's Conducted Aboard the Space Shuttle 1981-1986*. Bungo M, Bagian T, Bowman M, Levitan B (eds) Space Biomedical Research Institute and NASA, Houston, pp 145-152

Parker DE, Reschke MF, Aldrich N (1989) Performance. In: *Space Biology and Medicine*. 2nd Edition. Nicogossian A, Huntoon C, Pool S (eds) Lea & Febiger, Philadelphia, pp 167-178

Parker JF, Jones WL (1975) *Biomedical Results from Apollo*. NASA, Washington, DC, NASA SP-368

Parsons LM (1990) Body image. In: *The Blackwell Dictionary of Cognitive Psychology*. Eysenck MW (ed) Blackwell, Oxford, pp 46-47

Pavy-LeTraon A, Saivin S, Soulez-LaRiviere C, *et al.* (1997) Pharmacology in space: Pharmacotherapy. In: *Advances in Space Biology and Medicine*. Bonting SL (ed) JAI Press, Greenwich, Volume 6

Pelt SV, Van Gisbergen JA, Medendorp WP (2005) Visuospatial memory computations during whole-body rotations in roll. *J Neurophysiol* 94: 1432-1442

Persterer A, Berger M, Koppensteiner C, *et al.* (1992) Audimir – Directional hearing at microgravity. In: *Health from Space Reseach. Austrian Accomplishments Austrian*. Society for Aerospace Medicine (ed) Springer-Verlag, Wien, pp 21-38

Perterka RJ, Benolken MS (1992) Relation between perception of vertical axis rotation and vestibulo-ocular reflex symmetry. *J Vestib Res* 2: 59-70

Pestov ID, Gerathewohl SJ (1975) Weightlessness. In: *Foundations of Space Biology and Medicine*. Calvin M, Gazenko OG (eds) NASA, Washington DC, Volume II, Book 1, Chapter 8, pp 305-354

Pettit D (2003) Expedition Six Space Chronicles. Retrieved 25 April 2007 from URL: http://space flight.nasa.gov/station/crew/exp6/spacechronicles.html

Piaget J (1955) *The Child's Construction of Reality*. Routledge and Kegan Paul, London

Planel H (2004) *Space and Life: An Introduction to Space Biology amd Medicine*. CRC Press, Boca Raton

Polyakov BI (1987) Discrete adaptation to sensory conflict. *Kosm Biol Aviakosm Med* 21: 82-86

Popov VA, Boyko NI. (1967) *Vision During Space Flight*. NASA Johnson Space Center, NASA Technical Memorandum X-60574

Popov NI, Solodovnik FA, Khlebnikov GF (1970) Vestibular training of test pilots by passive methods. In: *Physiology of the Vestibular Analyzer*. NASA, Washington DC, NASA TT-F-616, pp 173-176

Pozzo T, Berthoz A, Lefort L, *et al.* (1991) Head stabilization during various locomotor tasks in humans: II. Patients with bilateral peripheral vestibular deficits. *Exp Brain Res* 85: 208-217

Pozzo T, Papaxanthis C, Stapley P, *et al.* (1998) The sensorimotor and cognitive integration of gravity. *Brain Res Rev* 28: 92-101

Previc FH, Ercoline WR (2004) Spatial disorientation in aviation. In: *Progress in Astronautics and Aeronautics*. American Institute of Aeronautics and Astronautics Inc, Reston, Virginia, Volume 23

Prisk GK, Fuller CA (2001) *Sleep and Chronobiology*. Retrieved on 12 March 2007 from URL: www.dsls.usra.edu/meetings/bio2001/pdf/311.pdf

Prisk GK, Eliott AR, Paiva M, *et al.* (1993) Sleep and respiration in microgravity. In: *The Neurolab Spacelab Mission: Neuroscience Research in Space*. Buckey JC, Homick JL (eds) NASA, Washington DC, NASA SP-2003-535, pp 223-232

Pulaski PD, Zee DS, Robinson DA (1981) The behavior of the vestibuloocular reflex at high velocities of head rotation. *Brain Res* 222: 159-165

Pyle R (2005) *Destination Moon. The Apollo Missions in the Astronauts' Own Words*. Collins, New York

Ramachandran VS, Blakeslee S (1998) *Phantoms in the Brain: Probing the Mysteries of the Human Mind*. William Mollow, New York

Rambaut PC, Johnson PC (1989) Nutrition. In: *Space Physiology and Medicine*. Nicogossian AE, Huntoon CL, Pool SL (eds) Lea and Febiger, Philadelphia, pp 202-213

Raphan T, Matsuo V, Cohen B (1979) Velocity storage in the vestibulo-ocular reflex arc (VOR). *Exp Brain Res* 35: 229-248

Raphan T, Dai MJ, Cohen B (1992) The spatial orientation of the vestibular system. *Ann NY Acad Sci* 656: 140-157

Ratino DA, Repperger DW, Goodyear C, *et al.* (1988) Quantification of reaction time and time perception during Space Shuttle operations. *Aviat Space Environ Med* 59: 220-224

Raybeck D (1991) Proxemics and privacy: Managing the problems of life in confined environments. In: *From Antarctica to Outer Space*. Harrison AA, Clearwater YA, McKay CP (eds), Springer, New York

Raymond J, Dememes D, Blanc E, *et al.* (2003) Development of the vestibular system in microgravity. In: *The Neurolab Spacelab Mission: Neuroscience Research in Space*. Buckey JC, Homick JL (eds) NASA, Washington DC, NASA SP-2003-535, pp 143-149

Reason JT (1978) Motion sickness adaptation: A neural mismatch model. *J Royal Soc Med* 71: 819-829

Reason JT, Brand JJ (1975) *Motion Sickness*. Academic Press, London

Reason JT, Graybiel A (1970) Progressive adaptation to Coriolis accelerations associated with 1 rpm increments in the velocity of the slow rotating room. *Aerospace Med* 41: 73-79

Reschke MF (1988) Microgravity vestibular investigations: Experiments on vestibular and sensorimotor adaptation to space flight. In: *Basic and Applied aspects of*

Vestibular Function. Hwang JC, Daunton NG, Wilson VJ (eds) Hong Kong Press, Hong Kong, pp 105-121

Reschke MF (1989) Open Discussion. In: *Proceeding, Workshop on Nervous System Plasticity in Relation to Long-Term Exposure to Microgravity Environment, October 1987.* Igarashi M, Nute NG, MacDonald S (eds) NASA Space Biomedical Research Institute, USRA Division of Space Biomedicine, Houston, pp 135-136

Reschke MF (1990) Statistical prediction of space motion sickness. In: *Motion and Space Sickness.* Crampton GH (ed) CRC Press, Boca Raton, pp 263-316

Reschke MF, Parker DE (1987) Effects of prolonged weightlessness on self-motion perception and eye movements evoked by roll and pitch. *Aviat Space Environ Med* 58: A153-158

Reschke MF, Anderson DJ, Homick JL (1984) Vestibulospinal reflexes as a function of microgravity. *Science* 225: 212-214

Reschke MF, Parker DE, Skinner N (1985) Oculer counterrolling. In: *Results of the Life Sciences DSOs Conducted Aboard the Space Shuttle (1981-1986).* Bungo MW, Bagian TM, Bowman MA, Levitan BM (eds) NASA Space Biomedical Research Institute, Houston, pp 141-144

Reschke MF, Anderson DJ, Homick JL (1986) Vestibulo-spinal response modification as determined with the H-reflex during the Spacelab-1 flight. *Exp Brain Res* 64: 335-346

Reschke MF, Parker DE, Harm DL, *et al.* (1988) Ground-based training for the stimulus rearrangement encountered during space flight. *Acta Otolaryngol* (Suppl) 460: 87-93

Reschke MF, Harm DL, Parker DE, *et al.* (1991) DSO 459: Otolith tilt-translation reinterpretation. In: *Results of the Life Sciences DSOs Conducted Aboard the Space Shuttle, 1988-1990.* NASA, Houston, pp 33-50

Reschke MF, Harm DL, Parker DE, *et al.* (1994) Physiologic adaptation to space flight. Neurophysiologic aspects: Space motion sickness. In: *Space Physiology and Medicine.* Nicogossian AE, Leach CE, Pool SL (eds) Lea & Febiger, Philadelphia, pp 228-260

Reschke MF, Kornilova LM, Harm DL, *et al.* (1996) Neurosensory and sensory-motor function. In: *Space Biology and Medicine.* Nicogossian AE, Mohler SR, Gazenko AE, Grigoriev AI (eds) American Institute of Aeronautics and Astronautics, Reston, Volume 3, pp 135-193

Reschke MF, Bloomberg JJ, Harm DL, *et al.* (1998) Posture, locomotion, spatial orientation, and motion sickness as a function of space flight. *Brain Res Rev* 28: 102-117

Reschke M, Somers JT, Leigh RJ, *et al.* (2004) Sensorimotor recovery following space flight may be due to frequent square-wave saccadic intrusions. *Aviat Space Environ Med* 75: 700-704

Reschke MF, Krnavek JM, Somers JT, *et al.* (2007) *A brief history of space flight with a comprehensive compendiun of vestibular and sensorimotor research conducted across the various flight programs.* NASA Johson Space Center, Houson, NASA SP-2007-560

Riewe PC, Horn ER (2000) The Scorpion: An ideal animal model to study long-term microgravity effects on circadian rhythms. Space Technology and Applications

International Forum, 19 January 2000, *American Institute of Physics, Conference Proceedings* 504: 383-388

Robbins FD (1988) *Exploring the Living Universe*. A Report of the NASA Life Sciences Strategic Planning Study Committee. NASA, Washington DC

Roberts TDM (1968) Labyrinthine control of the postural muscles. In: *Third Symposium on the Role of the Vestibular Organs in the Exploration of Space*. Graybiel A (ed) NASA, Washington DC, NASA SP-152, pp 149-168

Rock I (1956) The orientation of forms on the retina and in the environment. *Am J Psychol* 69: 513-528

Rock I (1973) *Orientation and Form*. Academic Press, New York

Roll JP, Popov KE, Gurfinkel VS, *et al.* (1993) Sensorimotor and perceptual functions of muscle proprioception in microgravity. *J Vestib Res* 3: 259-273

Roller CA, Cohen HS, Kimball KT, *et al.* (2001) Variable practice with lenses improves visuo-motor plasticity. *Cog Brain Res* 12: 341-52

Ross HE, Brodie EE, Benson AJ (1986) Mass discrimination in weightlessness and readaptation to Earth's gravity. *Exp Brain Res* 64: 358-366

Ross HE, Schwartz E, Emmerson P (1987) The nature of sensorimotor adaptation to altered G-levels: Evidence from mass discrimination. *Aviat Space Environ Med* 58 (Suppl 9): A148-A152

Ross MD (1992) A study of the effects of space travel on mammalian gravity receptors. *Space Life Sciences-1 180-Day Experimental Reports*. NASA, Washington DC

Ross MD (1993) Morphological changes in rats vestibular system following weightlessness. *J Vestib Res* 3: 241-251

Ross MD (1994) A space flight study of synaptic plasticity in adult rat vestibular maculas. *Acta Otolaryngol* (Suppl) 516: 1-14

Ross MD, Tomko DL (1998) Effects of gravity on vestibular neural development. *Brain Res Rev* 28: 44-51

Rossetti Y, Rode G (2002) Reducing spatial neglect by visual and other sensory manipulations: noncognitive (physiological) routes to the rehabilitation of a cognitive disorder. In: *The Cognitive and Neural Bases of Spatial Neglect*. Karnath H-O, Milner AD, Vallar G (eds) Oxford University Press, Oxford, pp 375-396

Rupert A (2000) Tactile situation awareness system: Proprioceptive prostheses for sensory deficiencies. *Aviat Space Environ Med* 71: A92-A99

Saivin S, Pavy-LeTraon A, Soulez-LaRiviere C, *et al.* (1997) Pharmacology in space: Pharmacokinetics. In: *Advances in Space Biology and Medicine*. Bonting SL (eds) JAI Press, Greenwich, Connecticut, Volume 6

Salnitski VP, Myasnikov VI, Bobrov AF, *et al.* (1999) Integrated evaluation and prognosis of cosmonaut's professional reliability during space flight. *Aviakosm Ecolog Med* 33: 16-22

Sangals J, Heuer H, Manzey D, *et al.* (1999) Changed visuomotor transformations during and after prolonged microgravity. *Exp Brain Res* 129: 378-390

Santy PA (1994) *Choosing the Right Stuff: The Psychological Selection of Astronauts and Cosmonauts*. Praeger Scientific, Westport

Santy PA, Bungo MW (1991) Pharmacologic considerations for Shuttle astronauts. *J Clin Pharmacol* 31: 931-933

Santy PA, Kapanka H, Davis JR, *et al.* (1988) Analysis of sleep on Shuttle missions. *Aviat Space Environ Med* 59: 1094-1097

Sauer J, Hockey GRJ, Wastell DH (1999) Maintenance of complex performance during a 135- day spaceflight simulation. *Aviat Space Environ Med* 70: 236-244

Sawin CF, Baker E, Black FO (1998) Medical investigations and resulting countermeasures in support of 16-day Space Shuttle missions. *J Gravit Physiol* 5: 1-12

Schachter S (1975) Cognition and peripheralist-centralist controversies in motivation and emotion. In: *Handbook of Psychobiology*. Gazzaniga MS, Blakemore C (eds) Academic Press, New York, pp 234-253

Scherer H, Clarke A (1985) The caloric vestibular reaction in space. Physiological considerations. *Acta Otolaryngol (Stockh)* 100: 328-336

Scherer H, Brandt U, Clarke AH, *et al.* (1986) European vestibular experiments on the Spacelab-1 mission: 3. Caloric nystagmus in microgravity. *Exp Brain Res* 64: 255-263

Scherer H, Helling K, Hausmann S, *et al.* (1997) On the origin of interindividual susceptibility to motion sickness. *Acta Otolaryngol (Stockh)* 117: 149-153

Schmitt HH, Reid DJ (1985) *Anecdotal Information on Space Adaptation Syndrome.* NASA, Johnson Space Center, Houston

Schöne H (1964) On the role of gravity in human spatial orientation. *Aerospace Med* 35: 764-772

Sekitani T, Honjo S, Mitani N, *et al.* (1976) Square drawing test: a new quantitative ataxia test. *Agressologie* 17: 35-40

Selemon LD, Goldman-Rakic PS (1985) Longitudinal topography and interdigitation of corticostriatal projections in the Rhesus monkey. *J Neurosci* 5: 776-794

Séverac A (1992) Electrical vestibular stimulation and space motion sickness. *Acta Astronautica* 28: 401-408

Shaw RE, Kugler PN, Kinsella-Shaw J (1990) Reciprocities of intentional systems. In: *Perception and Control of Self-Motion*. Warren R, Wertheim AH (eds) Lawrence Erlbaum, Hillsdale, pp 579-620

Shelhamer MJ (2007) Context-Specific Adaptation of Gravity-Dependent Vestibular Reflex Responses. Retrieved 14 Novembre 2007 from URL: http://mvl.mit.edu/Neurovestibular/Pages/project1.html

Shepard RN, Metzler J (1971) Mental rotation of three-dimensional objects. *Science* 171: 701-703

Shepard RN, Hurwitz S (1984) Upward direction, mental rotation, and discrimination of left and right turns in maps. *Cognition* 18: 161-193

Shephard JM, Kho S, Chen J, *et al.* (2006) MiniCog: A method for administering psychological tests and experiments on a handheld personal digital assistant *Behav Res Methods* 38: 648-655

Soechting JF, Flanders M (1989) Sensorimotor representations for pointing to targets in three-dimensional space. *J Neurophysiol* 62: 582-594

Somers JT, Reschke MF, Berthoz A, *et al.* (2002) Smoot pursuit tracking – Saccade amplitude modulation during exposure to microgravity. *Ann NY Acad Sci* 956: 426-429

Space Studies Board (1991) *Assessment of Programs in Space Biology and Medicine 1991*. National Research Council. National Academy Press, Washington DC

Space Studies Board (1998) *A Strategy for Research in Space Biology and Medicine in the New Century*. National Research Council. National Academy Press, Washington DC

Speers RA, Paloski WH, Kuo AD (1998) Multivariate changes in coordination of postural control following spaceflight. *J Biomech* 31: 883-889

Stampi C (1994) Sleep and circadian rhythms in space. *J Clin Pharmacol* 34: 518-534

Stone RW (1973) An overview of artificial gravity. In: *Fifth Symposium on the Role of the Vestibular Organs in Space Exploration*. NASA, Washington DC, NASA SP-314, pp 23-33

Suzuki JI, Tokumasu K, Goto K (1969) Eye movements from single utricular nerve stimulation in the cat. *Acta Otolaryngol (Stockh)* 68: 350-362

Suzuki M, Harada Y, Sekitani T (1993) Vestibular endorgan of the frog after the space flight and postural alteration of the neurectomized frog--its morphological and functional resilience. *J Vestib Res* 3: 253-258

Swenson L, Grinwood J, Alexander C (1966) *This New Ocean: A History of Project Mercury*. NASA, Washington DC, NASA SP-4201

Tafforin C, Lambin M (1993) Analysis of the relationships between sensory-motor alterations and behavioral modifications of the human in space flight (Spacelab-1). *Aviat Space Environ Med* 64: 146-152

Task L, Genko L (1987) Effects of long-term space flight in several visual functions. In: *Results of Life Sciences DSO's Conducted Aboard the Space Shuttle 1981-1986*. Bungo M, Bagian T, Bowman M, Levitan B (eds) NASA Space Biomedical Research Institute, Houston, Texas, pp 173-178

Thompson P (1980) Margaret Thatcher: A new illusion. *Perception* 9: 483-484

Thornton WE (1978) Anthropometric changes in weightlessness. In: *Anthropometric Source Book*. National Technical Information Service, Springfield, Illinois, Volume 1, pp 102-140

Thornton WE, Biggers WP, Thomas WG, *et al.* (1985) Electronystagmography and audio potentials in space flight. *Laryngoscope* 95: 924-932

Thornton WE, Linder BJ, Moore TP, *et al.* (1987) Gastrointestinal motility in space motion sickness. *Aviat Space Environ Med* 58: A16-A21

Thornton WE, Moore TP, Pool SL, *et al.* (1987) Clinical characterization and etiology of space motion sickness. *Aviat Space Environ Med* 58: A1-A8

Thornton WE, Moore TP, Pool SL (1987) Space motion sickness: characterization and etiology. In: *Results of the Life Sciences DSOs Conducted Aboard the Space Shuttle 1981-1986*. Bungo M, Bagian T, Bowman M, Levitan B (eds) NASA Space Biomedical Research Institute, Houston, Texas, pp 159-170

Thornton WE, Uri JJ, Moore TP, *et al.* (1989) Studies of the horizontal vestibulo-ocular reflex in space flight. *Arch Otolaryngol Head Neck Surg* 15: 943-949

Todorovic A, Kasapovic J, Pejic S, et al. (2005) Differences in antioxidative response of rat hippocampus and cortex after exposure to clinical dose of gamma-rays. *Ann NY Acad Sci* 1048: 369-372

Tomko DL, Wall C, Robinson FR, *et al.* (1988) Influence of gravity on cat vertical vestibulo-ocular reflex. *Exp Brain Res* 69: 307-314

Tran Ba Huy P, Toupet M (2001) Otolith functions and disorders. *Adv Otorhinolaryngol*, Karger, Basel, Volume 58

Treisman M (1977) Motion sickness: An evolutionary hypothesis. *Science* 197: 493-495

Tsiolkovsky KE (1911) *Personal Letter to a Friend*. August 12, 1911

Udo de Haes and Schöne H (1970) Interaction between statolith organs and semicircular canals on apparent vertical and nystagmus. Investigations on the effectiveness of the statolith organs. *Acta Otolaryngol (Stockh)* 69: 25-31

Uri JJ, Linder BJ, Moore TD, *et al.* (1989) *Saccadic Eye Movements during Space Flight.* NASA Johnson Space Center, Houston, TM-100475

Vallar G (1998) Spatial hemineglect in humans. *Trends Cognit Sci* 2: 87-97

Vernikos J (1996) Human physiology in space. *Bioessays* 18: 1029-1037

Viéville T, Clément G, Lestienne F, *et al.* (1986) Adaptive modifications of the optokinetic vestibulo-ocular reflexes in microgravity. In: *Adaptive Processes in Visual and Oculomotor Systems.* Keller EL, Zee DS (eds) Pergamon Press, New York, pp 111-120

Villard E, Tintó Garcia-Moreno F, Peter N, *et al.* (2005) Geometric visual illusions in microgravity during parabolic flight. *NeuroReport* 16: 1395-1398

Vinnikov YA, Gazenko O, Lychakov D, *et al.* (1984) *The Development of the Vestibular Apparatus under Conditions of Weightlessness.* NASA, Washington DC, TM-77517

Viviani P, McCollum G (1983) The relation between linear extent and velocity in drawing movements. *Neuroscience* 10:211-218

Vogel H, Kass JR (1986) European vestibular experiments on the Spacelab-1 mission: 7. Ocular counterrolling measurements pre- and postflight. *Exp Brain Res* 64: 284-290

Vorobyev YI, Gazenko OG, Gurovskiy NN, *et al.* (1976) Preliminary results of medical investigations carried out during flight of the second expedition of the 'Salyut -4' orbital station. *Kosm Biol Aviakosm Med* 10: 3-18

Wade NJ (1970) Effect of instructions on visual orientation. *J Exp Psychol* 83: 331-332

Waespe W, Henn V (1979) The velocity response of vestibular neurons during vestibular, visual and combined angular accelerations. *Exp Brain Res* 37: 337-347

Walk R, Gibson E (1961) A comparative analytical study of visual depth perception. *Psychological Monographs* 75: 2-34

Wassersug RJ (1992) The basic mechanisms of ascent and descent by anuran larvae (Xenopus laevis). *Copeia* 3: 890-894

Wassersug RJ, Souza KA (1990) The bronchial diverticula of Xenopus larvae. Are they essential for hydrostatic assessment? *Naturwissenschaften* 77: 443-445

Wassersug RJ, Pronych SP, Schofoeld SM (1991) On the way of life of aquatic vertebrates and their behavioral responses to reduced gravity. *ASGSB Bulletin* 5: 42

Wassersug RJ, Pronych S, Izumi-Kurotani A (1993) The behavioural responses of vertebrates to microgravity: a comparative approach. *Proceedings of the Spacebound'93 Symposium, Ottawa, May 16-18, 1993.* Canadian Space Agency paper number SB 93-001

Wassersug RJ, Izumi-Kurotani A (1993) The behavioral reactions of a snake and a turtle to abrupt decreases in gravity. *Zool Sci* 10: 505-509

Wassersug RJ, Izumi-Kurotani A, Yamashita M, *et al.* (1993) Motion sickness in amphibians. *Behav Neural Biol* 60: 42-51

Watanabe S, Takabayashi A, Tanaka M, *et al.* (1991) Neurovestibular physiology in fish. In: *Advances in Space Biology and Medicine.* Bonting S (ed) JAI, Tokyo, Volume 1

Watt DGD (1997) Pointing at memorized targets during prolonged microgravity. *Aviat Space Environ Med* 68: 99-103

Watt DGD (2003) Effects of prolonged exposure to microgravity on H-reflex loop excitability. *Proceedings of the 14th IAA Humans in Space Symposium*, Banff, Alberta. 18-24 May 2003

Watt DGD (2003) Effects of altered gravity on spinal cord excitability (final results). *Proceedings of the Bioastronautics Investigators' Workshop*, Galveston, 7-11 January 2003

Watt DGD (2007) *Effects of Altered Gravity on Spinal Cord Excitability (H-Reflex).* Retrieved 24 April 2007 from URL:
http://exploration.nasa.gov/programs/station/H-Reflex.html

Watt DGD, Lefebvre L (2001) Effects of altered gravity on spinal cord excitability. *First Research on the International Space Station*. Conference and Exhibit on International Space Station Utilization, Cape Canaveral, FL, 15-18 October 2001. AIAA 2001-4939

Watt DGD, Money KE, Bondar RL, *et al.* (1985) Canadian medical experiments on Shuttle flight 41-G. *Can Aeron Space J* 31: 215-226

Watt DGD, Money KE, Tomi LM (1986) M.I.T./Canadian vestibular experiments on the Spacelab-1 mission: 3. Effects of prolonged weightlessness on a human otolith-spinal reflex. *Exp Brain Res* 64: 367-379

Wei G, Lafortune-Kahane S, Ireland D, *et al.* (1997) Modification of vertical OKN and vertical OKAN asymmetry in humans during parabolic flight. *J Vestib Res* 7: 21-34

Welch RB (1986) Adaptation of space perception. In: *Handbook of Perception and Human Performance*. Boff KR, Kaufman L, Thomas JP (eds) John Wiley & Sons, New York, Volume 1, pp 24-1 to 24-45

Welch RB, Bridgeman B, Anand S, *et al.* (1993) Alternating prism exposure causes dual adaptation and generalization to a novel displacement. *Percept Psychophys* 54: 195-204

Welch RB, Cohen MM, DeRoshia CW (1996) Reduction of the elevator illusion from continued hypergravity exposure and visual error-corrective feedback. *Percept Psychophys* 58: 22-30

Welch RB (2005) Risks to astronaut performance readiness from acute and chronic exposure to altered gravity. *Space Human Factors Engineering*. Sub-Element 6.3.2.3. NASA Ames Research Center, Moffet Field

Wetzig J (1987) Rotation speed of labyrinthectomized fish during short-duration weightlessness. *Aviat Space Environ Med* 58: A257-A261

Wetzig J, Hoffstetter-Degen K, Von Baumgarten RJ (1993) Responses to eccentric rotation in two space-bound subjects. *Clin Investig* 71: 757-760

White O, McIntyre J, Augurelle AS, *et al.* (2005) Do novel gravitational environments alter the grip-force/load-force coupling at the fingertips? *Exp Brain Res* 163: 324-334

White RJ (2007) *Human Health and Performance in Space as Exploration Begins.* Paper presented at the 16th IAA Humans in Space Symposium, 21-24 May 2007, Beijing, China

White RJ, Averner M (2001) Humans in space. *Nature* 409: 1115-1118

Wiederhold ML, Gao W, Harrison JL, *et al.* (2003) Early development of gravity-sensing organs in microgravity. In: *The Neurolab Spacelab Mission: Neuroscience Research in Space*. Buckey JC, Homick JL (eds) NASA, Washington DC, NASA SP-2003-535, pp 123-132

Wilson VJ, Melvill Jones G (1979) *The Mammalian Vestibular System*. Plenum Press, New York

Wolfe JM, Held R (1979) Eye torsion and visual tilt are mediated by different binocular processes. *Vision Res* 19: 917-920

Wood CD (1990) Pharmacological countermeasures against motion sickness. In: *Motion and Space Sickness*. Crampton GH (ed) CRC Press, Boca Raton, pp 343-351

Wood CD, Graybiel A (1968) Evaluation of sixteen anti-motion sickness drugs under controlled laboratory conditions. *Aerospace Med* 39: 1341-1344

Wood CD, Graybiel A (1972) Theory of antimotion-sickness drug mechanisms. *Aerospace Med* 43: 249-252

Wood CD, Manno JE, Manno BR, *et al.* (1986) The effect of antimotion sickness drugs on habituation to motion. *Aviat Space Environ Med* 57: 539-542

Wood SJ, Paloski WH, Reschke MF (1998) Spatially-directed eye movements during roll-tilt relative to perceived head and earth orientations. *Exp Brain Res* 121:51-58

Wood SJ, Ramsdell CD, Mullen TJ, *et al.* (2000) Transient cardio-respiratory responses to visually induced tilt illusions. *Brain Res Bull* 53: 25-31

Wood SJ (2002) Human otolith-ocular reflexes during off-vertical axis rotation: Effect of frequency on tilt-translation ambiguity and motion sickness. *Neurosci Lett* 323: 41-44

Wood SJ, Reschke MF, Sarmiento L, *et al.* (2007) Tilt and translation motion perception during off-vertical axis rotation. *Exp Brain Res* 182: 365-377

Yakovleva IY, Kornilova LN, Tarasov IK (1982) Results of studies of cosmonauts' vestibular function and spatial perception. *Kosm Biol Aviakosm Med* 16: 20-26

Yates BJ (1992) Vestibular influences on the sympathetic nervous system. *Brain Res Rev* 17: 51 59

Yates BJ, Miller AD (1996) *Vestibular-Autonomic Regulation*. CRC Press, Boca Raton

Yates BJ, Aoki M, Burchill P, *et al.* (1999) Cardiovascular responses elicited by linear acceleration in humans. *Exp Brain Res* 125: 476-484

Yates BJ, Jian LA, Cotter SP (2000) Responses of vestibular nucleus neurons to tilt following chronic bilateral removal of vestibular inputs. *Exp Brain Res* 130: 151-158

Yegorov AD (1979) *Results of Medical Studies during Long-Term Manned Flights on the Orbital Salyut-6 and Soyuz Complex*. U.S.S.R. Academy of Sciences. NASA TM-76014

Young LR (1984) Perception of the body in space: Mechanisms. In: *Handbook of Physiology*. Section 1, *The Nervous System*, Volume III, *Sensory Processes*, Part 2. Darian-Smith I (ed) American Physiological Society, Bethesda, pp 1023-1066

Young LR (1993) Space and the vestibular system: What has been learned? Guest editorial. *J Vestib Res* 3: 203-206

Young LR (2000) Vestibular reactions to space flight: Human factors issues. *Aviat Space Environ Med* 71: A100-A104

Young LR (2006) Neurovestibular aspects of short-radius artificial gravity: Toward a comprehensive countermeasure. *NSBRI Sensorimotor Adaptation Project Technical Summary*. Retrieved 22 May 2006 from URL: http://www.nsbri.org/Research/Projects/viewsummary.epl?pid=184

Young LR, Shelhamer M (1990) Microgravity enhances the relative contributions of visually-induced motion sensation. *Aviat Space Environ Med* 61: 225-230

Young LR, Sinha P (1998) Spaceflight influences on ocular counterrolling and other neurovestibular reactions. *Otolaryngol Head Neck Surg* 118: 31-34

Young LR, Oman CM, Watt DGD, *et al.* (1984) Spatial orientation in weightlessness and readaptation to Earth's gravity. *Science* 225: 205-208

Young LR, Oman CM, Merfeld D, *et al.* (1993) Spatial orientation and posture during and following weightlessness: Human experiments on Spacelab Life Sciences 1. *J Vestib Res* 3: 231-239

Young LR, Mendoza JC, Groleau N, *et al.* (1996) Tactile influences on astronaut visual spatial orientation: Human neurovestibular experiments on Spacelab Life Sciences 2. *J Appl Physiol* 81: 44-49

Young LR, Hecht H, Lyne L, *et al.* (2001) Artificial gravity: Head movements during short-radius centrifugation. *Acta Astronautica* 49: 215-226

Yuganov YM, Kopanec VI (1975) Physiology of the sensory sphere under spaceflight conditions. In: *Foundations of Space Biology and Medicine*. Calvin M, Gazenko OG (eds) NASA, Washington DC, Volume II, Book 2, Chapter 15, pp 571-599

Zee DS, Yamazaki A, Butler PH, *et al.* (1981) Effects of ablation of flocculus and paraflocculus on eye movements in primate. *J Neurophysiol* 46: 878-899

Zubrin R (1996) *The Case for Mars: The Plan to Settle the Red Planet and Why We Must.* The Free Press, New York

INDEX

A

Acceleration
    Angular Acceleration 14-15, 117, 165, 205
    Centripetal Acceleration 7, 127-130, 256
    Cross-Coupled Accelerations 75, 236, 238
    Gravitational Acceleration 23-24, 193, 267
    Linear Acceleration 7, 14, 23, 114, 116-117, 123, 174, 177, 204, 209
Acrophobia 197
Acuity 44-50, 103-105, 131, 154, 156-157, 160-161, 182, 187, 275, 288
AFT (Autogenic Feedback Training) 241-242
Alexander's Law 178
ALFA (Air Lubricated Free Axis) 248
Ambiguity 15, 23, 83-84, 125, 207-209
Amphetamine 242-244
Ampulla 115-116
AR (Augmented Reality) 97
Artificial Gravity 7, 91, 99, 128-129, 131, 188, 211, 235, 252-259, 261, 265
Asymmetry 80-81, 153, 167, 169, 176-178, 187, 191
Ataxia 61-64, 69, 144-145, 151, 160
Attention 32, 46-47, 66-67, 93-96, 212, 220
Autonomic Nervous System 16, 38, 84, 240, 266-267
Awareness 23, 72, 97, 99, 101, 110, 134, 142, 149, 189, 198, 210-212, 220

B

Balance 13, 15, 31, 69, 71, 101, 114, 117, 135, 139, 142, 145, 147-149, 151, 158, 160, 253, 280
Bàràny Robert 164, 236
Behavior 1, 6, 12-14, 16, 23, 25, 27, 89-92, 95-100, 109, 115, 117, 121, 135-137, 139, 149, 153, 163, 171, 186, 212-213, 241, 248, 251, 258, 267-268, 272-273
Bioastronautics Critical Path Roadmap 233
Biofeedback Training *(see AFT)*
Binaural Hearing 106
Biosatellite 44, 120, 169
Blood 1, 30-31, 87, 245, 263
BRS (Body Restraint System) 123-124, 131, 168

C

CDT (Clock Drawing Test) 220-221
Central Vestibular System 120, 165, 224
Centrifuge 11, 17, 82, 120, 126, 129-131, 163, 168, 174, 180, 206-207, 220, 236, 249, 254, 257, 259, 266, 274
Cilia 115
Circadian Rhythm 5, 30, 54, 58, 86, 88, 96, 99, 258, 275
Cognilab 93
CNS (Central Nervous System) 1
Confinement 2, 5, 87, 89-90, 108, 230, 251, 254, 273
Conjunction-Class Mission 9-10
Coordinate System 26, 30, 181, 190-192, 206, 211-212, 218-219, 221, 227, 231
Coriolis 17, 74-75, 236-238, 242, 257
Cortex 16, 22, 29-31, 83, 102, 109, 133, 211, 271
Couch 1, 4, 202, 249
Countermeasure (definition) 17